Design aids for EC2

Design aids for EC2
Design of concrete structures

Design aids for ENV 1992-1-1
Eurocode 2, part 1

Betonvereniging
The Concrete Society
Deutscher Beton-Verein

E & FN SPON
An Imprint of Chapman & Hall
London · Weinheim · New York · Tokyo · Melbourne · Madras

Published by E & FN Spon, an imprint of
Chapman & Hall, 2-6 Boundary Row, London SE1 8HN, UK

Chapman & Hall, 2–6 Boundary Row, London SE1 8HN, UK

Chapman & Hall GmbH, Pappelallee 3, 69469 Weinheim, Germany

Chapman & Hall USA, 115 Fifth Avenue, New York, NY 10003, USA

Chapman & Hall Japan, ITP-Japan, Kyowa Building, 3F, 2-2-1 Hirakawacho, Chiyoda-ku, Tokyo 102, Japan

DA Book (Aust.) Pty Ltd, 648 Whitehorse Road, Mitcham 3132, Victoria, Australia

Chapman & Hall India, R. Seshadri, 32 Second Main Road, CIT East, Madras 600 035, India

First edition 1997

Printed in Great Britain by TJ International Ltd, Cornwall

ISBN 0 419 21190 X

A catalogue record for this book is available from the British Library

Publisher's Note This book has been prepared from camera ready copy provided by Betonvereniging, The Concrete Society and Deutscher Beton-Verein E.V.

⊜ Printed on acid-free text paper, manufactured in accordance with ANSI/NISO Z39.48-1992 (Permanence of Paper).

Contents

Preface

The European concrete standards in practice

The German, UK and Netherlands Concrete Societies are working together on a SPRINT project for the development of supporting tools for use with the European Structural Concrete Code. The project is in three parts essentially covering:

1. An investigation of what tools the industry needs and prefers to enable it to work with the new code.

2. The development of preferred tools.

3. Publication and dissemination of the tools developed and consideration of the possible development of further aids to the use of the code.

In the first phase, the societies questioned a wide range of practitioners about their needs and preferences for design tools. It was found that, although there is considerable interest in developing information systems through computer processes, the immediate need and preference was for a traditional "hard copy" Technical Document containing information, guidance and examples of the use of the Code.

In response, the societies concentrated efforts in the second phase into the production of such a document, which this now is. During the development of the material, an important meeting was held in Amsterdam in October 1994 when the societies were able to present draft material for examination and comment and to seek views on the direction of their work. Discussion at this meeting confirmed the earlier analysis of the industry's immediate needs and interest in the development of other information systems for the future. Comments made on the draft at and after the meeting were subsequently considered by the societies and, where appropriate, material was modified or added.

The publication of this document marks the completion of the second phase and forms part of the final phase which will concentrate on the dissemination of the information in this document. This last phase will also involve a further examination of other methods to highlight the material that has been prepared and to consider how other tools and systems may be developed to aid industry.

Finally, it must be stressed that this document is not an alternative to the European Structural Concrete Code. It is an aid to use in conjunction with the Code to help designers in their work.

March 1996

1 General information

Dr.-Ing. H.-U. Litzner, Wiesbaden: Chairman of CEN/TC250/SC2

1.1 Construction products directive and European harmonized standards for concrete structures

The European construction market was officially established in January 1993. This means that in this market, as in other areas of the economy, goods, services, people and capital are able to move freely within the European Union (EU). An important instrument in this connection is the "Construction products directive" [1], adopted by the EU-Commission in December 1988. This directive sets out the conditions under which a construction product (e.g. cement, ready-mixed concrete, reinforcement, precast element) can be imported and exported and used for its intended purposes without impediment in EU countries. This directive has been integrated into the national legislation of most EU Member States.

"Technical specifications" - i.e. harmonized European standards, or, where these are lacking, European technical approvals - are necessary for the practical application of this directive. Figure 1.1 shows the European code of practice system for concrete structures that is currently being elaborated at different levels on the basis of the Directive. This standards system will quantify requirements for concept, design, detailing and execution of structures.

According to Article 6 of the directive, a construction product may move freely within the EU provided it meets certain basic requirements. These criteria, denoted in the Directive as "Essential requirements", primarily relate, however, to the structure into which the construction product is to be incorporated. The "Essential requirements" concern:

- mechanical resistance and stability
- safety in case of fire
- hygiene, health and the environment
- safety in use
- protection against noise
- energy economy and heat retention.

This establishes the framework for further consideration.

The "Essential requirements" are only qualitatively described in the directive text. Further European documents are needed for practical application. These include the so-called "Interpretative documents", in which the essential requirements are defined, the previously mentioned "Technical specifications" (European harmonized standards and European guidelines for technical approval), as well as regulations for the positive assessment of the conformity of a construction product ("Certification").

1.2 Future European code of practice for concrete structures

On the basis of provisional mandates of the EU, a code of practice for concrete structures is being established by the European Committee for Standardization (abbreviated CEN) which, in the longer term, will replace national standards. Its structure is comparable to that of existing national standards systems (Figure 1.1).

It comprises:

- a safety concept relevant to any type of construction (ENV 1991-1);
- Eurocode 1 concerning actions on structures (including traffic loads in ENV 1991-3);
- codes of practice for design and execution of structures;
- construction material standards (concrete, reinforcement, prestressing steel);
- standards for the testing of construction materials (ISO or CEN standards).

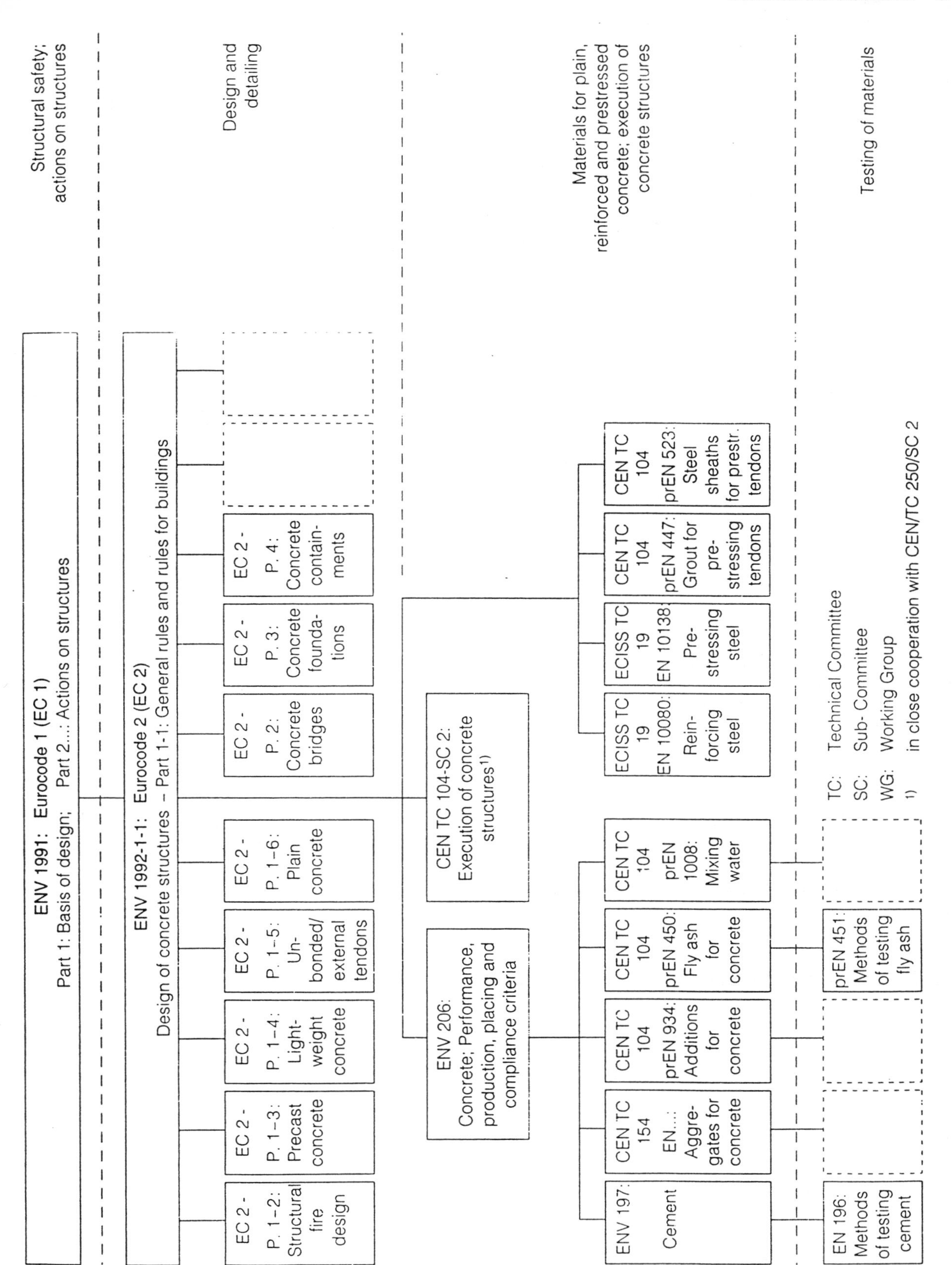

Figure 1.1 Structure of the future European harmonized standards for concrete.

From this it becomes clear that the future European standards for concrete structures are aimed at the "essential requirements", particularly at the mechanical resistance and stability, structural fire design and safety in use, whereby the initially mentioned requirement also incorporates criteria regarding durability. This objective is also expressed in the foreword to Eurocode 2 [2] which states, among other things, the following:

"0.1 Objectives of the Eurocodes

(1) The Structural Eurocodes comprise a group of standards for the structural and geotechnical design of buildings and civil engineering works.

(2) They are intended to serve as reference documents for the following purposes:

 (a) As a means to prove compliance of building and civil engineering works with the essential requirements of the Construction Products Directive (CPD)

 (b) As a framework for drawing up harmonized technical specifications for construction products.

(3) They cover execution control only to the extent that is necessary to indicate the quality of the construction products, and the standard of the workmanship, needed to comply with the assumptions of the design rules.

(4) Until the necessary set of harmonized technical specifications for products and for methods of testing their performance is available, some of the Structural Eurocodes cover some of these aspects in informative annexes."

"0.2 Background to the Eurocode programme

(1) The Commission of the European Communities (CEC) initiated the work of establishing a set of harmonized technical rules for the design of building and civil engineering works which would initially serve as an alternative to the different rules in force in the various Member States and would ultimately replace them. These technical rules became known as the 'Structural Eurocodes'.

(2) In 1990, after consulting their respective Member States, the CEC transferred work of further development, issue and updates of the Structural Eurocodes to CEN and the EFTA Secretariat agreed to support the CEN work.

(3) CEN Technical Committee CEN/TC250 is responsible for all Structural Eurocodes."

Paragraph 0.1 (2)(b) quoted above applies in particular to precast structural elements for which the CEN Technical Committee (TC) 229 is currently elaborating product standards in accordance with the 1988 Directive. These products include, for example, prestressed concrete hollow slabs and factory produced concrete masts and piles. As far as possible, the design concept is based on Eurocode 2 [2].

1.3 Safety concept relevant to any type of construction material

The outlines of the safety concept for any type of construction material in the Eurocodes are defined in the interpretative document "Mechanical resistance and stability". [3] Based on this, ENV 1991-1 [4] explains how the satisfaction of these "Essential requirements" in accordance with the Construction products directive [1] may be verified and provides as models the ultimate limit states concept as well as serviceability limit states.

The ultimate limit states concern the danger potential associated with collapse of the structure or other forms of structural failure. Among other criteria, these include the loss of global equilibrium (transformation into a mechanism, sliding, overturning), the failure or a state before failure of parts of the structure (failure of cross-section, states of deformation, exceeding the bearing capacity), loss of stability (buckling, lateral buckling of slender beams, local buckling of plates) as well as material fatigue.

These ultimate limit states are modelled mathematically in EC2. In its chapter 4.3, the ultimate limit states are distinguished as:

4.3.1 ultimate limit states for bending and longitudinal force;
4.3.2 ultimate limit states for shear;
4.3.3 ultimate limit states for torsion;
4.3.4 ultimate limit states of punching;
4.3.5 ultimate limit states induced by structural deformation (buckling).

The serviceability limit states in EC2 correspond to a structural state beyond which the specified service requirements are no longer met. The corresponding models in its chapter 4.4 are:

4.4.2 limit states of cracking;
4.4.3 limit states of deformation;

as well as excessive stresses in the concrete, reinforcing or prestressing steel under serviceability conditions, which likewise can adversely affect proper functioning of a member (section 4.4.1).

1.4 Eurocode 2 for the design and execution of concrete structures

1.4.1 General

Eurocode 2 "Design of concrete structures; Part 1-1: General rules and rules for buildings" was issued as European Prestandard ENV 1992-1-1 [2] by the European Committee for Standardisation (CEN). There is no obligation to implement this Prestandard into national standard systems or to withdraw conflicting national standards.

Consequently, the first parts of the future European system of harmonized standards for concrete structures (Figure 1.1) are available in the form of ENV 1992-1-1 (EC2) and the Prestandard ENV 206 for concrete technology. The gaps, which are due to the current lack of further ENV standards, e.g. covering constituent materials for concrete, reinforcement, prestressing steel, quality control, are covered by National Application Documents (NAD). This is to enable the provisional application of the new European standards as recommended by the EU. Approval ("notification") as a technical building regulation (guideline) by the relevant supervisory authorities has been carried out in most Member States.

1.4.2 Contents of Eurocode 2: principles and application rules: indicative numerical values

The design concept of EC2 does not differentiate between prestressed and non-prestressed structural members. Likewise, no distinction is made between full, limited or partial prestressing.

EC2 is divided into "Principles" and "Application rules". "Principles" comprise verbally defined general requirements (e.g. regarding structural safety), to which no alternative is permitted. On the whole, these are definitions and obvious requirements which can be adopted by all EU countries. The "Application rules" are generally recognized rules (for example detailing rules) that follow the "Principles" and satisfy their requirements.

It is permissible to use alternative design rules provided that it is shown that these rules accord with the relevant "Principles" and that they are at least equivalent to those in EC2. Similar questions regarding methods have yet to be resolved. However, the principle of interchangeability of rules is generally anchored in the national codes of practice. A further characteristic of EC2 is the so-called "indicative" values, i.e. figures given as an indication (e.g. the partial factors of safety) and identified in the text by a "box".

During an interim period, at least, they can be determined nationally by the individual EU countries. Where necessary, such modifications are given in special cases in the National Application Documents (NAD) during provisional application of EC2.

1.4.3 Essential requirements for design and execution

The essential requirements in chapter 2.1 of EC2 for design and construction stipulate among other things:

"P(1) A structure shall be designed and constructed in such a way that:

- with acceptable probability, it will remain fit for the use for which it is required, having due regard to its intended life and its cost, and
- with appropriate degrees of reliability, it will sustain all actions and influences likely to occur during execution and use and have adequate durability in relation to maintenance costs."

"P(2) A structure shall also be designed in such a way that it will not be damaged by events like explosions, impact or consequence of human errors, to an extent disproportionate to the original cause..."

"P(4) The above requirements shall be met by the choice of suitable materials, by appropriate design and detailing and by specifying control procedures for production, construction and use as relevant to the particular project."

With these requirements the overall framework is clearly defined into which the subsequent EC2 chapters 2.2 to 2.5 and 3 to 7 fit with their technical content (Table 1.1). Worthy of note is the fact that the durability requirement ranks high. This was one of the main reasons for the drafting of chapter 4.1 "Durability requirements" which, in the form of a sort of "checklist", specifies the essential parameters which are to be seen in connection with durability. Attention is also drawn here to the CEN standard ENV 206 which includes important requirements for the choice of constituent materials for concrete and for the composition of concrete.

Table 1.1 Contents of Eurocode 2

Chapter	Title
1	Introduction
2	Basis of design
2.1	Fundamental requirements
2.2	Definitions and classifications
2.3	Design requirements
2.4	Durability
2.5	Analysis
3	Material properties
4	Section and member design
4.1	Durability requirements
4.2	Design data
4.3	Ultimate limit states
4.4	Serviceability limit states
5	Detailing provisions
6	Construction and workmanship
7	Quality control

1.5 References

1. The Council of the European Communities: Council Directive of 21 December 1988 on the approximation of laws, regulations and administrative provisions of the Member States relating to construction products (89/106/EEC).
2. ENV 1992-1-1: 1991: Eurocode 2: Design of Concrete Structures. Part 1: General Rules and Rules for Buildings; European Prestandard. December 1991.
3. Commission of the European Communities: Interpretative Document for the Essential Requirement No. 1 - Mechanical Resistance and Stability. Last version complete, July 1993.
4. ENV 1991-1-Eurocode 1: Basis of design and actions on structures. Part 1: Basis of design. Edition 1994.

2 Main symbols used in EC2

A_c	Total cross-sectional area of a concrete section
A_{cl}	Maximum area corresponding geometrically to A_{co}, and having the same centre of gravity
A_{co}	Loaded area
$A_{ct,ext}$	Area of concrete external to stirrups
$A_{c.eff}$	Effective area of concrete in tension
A_k	Area enclosed within the centre-line of the idealized thin-walled cross-section including inner hollow areas
A_{ct}	Area of concrete within the tension zone
A_p	Area of a prestressing tendon or tendons
A_s	Area of reinforcement within the tension zone
A_{s2}	Area of reinforcement in the compression zone at the ultimate limit state
A_{sf}	Area of reinforcement across the flange of a flanged beam
$A_{s,min}$	Minimum area of longitudinal tensile reinforcement
$A_{s,prov}$	Area of steel provided
$A_{s,req}$	Area of steel required
$A_{s,surf}$	Area of surface reinforcement
A_{st}	Area of additional transverse reinforcement parallel to the lower face
A_{sv}	Area of additional transverse reinforcement perpendicular to the lower face
A_{sw}	Cross-sectional area of shear reinforcement
E_{cd}	Design value of the secant modulus of elasticity
$E_{c(t)}$	Tangent modulus of elasticity of normal weight concrete at a stress of $\sigma_c = 0$ and at time t
$E_{c(28)}$	Tangent modulus of elasticity of normal weight concrete at a stress of $\sigma_c = 0$ and at 28 days
E_{cm}	Secant modulus of elasticity of normal weight concrete
$E_{c,nom}$	Either the mean value of E_{cm} **or** The corresponding design value E_{cd}
$E_{d,dst}$	Design effects of destabilising actions
$E_{d,stb}$	Design effects of stabilising actions
E_s	Modulus of elasticity of reinforcement or prestressing steel
F_c	Force due to the compression block at a critical section at the ultimate limit state
ΔF_d	Variation of the longitudinal force acting in a section of flange within distance a_v
F_{px}	Ultimate resisting force provided by the prestressing tendons in a cracked anchorage zone
$F_{sd,sup}$	Design support reaction
F_s	Force in the tension reinforcement at a critical section at the ultimate limit state
F_s	Tensile force in longitudinal reinforcement
F_v	Vertical force acting on a corbel
$G_{d,inf}$	Lower design value of a permanent action
$G_{d,sup}$	Upper design value of a permanent action
G_{ind}	Indirect permanent action
$G_{k,inf}$	Lower characteristic value of a permanent action
$G_{k,sup}$	Upper characteristic value of a permanent action
$G_{k,j}$	Characteristic values of permanent actions
H_c	Horizontal force acting at the bearing on a corbel
H_{fd}	Additional horizontal force to be considered in the design of horizontal structural elements, when taking account of imperfections
ΔH_j	Increase in the horizontal force acting on the floor of a frame structure, due to imperfections
ΔM_{Sd}	Reduction in the design support moment for continuous beams or slabs, due to the

ρ_w	Reinforcement ratio for shear reinforcement
σ_c	Compressive stress in the concrete
σ_{cu}	Compressive stress in the concrete at the ultimate compressive strain
σ_{cg}	Stress in the concrete adjacent to the tendons, due to self-weight and any other permanent actions
σ_{cpo}	Initial stress in the concrete adjacent to the tendons, due to prestress
$\sigma_{o,\,max}$	Maximum stress applied to a tendon
σ_{pmo}	Stress in the tendon immediately after stressing or transfer
σ_{pgo}	Initial stress in the tendons due to prestress and permanent actions
σ_s	Stress in the tension reinforcement calculated on the basis of a cracked section
σ_{sr}	Stress in the tension reinforcement calculated on the basis of a cracked section under conditions of loading leading to formation of the first crack
τ_{Rd}	Basic shear strength of members without shear reinforcement
$\varphi(\infty, t_o)$	Final value of creep coefficient
$\varnothing$	Diameter of a reinforcing bar or of a prestressing duct
$\varnothing_n$	Equivalent diameter of a bundle of reinforcing bars
$\varnothing_s$	Adjusted maximum bar diameter
$\varnothing^*_s$	Unadjusted maximum bar diameter (Table 4.11)
ψ	Factors defining representative values of variable actions
ψ_0	Used for combination values
ψ_1	Used for frequent values
ψ_2	Used for quasi-permanent values

3 Overview of flow charts

The flow charts function as a guide through Eurocode 2. The cross-references used in the flow charts therefore refer to Eurocode 2.

There are three main levels of flow charts.

Level 1	Basis of design		2.
	Flow chart 3.0	Overview	
Level 2	Section and member design		4.
	Flow chart 3.0.1	General	
	Flow chart 3.0.2	Ultimate limit states (ULS)	
	Flow chart 3.0.3	Serviceability limit states (SLS)	
Level 3	Detailed calculations and detailing provisions		4.
Level 3.1	ULS		4.3
Level 3.1.1	Bending		4.3.1
	Flow chart 3.1.1.1	Bending and longitudinal force	
Level 3.1.2	Shear		4.3.2
	Flow chart 3.1.2.1	Design method	
	Flow chart 3.1.2.2	Elements with shear reinforcement	
Level 3.1.3	Torsion		4.3.3
	Flow chart 3.1.3.1	Pure torsion	
	Flow chart 3.1.3.2	Torsion, combined effects of actions	
	Flow chart 3.1.3.3	Torsion and flexure	
	Flow chart 3.1.3.4	Torsion and shear	
Level 3.1.4	Punching		4.3.4
	Flow chart 3.1.4.1	Punching	
	Flow chart 3.1.4.2	Punching shear reinforcement	
Level 3.1.5	Buckling		4.3.5
	Flow chart 3.1.5.1	General guide	
	Flow chart 3.1.5.2	Structure as a whole	
	Flow chart 3.1.5.3	Isolated columns	

Flowchart 3.0

Basis of design: overview

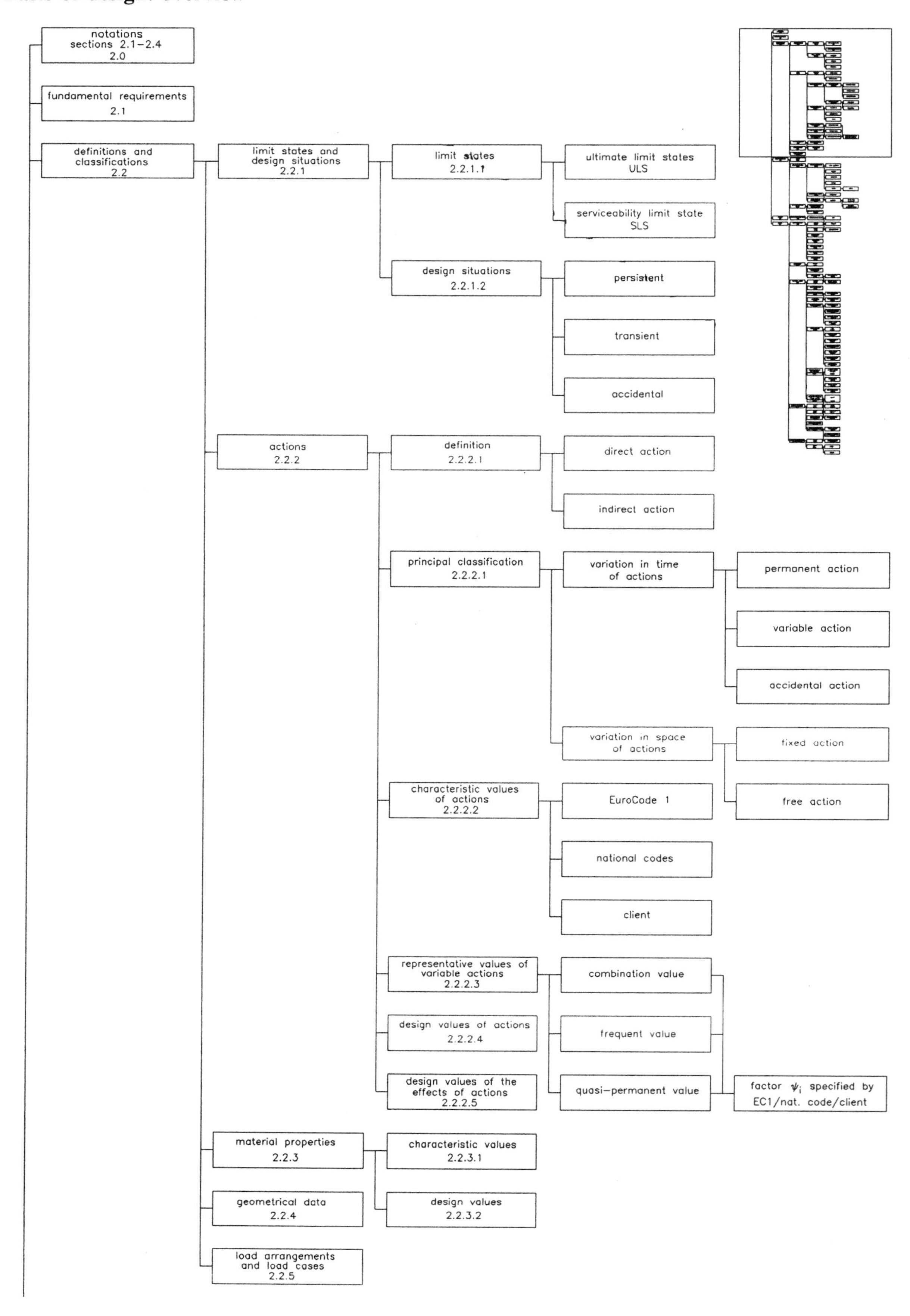

Flow chart 3.0.2

Section and member design: ultimate limit state (ULS)

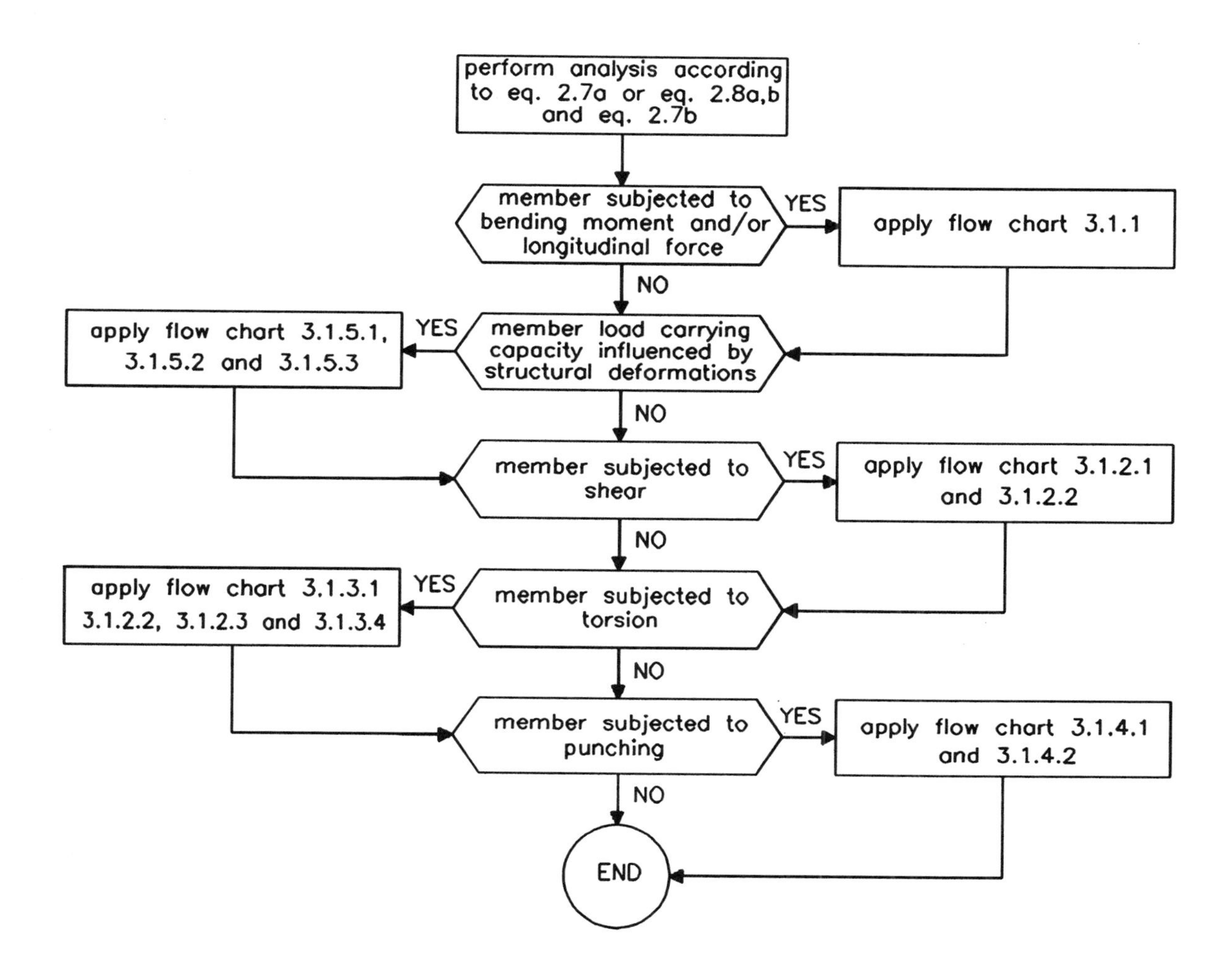

Flow chart 3.0.3

Section and member design: serviceability limit state (SLS)

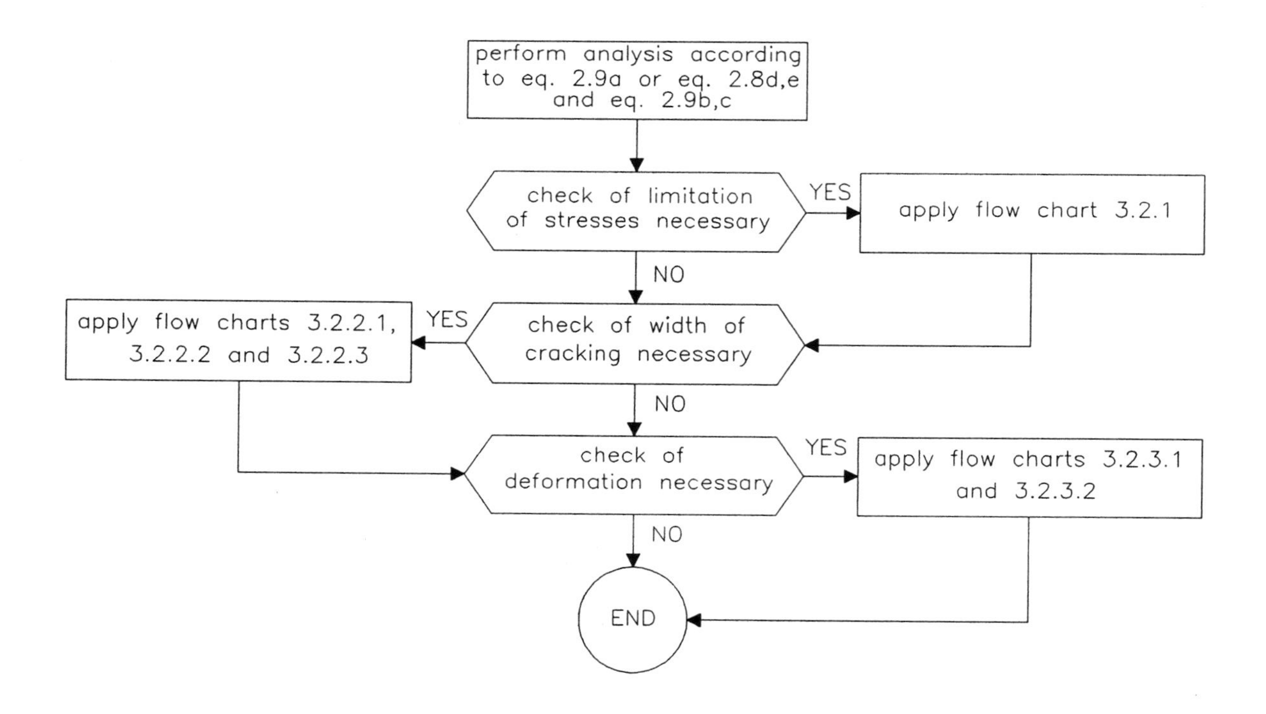

Flow chart 3.1.1.1

Bending: bending and longitudinal force

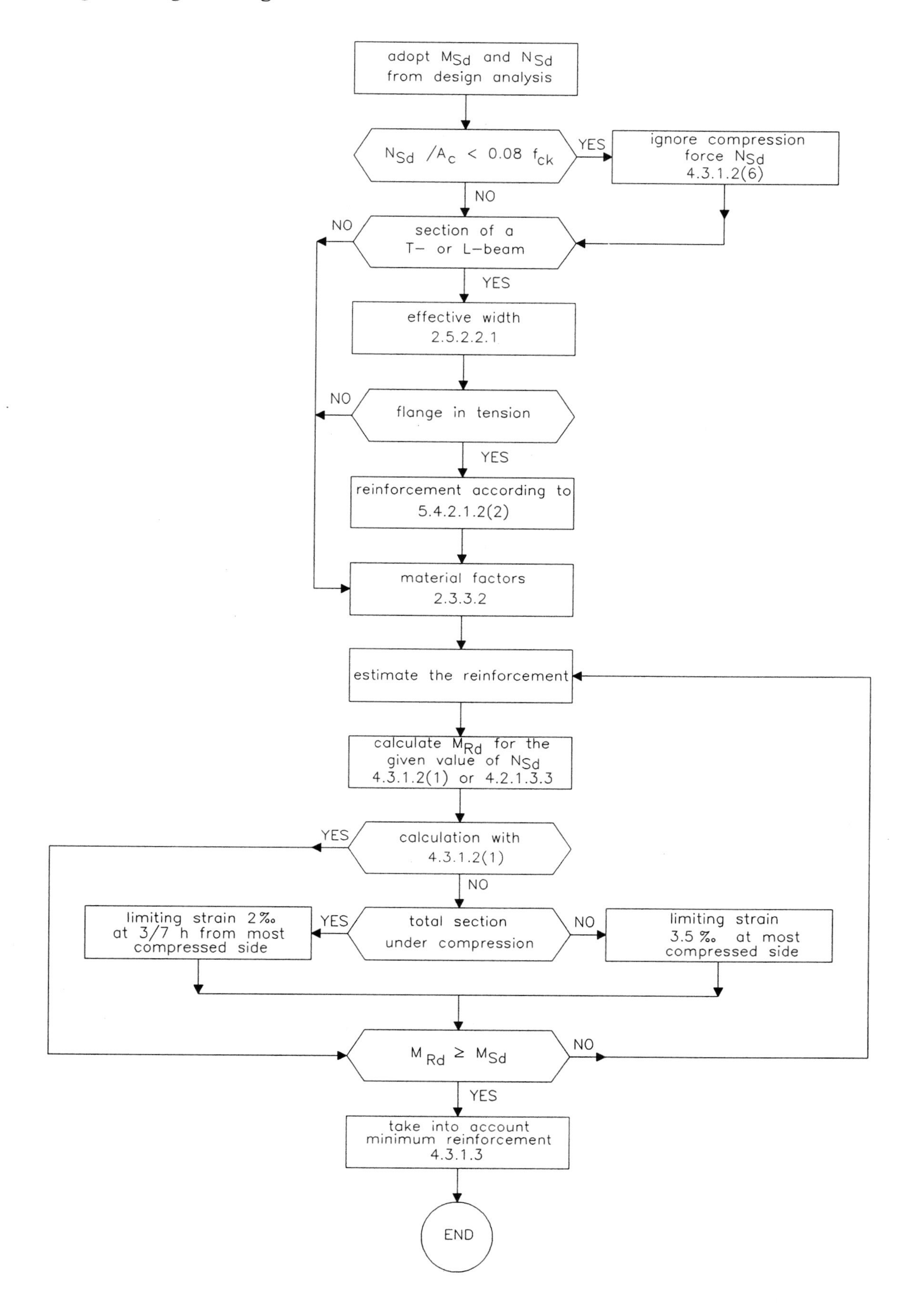

Flow chart 3.1.2.1

Shear: design method

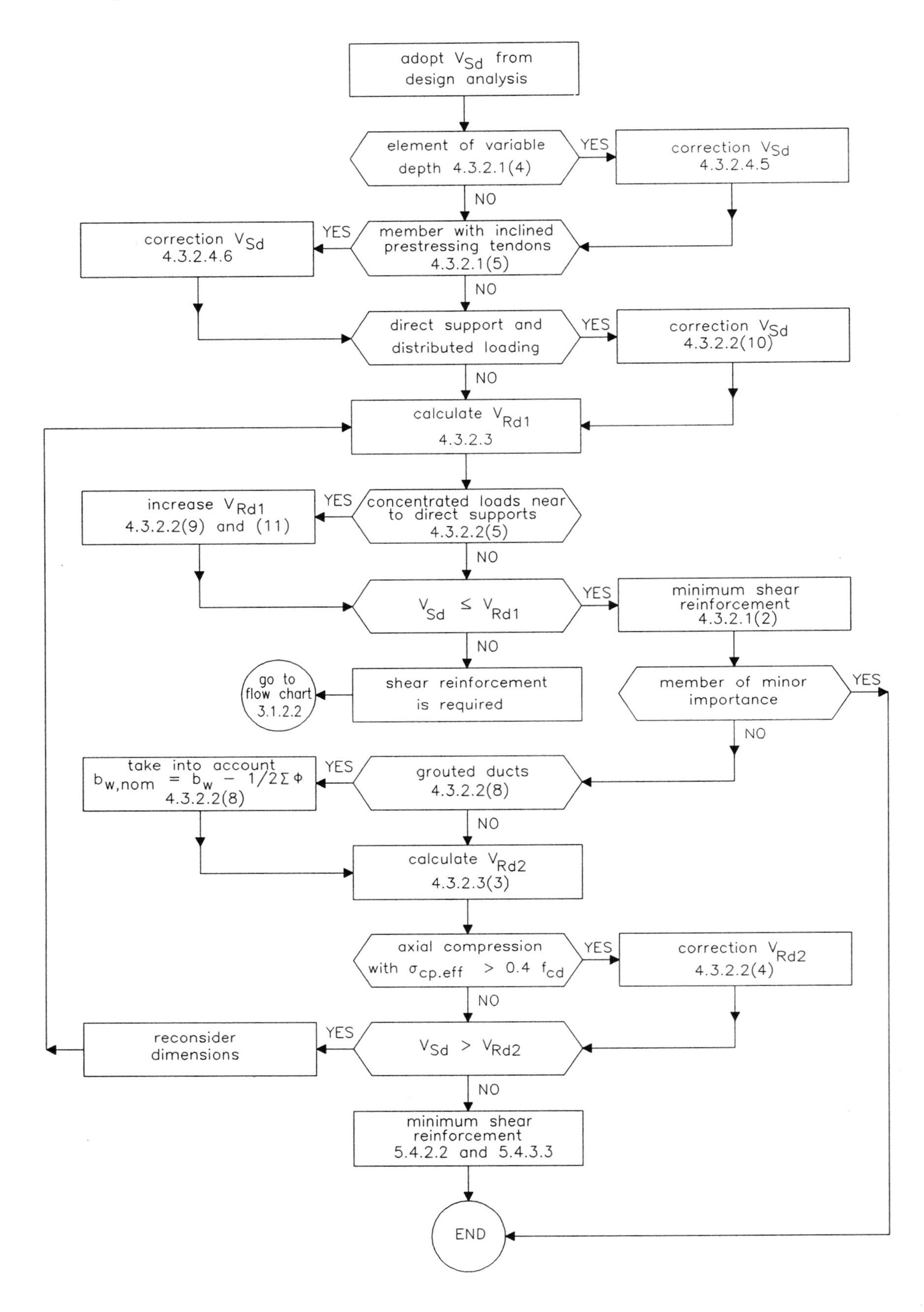

Flow chart 3.1.3.4

Torsion: torsion and shear

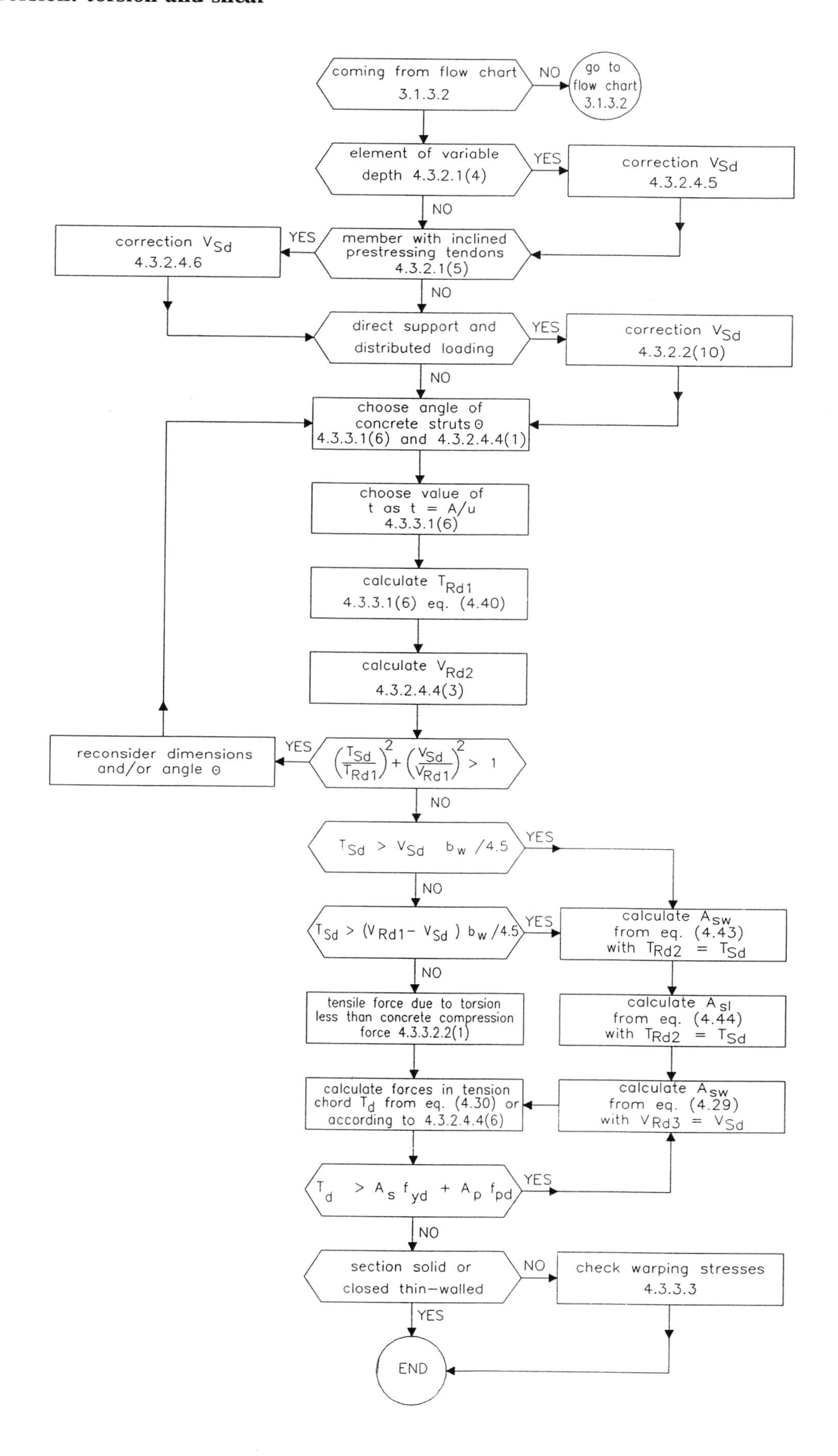

Flow chart 3.1.4.1

Punching: punching

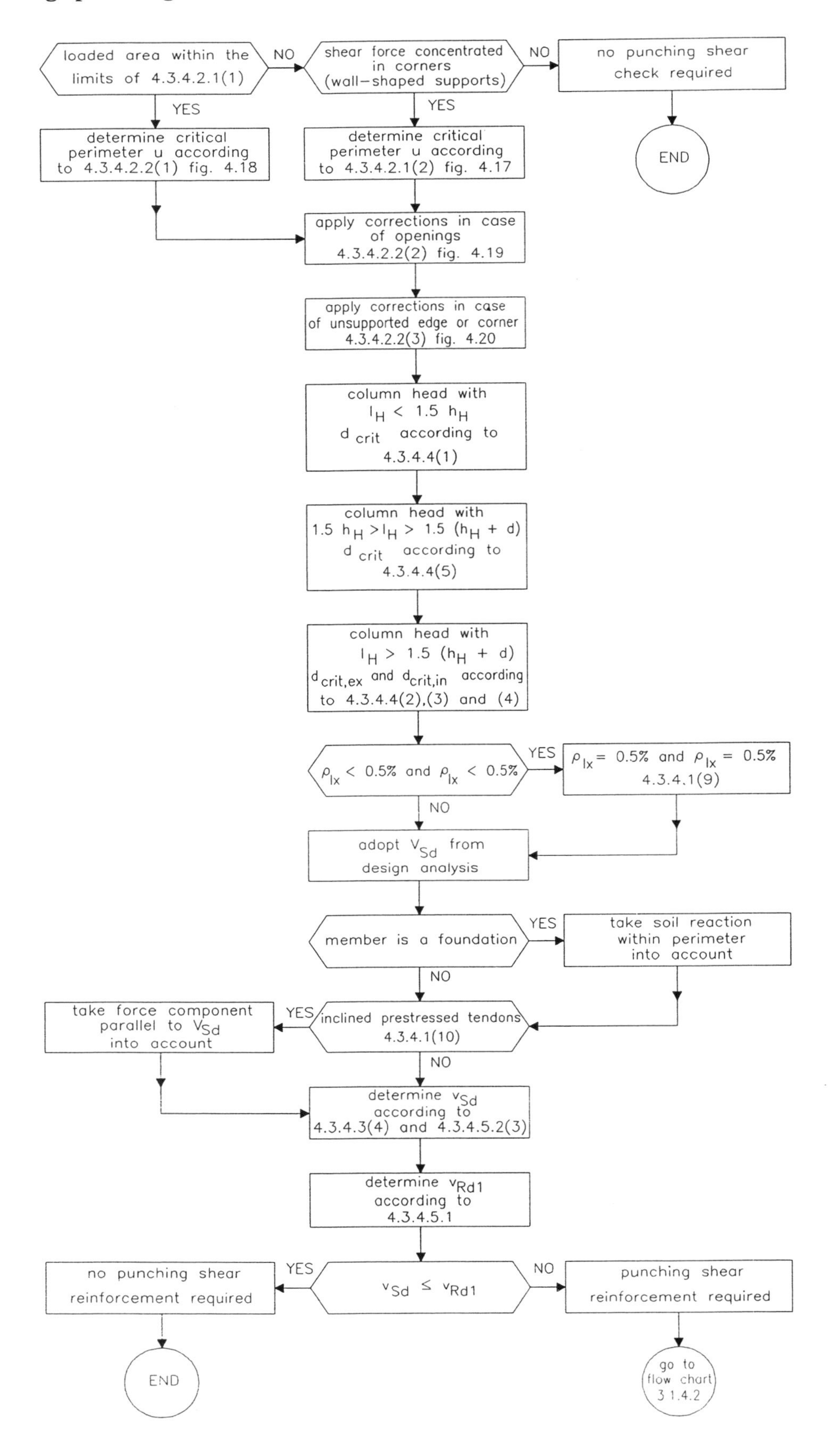

Flow chart 3.1.4.2

Punching: punching shear reinforcement

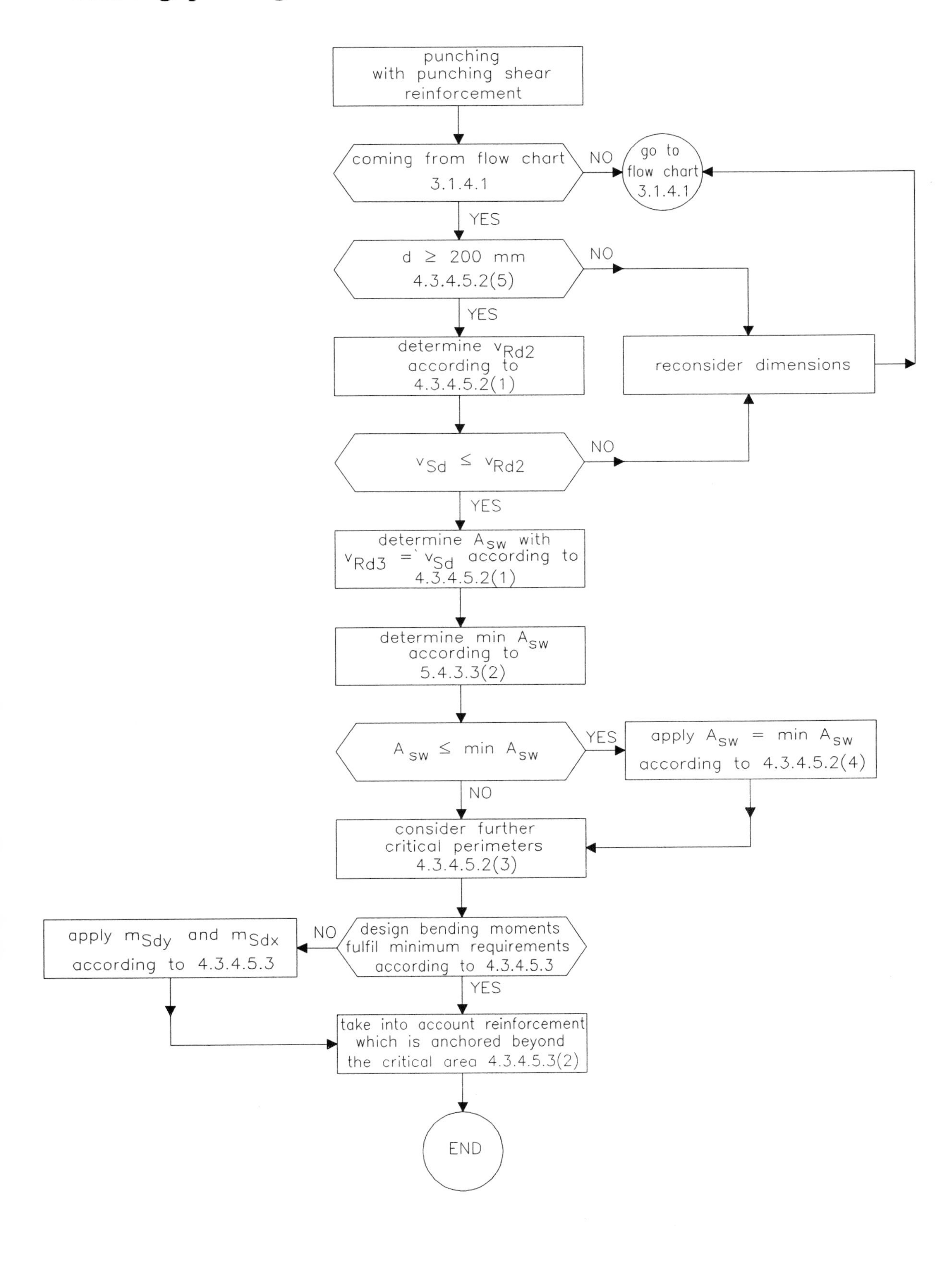

Flow chart 3.1.5.1

Buckling: general guide

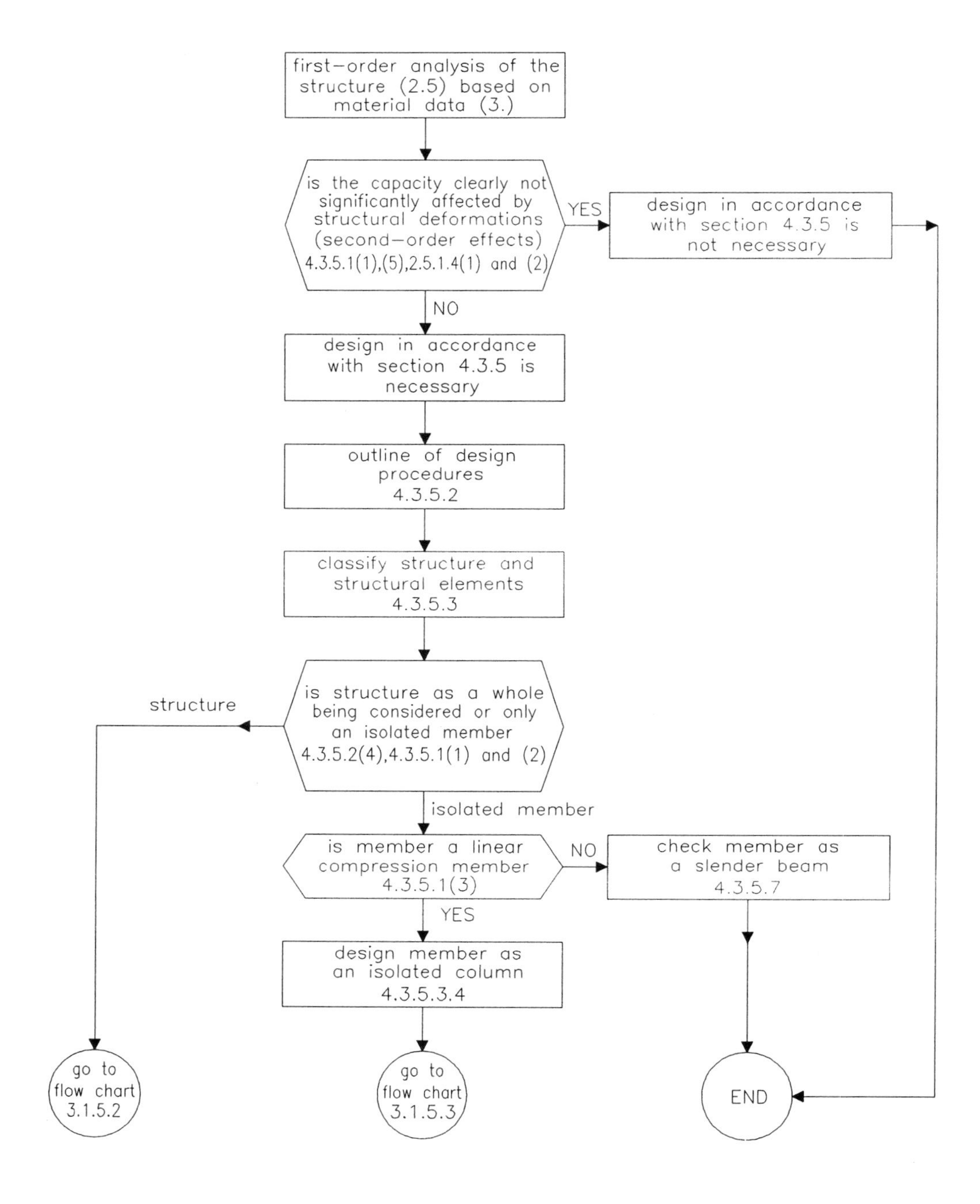

Flow chart 3.3.2.1

Splices: splices for bars or wires

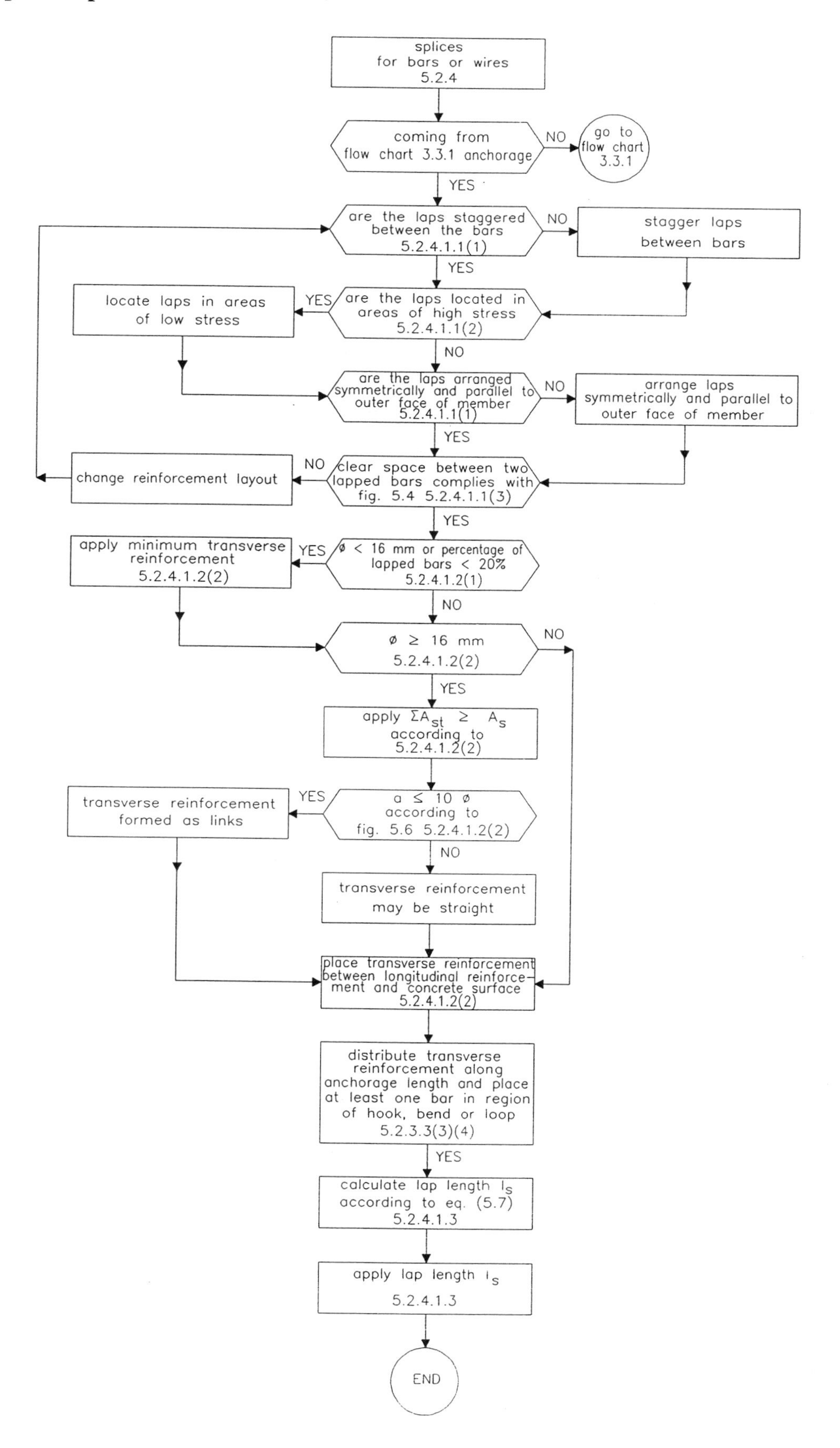

Flow chart 3.3.2.2

Splices: splices for welded mesh fabrics

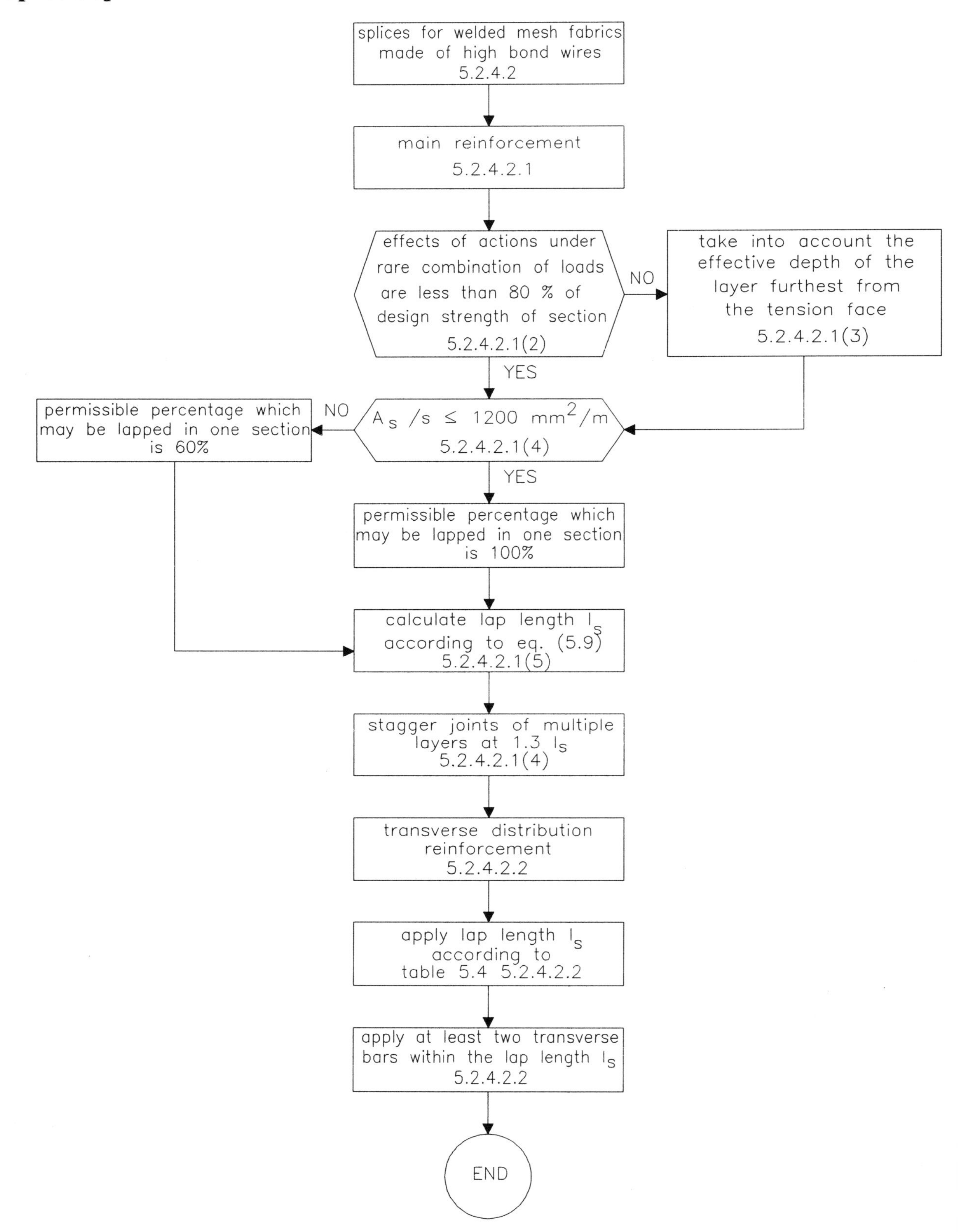

4 Design requirements

Throughout the following, the numbers on the right refer to relevant clauses of EC2 and square brackets - [ ] - in these references refer to relevant formulae in EC2. Please note, however, that square brackets in text indicate boxed values in the appropriate NAD.

4.1 Combinations of actions

Ultimate limit states 2.3.2.2

Fundamental combinations

$$\sum(\gamma_{G,j} G_{k,j}) + \gamma_{Q,1} Q_{k,1} + \sum_{i>1}(\gamma_{Q,i} \psi_{0,i} Q_{k,i}) \qquad [2.7(a)]$$

Accidental combinations

$$\sum(\gamma_{GA,j} G_{k,j}) + A_d + \psi_{1,1} Q_{k,1} + \sum_{i>1}(\psi_{2,i} Q_{k,i}) \qquad [2.7(b)]$$

$G_{k,j}$ = characteristic values of permanent actions
$Q_{k,1}$ = characteristic value of one of the variable actions
$Q_{k,i}$ = characteristic values of the other variable actions
A_d = design value (specified value) of the accidental actions
$\gamma_{G,j}$ = partial safety factors for any permanent action j
$\gamma_{GA,j}$ as $\gamma_{G,j}$ but for accidental design situations
$\gamma_{Q,i}$ = partial safety factors for any variable action i
ψ_0, ψ_1, ψ_2 combination coefficients to determine the combination, frequent and quasi-permanent values of variable actions

In expressions [2.7(a)] and [2.7(b)], prestressing shall be introduced where relevant.

Simplified method for fundamental combinations 2.3.3.1(8)

One variable action

$$\sum(\gamma_{G,j} G_{k,j}) + [1.5]\, Q_{k,1} \qquad [2.8(a)]$$

Two or more variable actions

$$\sum(\gamma_{G,j} G_{k,j}) + [1.35] \sum_{i\geq 1} Q_{k,i} \qquad [2.8(b)]$$

whichever gives the larger value

For the boxed values, apply the values given in the appropriate NAD.

Serviceability limit states 2.3.4

Rare combinations

$$\sum G_{k,j}\ (+P) + Q_{k,1} + \sum_{i>1}(\psi_{0,i} Q_{k,i}) \qquad [2.9(a)]$$

Frequent combinations

$$\sum G_{k,j}\ (+P) + \psi_{1,1} Q_{k,1} + \sum_{i>1}(\psi_{2,i} Q_{k,i}) \qquad [2.9(b)]$$

Quasi-permanent combinations

$$\sum G_{k,j} \ (+\ P) + \sum_{i\geq 1} (\psi_{2,i} Q_{k,i})$$ [2.9(c)]

P = prestressing force

Simplified method for rare combinations 2.3.4(6)

One variable action

$$\sum G_{k,j} + Q_{k,1}$$ [2.9(d)]

Two or more variable actions

$$\sum G_{k,j} + 0.9 \sum_{i\geq 1} Q_{k,i}$$ [2.9(e)]

whichever gives the larger value.

Permanent actions

Where the results of a verification may be very sensitive to variations of the magnitude of a permanent action from place to place in the structure, the unfavourable and the favourable parts of this action shall be considered as individual actions in ULS (2.3.2.3(3)).

For beams and slabs in buildings with cantilevers subjected to dominantly uniformly distributed loads, this requirement leads to the following decisive combinations of actions (see Figures 4.1 and 4.2):

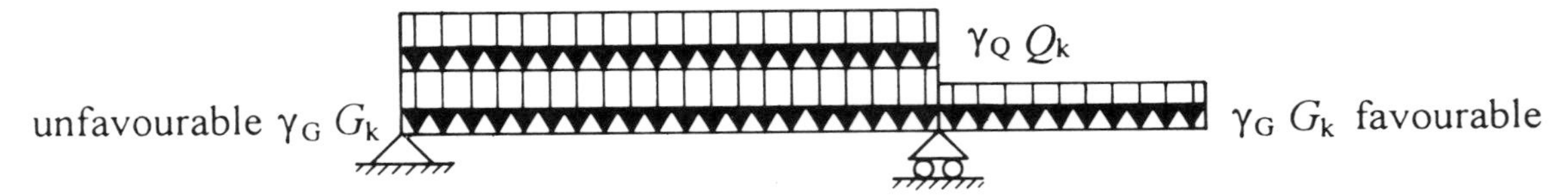

Figure 4.1 Maximum (positive) bending moment in middle of span and maximum shear at bearings of span.

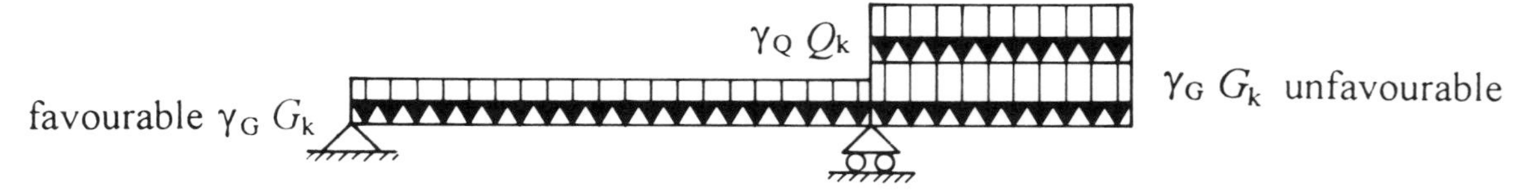

Figure 4.2 Minimum (positive or negative) bending moment in middle of span and maximum (negative) bending moment and maximum shear at bearing of cantilever.

For continuous beams and slabs in buildings without cantilevers subjected to dominantly uniformly distributed loads, it will generally be sufficient to consider only the two load cases in ULS (2.5.1.2(4)): alternate spans carrying the design variable and permanent loads ($\gamma_Q Q_k + \gamma_G G_k$), other spans carrying only the design permanent load ($\gamma_G G_k$) (2.5.1.2(4)(a)) (see Figure 4.3); any two adjacent spans carrying the design variable and permanent loads ($\gamma_Q Q_k + \gamma_G G_k$), other spans carrying only the design permanent load ($\gamma_G G_k$) (2.5.1.2(4)(b)) (see Figure 4.4).

4.4 Partial safety factors for actions

Table 4.3 Partial safety factors for actions (Eurocode 1, part 1 (ENV 1991-1:1993))

Case [1]	Action	Symbol	Situations	
			P/T	A
Case A Loss of static equilibrium; strength of structural material or ground insignificant (see 9.4.1)	Permanent actions: self-weight of structural and non-structural components, permanent actions caused by ground-water and free water - unfavourable - favourable	 γ_{Gsup} [2,4] γ_{Ginf} [2,4]	 [1.10] [0.90]	 [1.00] [1.00]
	Variable actions - unfavourable	γ_Q	[1.50]	[1.00]
	Accidental actions	γ_A		[1.00]
Case B[5] Failure of structure or structural elements, including those of the footing, piles, basement walls, etc., governed by strength of structural materials (see 9.4.1)	Permanent actions[6] (see above) - unfavourable - favourable	 γ_{Gsup} [3,4] γ_{Ginf} [3,4]	 [1.35] [1.00]	 [1.00] [1.00]
	Variable actions - unfavourable	γ_Q	[1.50]	[1.00]
	Accidental actions	γ_A		[1.00]
Case C[5] Failure in the ground	Permanent actions (see above) - unfavourable - favourable	 γ_{Gsup} [4] γ_{Ginf} [4]	 [1.00] [1.00]	 [1.00] [1.00]
	Variable actions - unfavourable	γ_Q	[1.30]	[1.00]
	Accidental actions	γ_A		[1.00]

P: Persistent situation   T: Transient situation   A: Accidental situation

NOTES

1. The design should be separately verified for each case A, B and C as relevant.
2. In this verification, the characteristic value of the unfavourable part of the permanent action is multiplied by the factor 1.1 and the favourable part by 0.9. More refined rules are given in ENV 1993 and ENV 1994.
3. In this verification, the characteristic values of all permanent actions from one source are multiplied by 1.35 if the total effect of the resulting action is unfavourable and by 1.0 if the total effect of the resulting action is favourable.
4. When the limit state is sensitive to variations of permanent actions, the upper and lower characteristic values of these actions should be taken according to 4.2 (3).
5. For cases B and C, the design ground properties may be different: see ENV 1997-1-1.
6. Instead of using γ_G (1.35) and γ_Q (1.50) for lateral earth pressure actions, the design ground properties may be introduced in accordance with ENV 1997 and a model factor γ_{Sd} applied.

For the boxed values, apply the values given in the appropriate NAD.

Table 4.4 Partial safety factors for actions (Eurocode 2, part 1 (ENV 1992-1-1:1991))

	Permanent actions (γ_G)	Variable actions (γ_Q)		Prestressing (γ_P)
		One with its characteristic value	Others with their combination value	
Favourable effect	[1.00]	-	-	[0.9] or [1.0]
Unfavourable effect	[1.35]	[1.50]	[1.50]	[1.2] or [1.0]

4.5 Partial safety factors for materials

Table 4.5 Partial safety factors for materials (Eurocode 2, part 1 (ENV 1992-1-1:1991))

Combination	Concrete (γ_c)	Steel reinforcement or prestressing tendons (γ_s)
Fundamental	[1.50]	[1.15]
Accidental (except earthquakes)	[1.30]	[1.00]

For the boxed values, apply the values given in the appropriate NAD.

5 Calculation methods

5.1 Flat slabs

5.1.1 Introduction

Slabs are classified as flat slabs when they transfer loads to columns directly without any beam supports. Slabs may be solid or coffered (ribbed in two directions). Unlike two-way spanning slabs, flat slabs can fail by yield lines in either of the two orthogonal directions. Flat slabs should therefore be designed to carry the total load on the panel in each direction.

EC2 does not provide any specific guidance for the analysis of the flat slabs. The methods given are based on common practice in a number of countries in Europe. General methods of analysis include: (a) equivalent frame method; (b) use of simplified coefficients; (c) yield-line analysis; and (d) grillage analysis.

5.1.2 Equivalent frame method

The structure is divided in two orthogonal directions into frames consisting of columns and strips of slab acting as "beams". The width of the slab to be used as "beams" is determined as follows:

For vertical loading,
when $l_y < 2l_x$,

width in x-direction $= 0.5(l_{x1} + l_{x2})$
width in y-direction $= 0.5(l_{x1} + l_{y2})$

when $l_y > 2l_x$,

width in x-direction $= 0.5\ (l_{x1} + l_{x2})$
width in y-direction $= (l_{x1} + l_{x2})$

In these expressions, l_x and l_y are the shorter and longer spans respectively and l_{x1} and l_{x2} refer to the lengths of adjacent spans in x-direction. The stiffness of the "beams" for analysis should be based on the widths calculated above. When the loading is horizontal, the stiffness used in analysis should be taken as half that derived for vertical loading, to allow for uncertainties associated with the slab-column joints.

Analysis

A braced structure may be analysed using any of the standard linear elastic methods such as moment distribution method. The structure may be analysed as a whole or split into sub-frames consisting of the slab at any one level and the columns. The remote ends of the columns are normally treated as fixed unless they are obviously not.

Lateral distribution of moments

The slab should be divided into column and middle strips as shown in Figure 5.1. The slab bending moments obtained from analysis should be apportioned across the width of the slab as follows:

	Column strip	Middle strip
Negative moments	75%	25%
Positive moments	55%	45%

These figures are percentages of the total positive or negative moments obtained in analysis. Where the width of the column strip is taken as equal to that of a drop and thereby the width of the middle strip is increased, the design moments to be resisted by the middle strip should be increased in proportion to the increased width. The design moments in the column strip may be reduced accordingly.

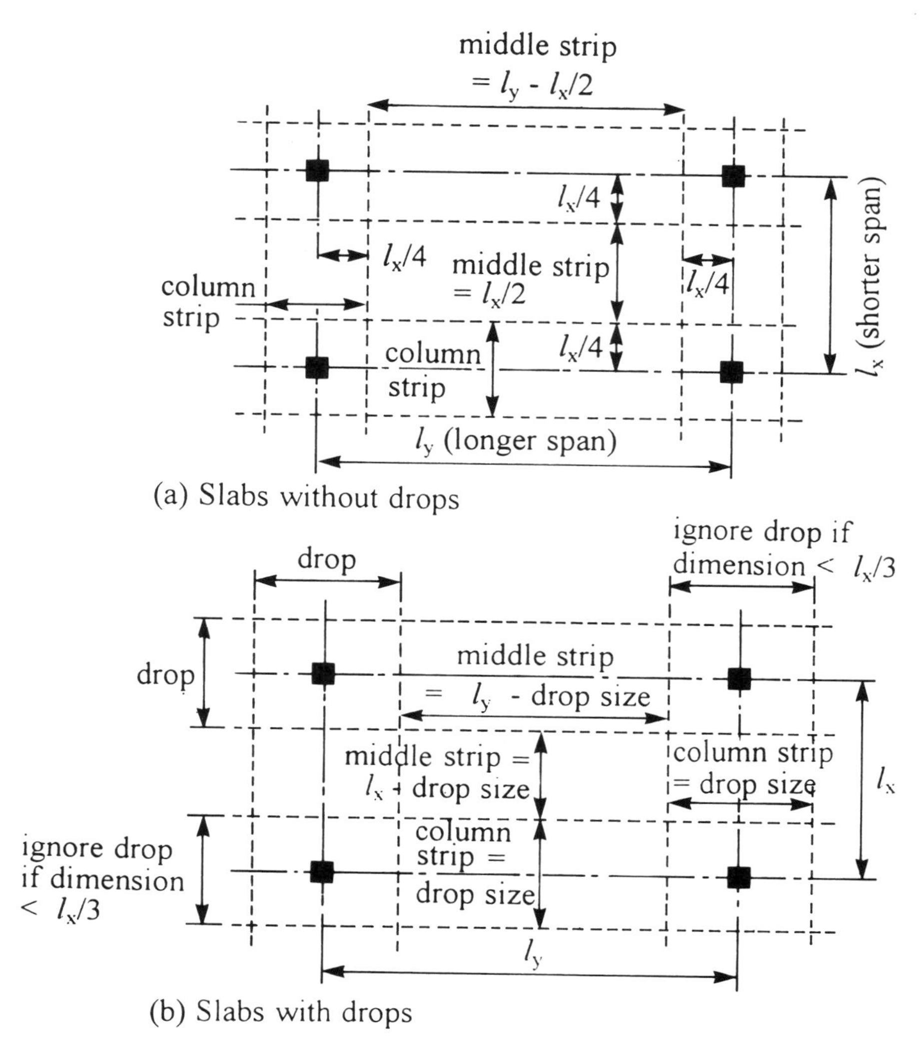

Figure 5.1: Division of slab into strips.

Moment transfer at edge columns

The effective width to the slab through which moments are transferred between the edge (or corner) columns and slab should be calculated as shown in Figure 5.2. The maximum moment that can be transferred to the column is

$M_{max} = 0.167 b_e d^2 f_{ck}$ for concrete grades C35/45 or less;

$M_{max} = 0.136 b_e d^2 f_{ck}$ for concrete grades C40/50 or greater.

The structure should be sized so that M_{max} is at least 50% of the moment obtained from an elastic analysis.

When the bending moment at the outer support obtained from the analysis exceeds M_{max}, the moment at this support should be limited to M_{max} and the moment in the span should be increased accordingly.

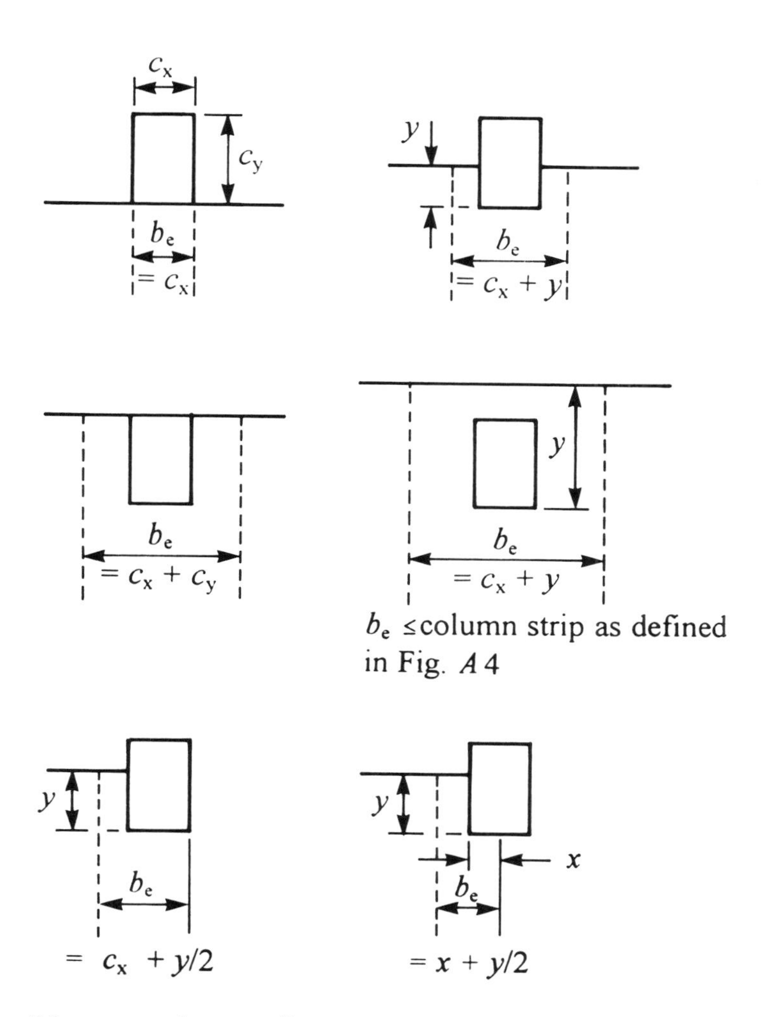

Figure 5.2

5.1.3 Use of simplified coefficients

Bending moments using the coefficients given below may be used for flat slabs where:
(a) the structure consists of at least three spans; and
(b) the ratio of the longest to the shortest span does not exceed 1.2; and
(c) the loading is predominantly uniformly distributed

At outer support	Near middle of end span	At first interior support	At middle of interior spans	At interior supports
0	$0.09Fl$	$0.11Fl$	$0.07Fl$	$0.10Fl$

NOTES
l is the effective span. F is the total ultimate load on the span $= 1.35G_k + 1.5Q_k$. No redistribution should be carried out on the moments.

5.1.4 Reinforcement

Reinforcement should be sufficient to resist the minimum bending moment specified in Table 4.9 of EC2. The reinforcement required in each column and middle strip should be distributed uniformly. In slabs without drops, the reinforcement required to resist the negative moment in the column strips should be placed with 66% of the reinforcement within the middle half of the strip.

5.2 Strut-and-tie models

Strut-and-tie models may be used for structural analysis, where the assumption of linear strain distribution through the structure is not valid. This powerful plastic method is useful in a number of instances, including anchorage zones of prestressed members, members with holes, pile caps, deep beams and beam-column junctions. Typical models are shown in Figure 5.3.

The structure is divided into struts (concrete) and ties (reinforcement bars). The model should reflect closely the elastic stress trajectories. In general, the angle between the struts and ties should not be less than 30°. Internal stresses are calculated so that equilibrium with external loads is achieved.

Limiting permissible stresses are as follows.

Reinforcement ties	f_{yd}
Struts under uniaxial stress	$0.6f_{cd}$
Struts under triaxial stress	$1.0f_{cd}$

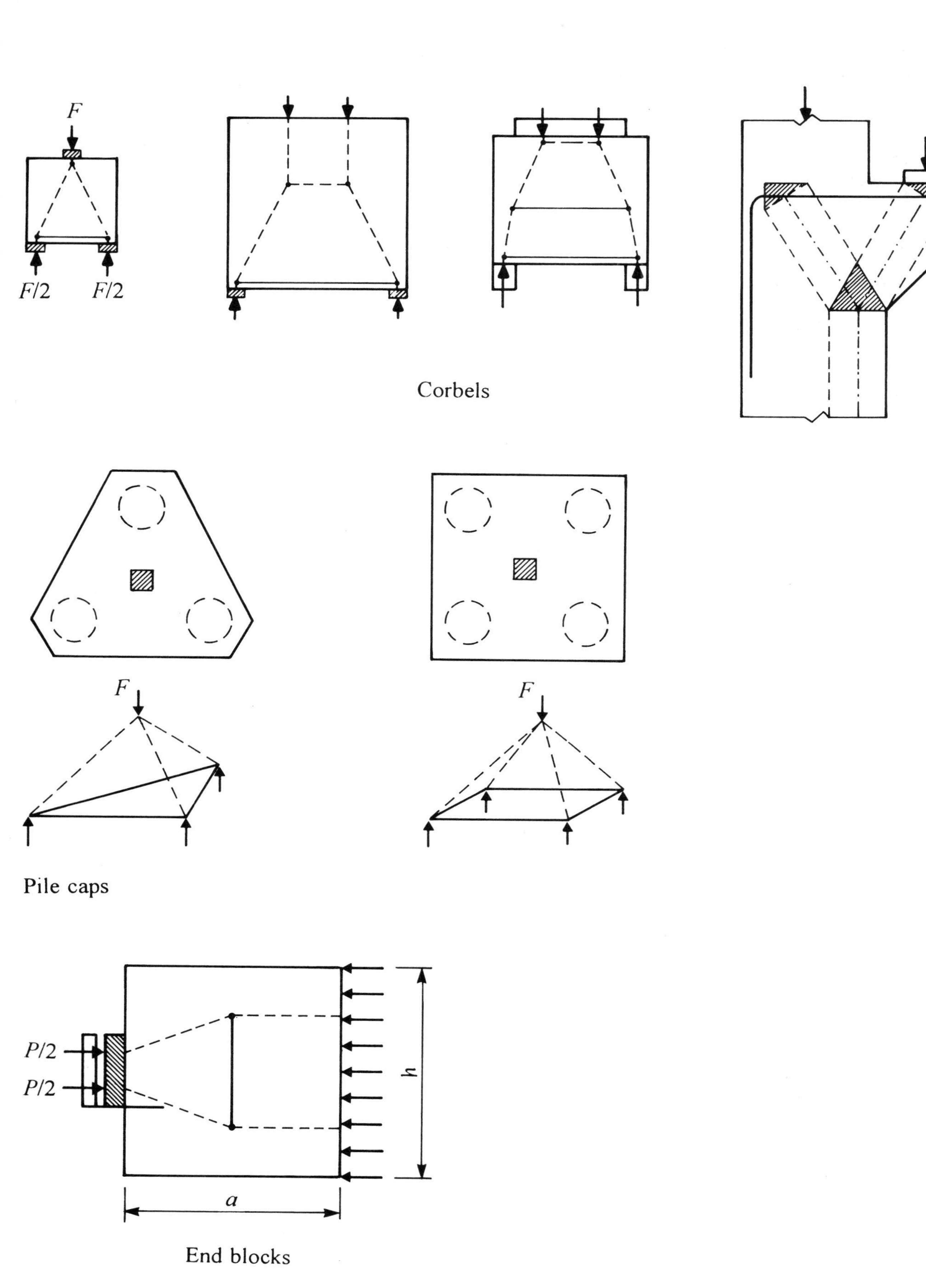

Figure 5.3 Typical strut-and-tie models.

6 Material properties

6.1 Concrete

Material properties of concrete (Eurocode 2, part 1 (ENV 1992-1-1:1993))

Strength class	f_{ck}	f_{cm} (1)	f_{cd}	$\alpha f_{ck}/\gamma_c$ (2)	f_{ctm}	$f_{ctk\,0.05}$	$f_{ctk\,0.95}$	τ_{Rd}	E_{cm} (1)	E_{cd} (1)	ϵ_{cu} (1) (‰)	ϵ_{cu} (2) (‰)
	(N/mm^2)											
C12/15	12	20	8.0	6.4	1.6	1.1	2.0	0.18	26000	17300	3.6	3.5
C16/20	16	24	10.7	9.1	1.9	1.3	2.5	0.22	27500	18300	3.5	3.5
C20/25	20	28	13.3	11.3	2.2	1.5	2.9	0.26	29000	19300	3.4	3.5
C25/30	25	33	16.7	14.2	2.6	1.8	3.3	0.30	30500	20300	3.3	3.5
C30/37	30	38	20.0	17.0	2.9	2.0	3.8	0.34	32000	21300	3.2	3.5
C35/45	35	43	23.3	19.8	3.2	2.2	4.2	0.37	33500	22300	3.1	3.5
C40/50	40	48	26.7	22.7	3.5	2.5	4.6	0.41	35000	23300	3.0	3.5
C45/55	45	53	30.0	25.5	3.8	2.7	4.9	0.44	36000	24000	2.9	3.5
C50/60	50	58	33.3	28.3	4.1	2.9	5.3	0.48	37000	24700	2.8	3.5

NOTES
1. Structural analysis of sections with a rectangular compression zone; take into account f_{cm} and E_{cm} or f_{cd} and E_{cd}
2. Cross-section design

f_{ck} = characteristic compressive cylinder strength of concrete at 28 days in N/mm^2

f_{cm} = mean value of compressive cylinder strength of concrete at 28 days in N/mm^2
= f_{ck} + [8] (N/mm^2)

f_{cd} = design value of compressive cylinder strength of concrete at 28 days in N/mm^2
= f_{ck}/γ_c where γ_c = partial safety factor for concrete =[1.5]; if $\gamma_c \neq 1.5$, multiply by $1.5/\gamma_c$

$\alpha \frac{f_{ck}}{\gamma_c}$ = reduced design compressive cylinder strength of concrete at 28 days in N/mm^2

α = coefficient taking account of long-term effects on the compressive cylinder strength of concrete and of unfavourable effects resulting from the way the load is applied
= [0.85]; if $\alpha \neq 0.85$, multiply by $\alpha/0.85$

γ_c = [1.5]; if $\gamma_c \neq 1.5$, multiply by $1.5/\gamma_c$

f_{ctm} = mean value of the axial tensile strength of concrete at 28 days in N/mm^2

$f_{ctk\,0.05}$ = lower characteristic axial tensile strength (5%-fractile) of concrete at 28 days in N/mm^2
= $0.7 f_{ctm}$

$f_{ctk\,0.95}$ = upper characteristic axial tensile strength (95%-fractile) of concrete at 28 days in N/mm^2
= $1.3 f_{ctm}$

τ_{Rd} = basic design shear strength of concrete at 28 days in $N/mm^2 = \frac{025 f_{ctk\,0.05}}{\gamma_c}$
with γ_c = [1.5]; if $\gamma_c \neq 1.5$, multiply by $1.5/\gamma_c$

E_{cm} = mean value of secant modulus of elasticity of concrete in N/mm^2
= $9.5 * 10^3 (f_{ck} + 8)^{1/3}$

E_{cd} = design value of secant modulus of elasticity of concrete in N/mm^2 = E_{cd}/γ_c
with γ_c = [1.5]; if $\gamma_c \neq 1.5$, multiply by $1.5/\gamma_c$

ϵ_{cu} = ultimate compressive strain in the concrete in ‰

For the boxed values, apply the values given in the appropriate NAD.

6.2 Reinforcing steel

Material properties of reinforcing steel (Eurocode 2, part 1 (ENV 1992-1-1:1993) and ENV 10080:1994)

Steel name	f_{tk} (N/mm^2)	f_{td} (N/mm^2)	f_{yk} (N/mm^2)	f_{yd} (N/mm^2)	ϵ_{uk} (%)
B500A	525	455	500	435	2.5[(1)]
B500B	540	470	500	435	5.0

NOTES
1. 2.0% for bars with d = 5.0 and 5.5 mm, where d is diameter of bar in mm

f_{tk} = characteristic tensile strength of reinforcing steel in N/mm^2
f_{td} = design tensile strength of reinforcing steel in N/mm^2 = f_{tk}/γ_s
γ_s = partial safety factor for reinforcing steel = [1.15]; if $\gamma_s \neq 1.15$, multiply by $1.15/\gamma_s$
f_{yk} = characteristic yield stress of reinforcing steel in N/mm^2
f_{yd} = design yield stress of reinforcing steel in N/mm^2 = f_{yk}/γ_s with γ_s = [1.15]; if $\gamma_s \neq 1.15$, multiply by $1.15/\gamma_s$
$f_{0.2k}$ = characteristic 0.2 % proof-stress of reinforcing steel in N/mm^2
$f_{0.2d}$ = design 0.2% proof-stress of reinforcing steel in N/mm^2 = $f_{0.2k}/\gamma_s$
ϵ_{uk} = characteristic elongation of reinforcing steel at maximum load in %
$(f_t/f_y)_k$ = characteristic ratio of tensile strength to yield stress
E_s = modulus of elasticity of reinforcing steel $E_s = 2 * 10^5$ N/mm^2
Density = 7850 kg/m^3.
Coefficient of thermal expansion = 10^{-5}/°C

Bond characteristics

Ribbed bars: resulting in high bond action (as specified in EN 10080)
Plain, smooth bars: resulting in low bond action

Ductility characteristics

High ductility: $\epsilon_{uk} > [5.0]\%$ and $(f_t/f_y)_k > [1.08]$
Normal ductility: $\epsilon_{uk} > [2.5]\%$ and $(f_t/f_y)_k > [1.05]$

For the boxed values, apply the values given in the appropriate NAD.

6.3 Prestressing steel

Material properties of prestressing steel (Eurocode 2, part 1 (ENV 1992-1-1:1993) and ENV 10138:1994)

Wires

Steel name	f_{pk} (N/mm²)	f_{pd} (N/mm²)	$f_{p0.1k}$ (N/mm²)	$f_{p0.1d}$ (N/mm²)	E_s (N/mm²)	ϵ_{uk} (%)
Y1860C	1860	1620	1600	1390	205000	3.5
Y1770C	1770	1540	1520	1320	205000	3.5
Y1670C	1670	1450	1440	1250	205000	3.5
Y1570C	1570	1370	1300	1130	205000	3.5

Strands

Steel name	f_{pk} (N/mm²)	f_{pd} (N/mm²)	$f_{p0.1k}$ (N/mm²)	$f_{p0.1d}$ (N/mm²)	E_s (N/mm²)	ϵ_{uk} (%)
Y2060S	2060	1790	1770	1540	195000	3.5
Y1960S	1960	1700	1680	1460	195000	3.5
Y1860S	1860	1620	1600	1639	195000	3.5
Y1770S	1770	1540	1520	1250	195000	3.5

Bars

Steel name	f_{pk} (N/mm²)	f_{pd} (N/mm²)	$f_{p0.1k}$ (N/mm²)	$f_{p0.1d}$ (N/mm²)	E_s (N/mm²)	ϵ_{uk} (%)
Y1030	1030	900	830	720	205000	4.0
Y1100	1100	960	900	780	205000	4.0
Y1230	1230	1070	1080	940	205000	4.0

f_{pk} = characteristic tensile strength of prestressing steel in N/mm²
f_{pd} = design tensile strength of prestressing steel in N/mm² = f_{pk}/γ_s
γ_s = partial safety factor for prestressing steel = [1.15]; if $\gamma_s \neq 1.15$, multiply by $1.15/\gamma_s$
$f_{p0.1k}$ = characteristic 0.1 % proof-stress of prestressing steel in N/mm²
$f_{p0.1d}$ = design 0.1% proof-stress of prestressing steel in N/mm² = $f_{p0.1k}/\gamma_s$ with γ_s = [1.15]; if $\gamma_s \neq 1.15$, multiply by $1.15/\gamma_s$
ϵ_{uk} = characteristic elongation of prestressing steel at maximum load in %
E_s = modulus of elasticity of reinforcement $E_s = 2 * 10^5$ N/mm² (taken into account in stress-strain diagram)
Density = 7850 kg/m³
Coefficient of thermal expansion = 10^{-5}/°C

Classes of relaxation

Class 1: for wires and strands, high relaxation
Class 2: for wires and strands, low relaxation
Class 3: for bars

For the boxed values, apply the values given in the appropriate NAD.

7 Basic design

Table 7.1 Exposure classes

Exposure class		Examples of environmental conditions
1 Dry environment		Interior of dwellings or offices
2 Humid environment	(a) Without frost	Interior of buildings with high humidity, e.g. laundries Exterior components Components in non-aggressive soil and/or water
	(b) With frost	Exterior components exposed to frost Components in non-aggressive soil and/or water and exposed to frost Interior components where the humidity is high and exposed to frost
3 Humid environment with frost and de-icing agents		Interior and exterior components exposed to frost and de-icing agents
4 Seawater environment	(a) Without frost	Components completely or partially submerged in seawater or in the splash zone Components in saturated salt air (coastal area)
	(b) With frost	Components partially submerged in seawater or in the splash zone and exposed to frost Components in saturated salt air and exposed to frost
The following classes may occur alone or in combination with the above		
5 Aggressive chemical environment(2)	(a)	Slightly aggressive chemical environment (gas, liquid or solid) Aggressive industrial atmosphere
	(b)	Moderately aggressive chemical environment (gas, liquid or solid)
	(c)	Highly aggressive chemical environment (gas, liquid or solid)

NOTES

1. This exposure class is valid as long as, during construction, the structure or some of its components are not exposed to more severe conditions over a prolonged period
2. Chemically aggressive environments are classified in ISO 9690. The following exposure conditions may be used:
   Exposure class 5a: ISO classification A1G, A1L, A1S
   Exposure class 5b: ISO classification A2G, A2L, A2S
   Exposure class 5c: ISO classification A3G, A3L, A3S

Table 7.2 Minimum cover requirements for normal weight concrete

		Exposure class according to Table 7.1								
		1	2a	2b	3	4a	4b	5a	5b	5c
Minimum cover (mm)	Reinforcement	15	20	25	40	40	40	25	30	40
	Prestressing steel	25	30	35	50	50	50	35	40	50

NOTES

1. For slab elements, a reduction of 5 mm may be made for exposure classes 2-5.
2. A reduction of 5 mm may be made where concrete of strength class C40/50 and above is used for reinforced concrete in exposure classes 2a-5b and for prestressed concrete in exposure classes 1-5b. However, the minimum cover should never be less than that for class 1.
3. For exposure class 5c, a protective barrier should be used to prevent direct contact with aggressive media.

Table 7.8 Minimum dimensions for fire resistance of continuous reinforced concrete (normal weight) beams

Standard fire resistance (mm)	Possible combinations of the average axis distance a and the beam width b (both in mm)			Web thickness b_w of I-beams (mm)
R 30	$a=12$ $b=80$		$a=20$ $b=200$	80
R 60	$a=25$ $b=120$	$a=12$ $b=200$	$a=25$ $b=300$	100
R 90	$a=35$ $b=150$	$a=45$ $b=250$	$a=25$ $b=400$	100
R 120	$a=45$ $b=200$	$a=35$ $b=300$	$a=35$ $b=500$	120
R 180	$a=50$ $b=240$		$a=50$ $b=600$	140
R 240	$a=60$ $b=280$		$a=60$ $b=700$	160
$a_{st} = a + 10$ mm (see note below)			$a_{st} = a$ (see note below)	
a_{st} = increased axis distance of the outermost bar (tendon, wire) from the side surface of the cross-section, where steel is in a single layer				

NOTES

1. For prestressed members, the axis distances should be increased by 10 mm for prestressing bars and by 5 mm for wires or strands.
2. The table applies to beams exposed to fire on three sides.
3. For beams exposed to fire on all four sides, the height should at least equal the minimum dimension b_{min} in the table for the required fire resistance and its cross-sectional area should be at least $2b_{min}^2$.
4. The minimum axis distance to any individual bars should not be less than that required for R 30 in the table nor less than half the average axis distance.
5. For R 90 and above, the top reinforcement over each intermediate support should extend at least $0.3l_{eff}$ from the centre of support, where the effective span $l_{eff} > 4$ metres and $l_{eff}/h > 20$, h being the beam depth. In other cases, this minimum may be reduced to $0.15l_{eff}$.
6. If the above detailing requirement is not met and the moment redistribution in the analysis exceeds 15%, each span of the continuous beam should be assessed as a simply supported beam.
7. In a continuous I-beam, b_w should not be less than b for a distance of $2h$ from an intermediate support unless a check for explosive spalling is carried out.
8. In two-span I-beam systems with no rotational restraint at the end, with predominantly concentrated loading with M_{sd}/V_{sd} between 2.5 and 3, and with $V_{sd} > \frac{2}{3}V_{rd2}$, the minimum width of the beam web between the concentrated loads should be: 220 mm for R 120, 400 mm for R 180 and 600 mm for R 240.

Table 7.9 Minimum dimensions for fire resistance for solid (normal weight) reinforced concrete slabs spanning one and two ways

Standard fire resistance	Slab thickness h_s (mm)	Average axis distance span a (mm)		
		One way	Two way	
			$l_y/l_x < 1.5$	$1.5 < l_y/l_x < 2$
REI 30	60	10	10	10
REI 60	80	20	10	15
REI 90	100	30	15	20
REI 120	120	40	20	25
REI 180	150	55	30	40
REI 240	175	65	40	50
l_x and l_y are the spans of a two-way slab (two directions at right-angles) where l_y is the longer span				

NOTES

1. For prestressed members, the axis distances should be increased by 10 mm for prestressing bars and by 15 mm for wires or strands.
2. The minimum cover to any bar should not be less than half the average axis distance.
3. The table values of axis distance for two-way slabs apply to slabs supported on all four edges. For all other support conditions, the values for one-way slabs should be used.
4. The table values of slab thickness and cover for two-way slabs with $l_y/l_x < 1.5$ should be used.
5. For R 90 and above, the top reinforcement over each intermediate support should extend at least $0.3l_{eff}$ from the centre of support, where the effective span $l_{eff} > 4$ metres and $l_{eff}/h > 20$, h being the beam depth. In other cases, this minimum may be reduced to $0.15l_{eff}$.
6. If the above detailing requirement is not met and the moment redistribution in the analysis exceeds 15%, each span of the continuous slab should be assessed as a simply supported slab.
7. Minimum top reinforcement of $0.005A_c$ should be used over intermediate supports when the reinforcement has "normal" ductility, when there is not rotational restraint at ends of two-span slabs, and when transverse redistribution of load effects cannot occur.

Table 7.10 Minimum dimensions for fire resistance of reinforced and prestressed (normal weight) concrete slabs

Standard fire resistance	Slab thickness h_s (mm), excluding finishes	Axis distance a (mm)
REI 30	150	10
REI 60	200	15
REI 90	200	25
REI 120	200	35
REI 180	200	45
REI 240	200	50

NOTES

1. For prestressed members, the axis distances should be increased by 10 mm for prestressing bars and by 15 mm for wires or strands.
2. It is assumed that the moment redistribution in this analysis does not exceed 15%. If it does exceed 15%, the axis distances in this table should be replaced by those for one-way slabs.
3. Over intermediate supports in each direction, at least 20% of the total top reinforcement calculated for cold design should extend over the full span, in the column strips.

8 Bending and longitudinal force

8.1 Conditions at failure

Figure 8.1 (taken from 4.11 in EC2) shows the strain conditions assumed at the ultimate limit state for reinforced concrete.

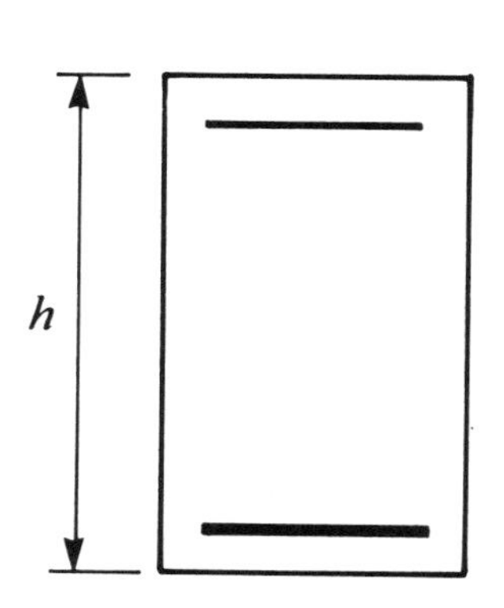

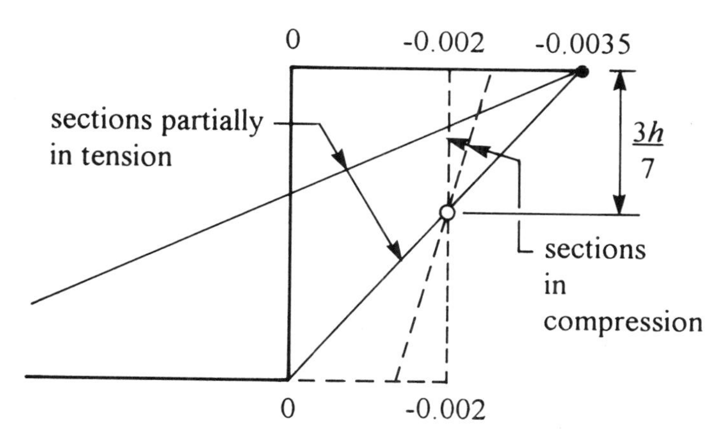

For cross-sections not fully in compression, the concrete is assumed to fail in compression when the strain reaches 0.0035. The strain in the tension reinforcement need not be limited where a horizontal top branch is assumed for the reinforcement stress-strain curve.

For cross-sections that are completely in compression, the strain is limited to 0.002 at a height of $3/7h$ from the most compressed face.

The strains in the reinforcement at ultimate are given by the formulae in Table 8.1.

Table 8.1 Strains in reinforcement at ultimate

(A) $x \leq h$

Compression reinforcement

$$\epsilon_s' = \frac{0.0035}{x}(x - d')$$

Reinforcement near tension or least compressed face

(B) $x > h$

Reinforcement near most compressed face

$$\epsilon_s = -\frac{0.002}{(x - 3h/7)}(x - d')$$

Reinforcement near least compressed face

$$\epsilon_s = \frac{0.002}{(x - 3h/7)}(d - x)$$

In general, it is satisfactory to assume that the reinforcement near to the most compressed face is yielding but there are cases when this may not be so. Table 8.2 sets out the conditions for the reinforcement to be yielding, assuming a bilinear stress-strain diagram.

Table 8.2 Conditions for yield of reinforcement

(A) $x \le h$

Compression steel

$$f_{yd} \le 700(1 - d'/x) \quad \text{or} \quad \frac{d'}{x} \le 1 - f_{yd}/700$$

Tension steel

$$\frac{x}{d} \le \frac{1}{f_{yd}/700 + 1}$$

(B) $x > h$

Compression steel

$$\frac{(1 - d'/x)}{(1 - 3h/7x)} > \frac{f_{yd}}{900} \quad \text{or} \quad \frac{x}{h} \le \frac{\left(\dfrac{3f_{yd}}{2800} - d'/h\right)}{\left(\dfrac{f_{yd}}{400} - 1\right)}$$

8.2 Design of rectangular sections subject to flexure only

I <u>Derivation of equations</u>

Stress-strain curves for reinforcement and concrete.

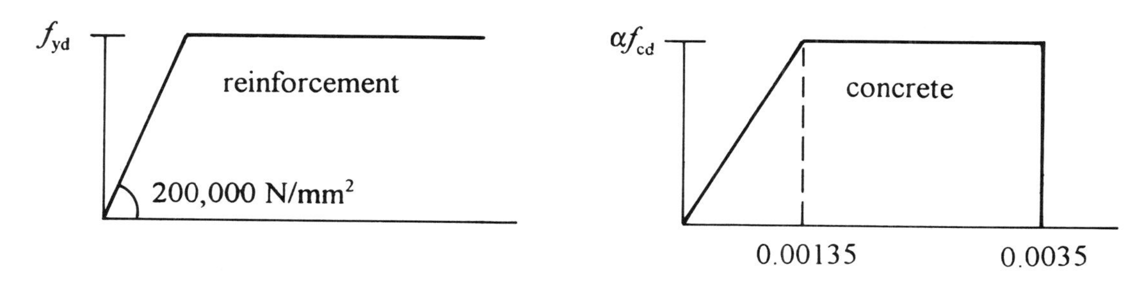

Conditions in section at ultimate in a singly reinforced section.

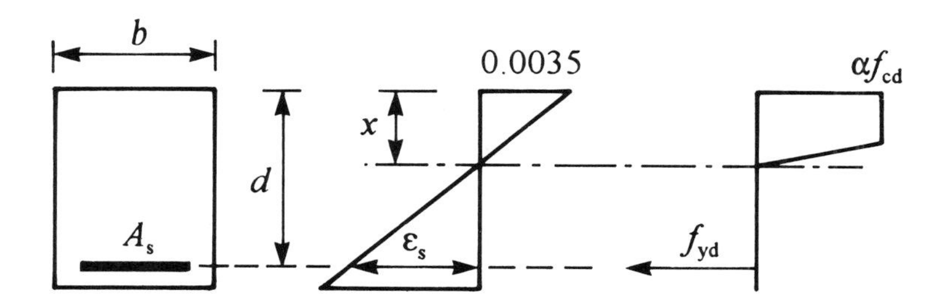

The limits to x/d will ensure that stress in steel is at yield. The average stress in compression zone is $0.807f_{cd}$. Distance from compression face to centre of concrete compression is $0.411x$.

Equilibrium of axial forces gives:

$$\frac{x}{d} = \frac{1.239 A_s f_{yd}}{\alpha f_{cd} b d}$$

Defining $\omega = \dfrac{A_s f_{yd}}{\alpha f_{cd} b d}$

$$\frac{x}{d} = 1.239\omega \qquad \text{I}$$

The lever arm, z, is given by:

$$\frac{z}{d} = 1 - 0.411\left(\frac{x}{d}\right) \qquad \text{IIa}$$

or

$$\frac{z}{d} = 1 - 0.5092\omega \qquad \text{IIb}$$

The moment is given by:

$$M = A_s f_{yd} z \qquad \text{IIIa}$$

hence $\dfrac{M}{bd^2 \alpha f_{cd}} = \omega(1 - 0.5092\omega)$

Defining

$m = \dfrac{M}{bd^2 \alpha f_{cd}}$ and solving for ω gives:

$$\omega = (1 - \sqrt{1 - 2.0368m})/1.0184 \qquad \text{IIIb}$$

or, approximately, $\omega = 1 - \sqrt{1 - 2m}$

Equation 2.17 in Eurocode 2 can be rewritten to give:

for $f_{cu} \leq 35$ $\quad \left(\dfrac{x}{d}\right)_{\text{lim}} = (\delta - 0.44)/1.25 \qquad$ IVa

for $f_{cu} > 35$ $\quad \left(\dfrac{x}{d}\right)_{\text{lim}} = (\delta - 0.56)/1.25 \qquad$ IVb

From I

$$\omega_{\text{lim}} = 0.807\left(\frac{x}{d}\right)_{\text{lim}} \qquad \text{V}$$

From III(a) and II(a)

$$m_{\text{lim}} = \left(\frac{x}{d}\right)_{\text{lim}}\left(0.411\left(\frac{x}{d}\right)_{\text{lim}}\right) \qquad \text{VI}$$

If $m > m_{\text{lim}}$, compression steel is needed to maintain the neutral axis at the limiting value. The moment capacity can then be calculated by assuming two superimposed sections.

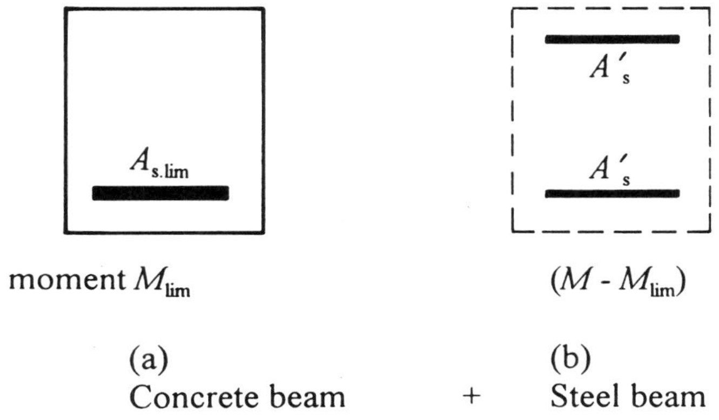

$A_{\text{s.lim}}$ A'_s A'_s

moment M_{lim} $(M - M_{\text{lim}})$

(a) Concrete beam + (b) Steel beam

The steel area required in the 'steel beam' is given by:

$$A_s = \frac{(M - M_{\text{lim}})}{f_{\text{yd}}(d - d')} \qquad \text{VIIa}$$

(Assuming reinforcement in compression is yielding)

or $$\omega = \frac{A_s f_{\text{yd}}}{\alpha f_{\text{cd}}} = \frac{(m - m')}{(1 - d'/d)} \qquad \text{VIIb}$$

The area of steel required for the 'concrete beam' is given by equation V.

Hence, total areas of reinforcement are given by:

$$\omega = \frac{(m - m')}{(1 - d'/d)}$$

and $$\omega = \omega_{\text{lim}} + \omega' \qquad \text{VIII}$$

The procedure for using these equations directly for calculating reinforcement areas is summarized below in Table 8.3.

Table 8.3 Design of rectangular beams

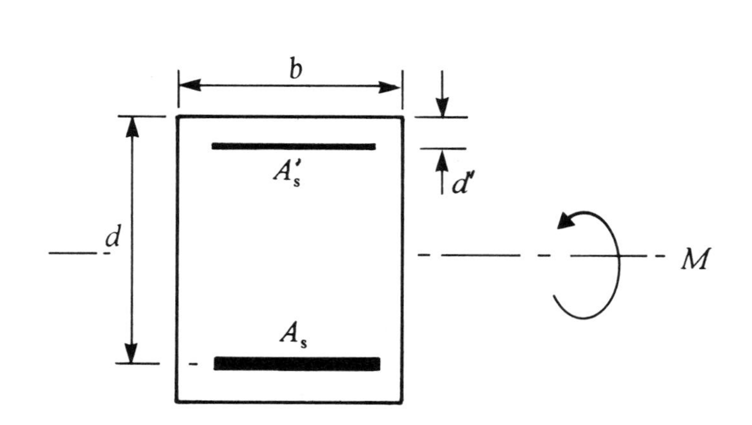

1. Calculate $m = \dfrac{M}{bd^2 \alpha f_{cd}}$

2. Calculate $\left(\dfrac{x}{d}\right)_{lim}$:

   if $f_{cu} \leq 35$, $\left(\dfrac{x}{d}\right)_{lim} = (\delta - 0.44)/1.25$

   if $f_{cu} > 35$, $\left(\dfrac{x}{d}\right)_{lim} = (\delta - 0.56)/1.25$

3. Calculate $M_{lim} = \left(\dfrac{x}{d}\right)_{lim} (1 - 0.411\left(\dfrac{x}{d}\right)_{lim})$

4. If $m < m_{lim}$, simply reinforced beam will suffice

   $$\omega = 1 - \sqrt{1 - 2m}$$

   Hence calculate A_s - END

5. If $m > m_{lim}$

   $$\omega' = \frac{(m - m_{lim})}{(1 - d'/d)}$$

   $$\omega = 0.807\left(\frac{x}{d}\right)_{lim} + \omega'$$

   Hence calculate A_s and A_s - END

Design Tables

The equations can be presented as design tables as shown below.

Table 8.4 gives values of x/d and ω for singly reinforced beams as a function of m.

Table 8.4 Values of x/d and ω for singly reinforced beams

$\frac{M}{bd^2\alpha f_{cd}}$	$\frac{A_s f_{yd}}{bd\alpha f_{cd}}$	$\frac{x}{d}$
0.01	0.010	0.012
0.02	0.020	0.025
0.03	0.030	0.038
0.04	0.041	0.052
0.05	0.051	0.064
0.06	0.062	0.077
0.07	0.073	0.090
0.08	0.084	0.104
0.09	0.095	0.117
0.1	0.106	0.131
0.11	0.117	0.145
0.12	0.128	0.159
0.13	0.140	0.173
0.14	0.152	0.188
0.15	0.164	0.203
0.16	0.176	0.218

$\frac{M}{bd^2\alpha f_{cd}}$	$\frac{A_s f_{yd}}{bd\alpha f_{cd}}$	$\frac{x}{d}$
0.17	0.188	0.233
0.18	0.200	0.248
0.19	0.213	0.264
0.2	0.226	0.280
0.21	0.239	0.296
0.22	0.252	0.313
0.23	0.266	0.330
0.24	0.280	0.347
0.25	0.294	0.364
0.26	0.308	0.382
0.27	0.323	0.400
0.28	0.338	0.419
0.29	0.354	0.438
0.3	0.370	0.458
0.31	0.386	0.478
0.32	0.402	0.499

Table 8.5 gives $\left(\frac{x}{d}\right)_{lim}$, ω_{lim} and m_{lim} as a function of the amount of redistribution.

Table 8.5 Limiting values of $\left(\frac{x}{d}\right)$, $\frac{M}{bd^2\alpha f_{ck}}$ and $\frac{A_s f_{yd}}{bd\alpha f_{ck}}$

Percentage redistribution	δ	$\left(\frac{x}{d}\right)_{lim}$		$\left(\frac{M}{bd^2\alpha f_{ck}}\right)_{lim}$		$\left(\frac{A_s f_{yd}}{bd\alpha f_{ck}}\right)_{lim}$	
		$f_{ck} \leq 35$	$f_{ck} > 35$	$f_{ck} \leq 35$	$f_{ck} > 35$	$f_{ck} \leq 35$	$f_{ck} > 35$
0	1.00	0.448	0.352	0.295	0.243	0.362	0.284
5	0.95	0.408	0.312	0.274	0.220	0.329	0.252
10	0.90	0.368	0.272	0.252	0.195	0.267	0.220
15	0.85	0.328	0.232	0.229	0.169	0.265	0.187
20	0.80	0.288	0.192	0.205	0.143	0.232	0.155
25	0.75	0.248	0.152	0.180	0.115	0.200	0.123
30	0.70	0.208	0.112	0.154	0.086	0.168	0.090

Tables 8.4 and 8.5 can be used to streamline the procedure set out in Table 8.3.

Flanged beams

Since concrete in tension is ignored, the design of a flanged beam is identical to that for a rectangular beam provided that the neutral axis at failure lies within the flange.

Thus the procedure for design can be:

1. Follow steps 1 to 4 in Table 8.3 using the overall flange breadth as b.
2. Calculate $\left(\frac{x}{d}\right) = \omega/0.807 \not> \left(\frac{x}{d}\right)_{lim}$

   If $\left(\frac{x}{d}\right) < \left(\frac{h_f}{d}\right)$, design is OK. This will normally be the case.

   If $\left(\frac{x}{d}\right) > \left(\frac{h_f}{d}\right)$, then further equations need to be derived. This can most easily be achieved by considering the base to be made up of two parts as shown below:

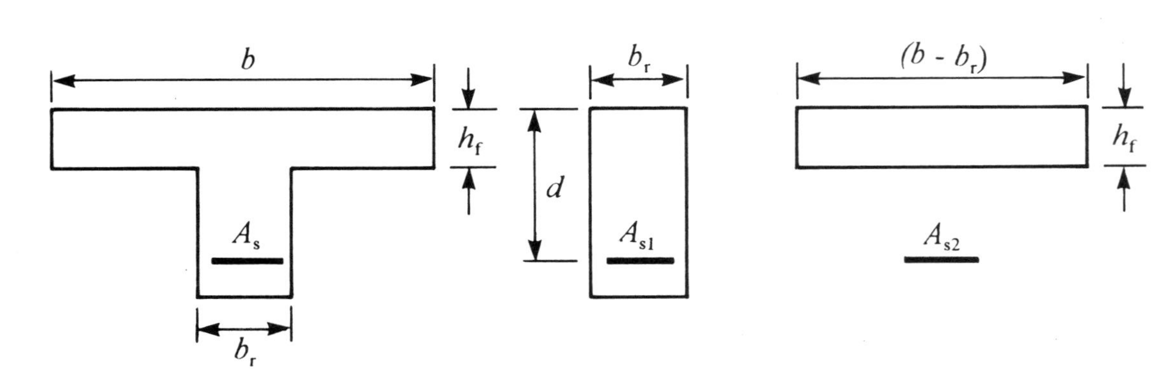

It will be assumed that the neutral axis is large enough for the whole flange to be at a stress of αf_{cd}. Hence, by equilibrium,

$$A_{s2} = (b - b_r)h_f \alpha f_{cd} / f_{yd} \qquad \text{IX}$$

$$M_2 = A_{s2} f_{yd} (d - h_f/2) \qquad \text{X}$$

The steel area required for the rectangular rib can now be obtained by using Table 8.3 to assess the reinforcement area needed for a rectangular beam of breadth b_r to support a moment of $M_1 = (M - M_2)$.

Although very unlikely to be exceeded, the limiting moment for a flanged beam where $(x/d)_{lim}$ exceeds (h_f/d) is given by:

$$M_{lim} = M_{1.lim} + b_{hf} \alpha f_{cd} (d - h_f/2)$$

$$\left[\left(\frac{x}{d}\right)_{lim}\left[1 - 0.0411\left(\frac{x}{d}\right)_{lim}\right]b_r d^2 + bh_f\left[d - \left(\frac{h_f}{z}\right)\right]\right]\alpha f_{cd}$$

The required steel areas can then be calculated using Equations VIIIa, XI, X and V.

The procedure for the design of flanged sections is summarized in Table 8.6.

Table 8.6 Design of flanged sections for flexure

1.	Calculate $$m = \frac{M}{bd\alpha f_{cd}}$$
2.	Follow Table 8.3 to obtain ϖ. Calculate $\frac{x}{d} = \frac{\omega}{0.807} \ngtr \left(\frac{x}{d}\right)_{lim}$ If $\left(\frac{x}{d}\right) \leq \left(\frac{h_f}{d}\right)$, calculate A_s from ω (END)
3.	If $\left(\frac{x}{d}\right) > \left(\frac{h_f}{d}\right)$ Calculate $A_{s2} = (b - b_r)h_f \alpha f_{cd} / f_{yd}$ $M_2 = A_{s2} f_{yd} (d - h_f/2)$
4.	Use Table 8.3 to calculate steel areas for rectangular sections of breadth b_r to resist moment of $(M - M_2)$.
5.	Areas of steel = sum of those obtained from steps 3 and 4.

Minimum reinforcement

There are two provisions defining minimum areas of flexural steel. These are:

(a) minimum for crack control 4.4.2.2.
(b) overall minimum 5.4.2.1.1.

The formula in 4.4.2.2 is:

$$A_s \geq k_c k f_{ct.eff} A_{ct} / \sigma_s$$

where, for bending, $k_c = 0.4$

$f_{ct.eff}$ is suggested as 3, k is 0.8 for sections with depths not greater than 300 mm and 0.5 for sections deeper than 800 mm, σ_s may be taken as f_{yk}. A_{ct}, the area of concrete in the tension zone immediately before cracking, will be $bh/2$ for rectangular sections and an approximate value for flanged beams could be taken as 0.75 $b_t h$ where b_t is the breadth of the tension zone. If h is assumed to be $1.15d$, the above equation thus reduces to:

for rectangular beams	$h \leq 300$mm	$0.55bd/f_{yk}$
	$h \geq 800$mm	$0.34bd/f_{yk}$
for flanged beams	$h \leq 300$mm	$0.83b_t d/f_{yk}$
	$h \geq 800$mm	$0.55b_t d/f_{yk}$

Interpolation is permitted for depths between 300 and 800 mm.

Clause 5.4.2.1.1 gives:

$$A_s \geq \frac{0.6 b_t d}{f_{yk}} \geq 0.0015 b_t d$$

Assuming $f_{yk} > 400$, $0.0015b_t d$ will govern.

It will be seen, in any case, that the rule in 5.4.2.1.1 will always govern except for shallow flanged beams and, for commonly used reinforcement, the limit of 0.0015 $b_t d$ will be the controlling factor in 5.4.2.1.1. The following general rule therefore seems adequate for normal beams.

Table 8.7: Minimum tension reinforcement

If	$f_{yk} = 500$ N/mm^2
or	$f_{yk} < 500$ N/mm^2 and beam is either rectangular or flanged with $h < 700$ mm
then	$A_s \geq 0.001\, b_t d$
else	$A_s \geq 0.083 - \frac{(h - 300)}{1786} \frac{b_t d}{f_{yk}} \geq 0.0015 b_t d$

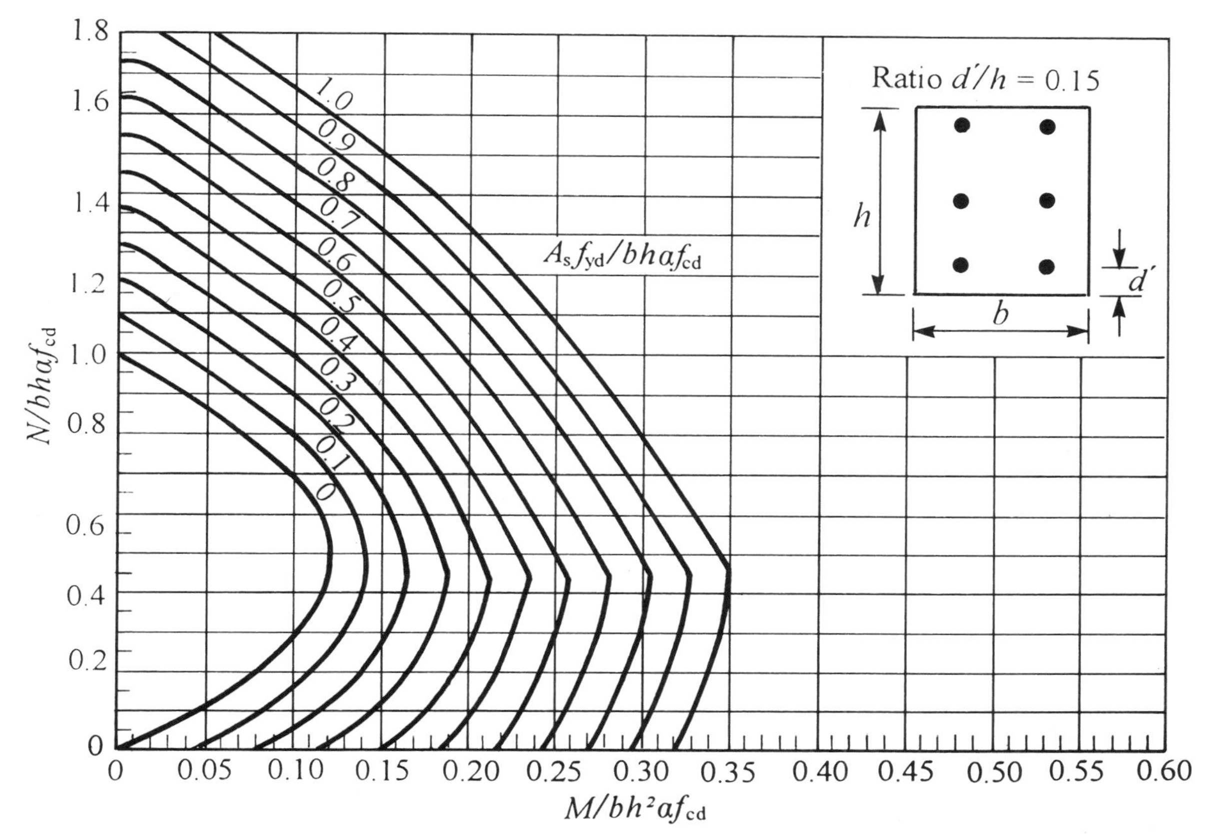

Chart 8.7

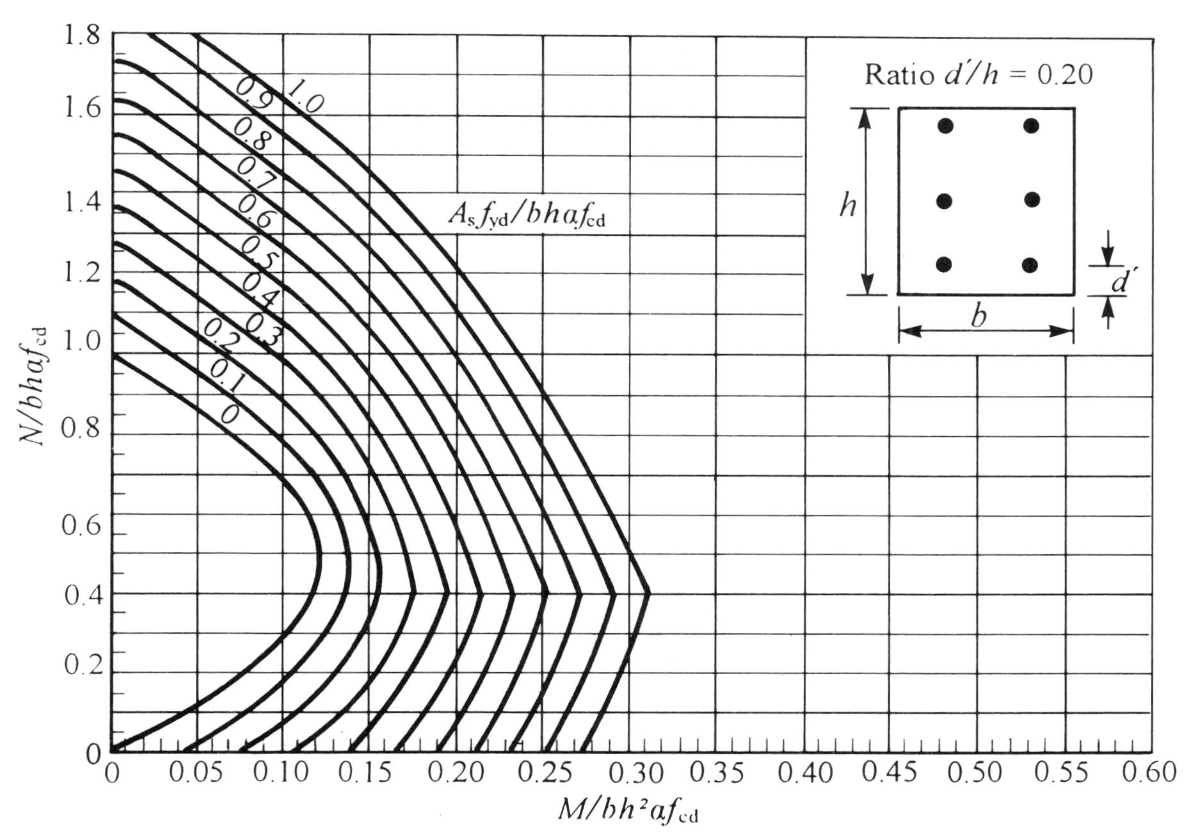

Chart 8.8

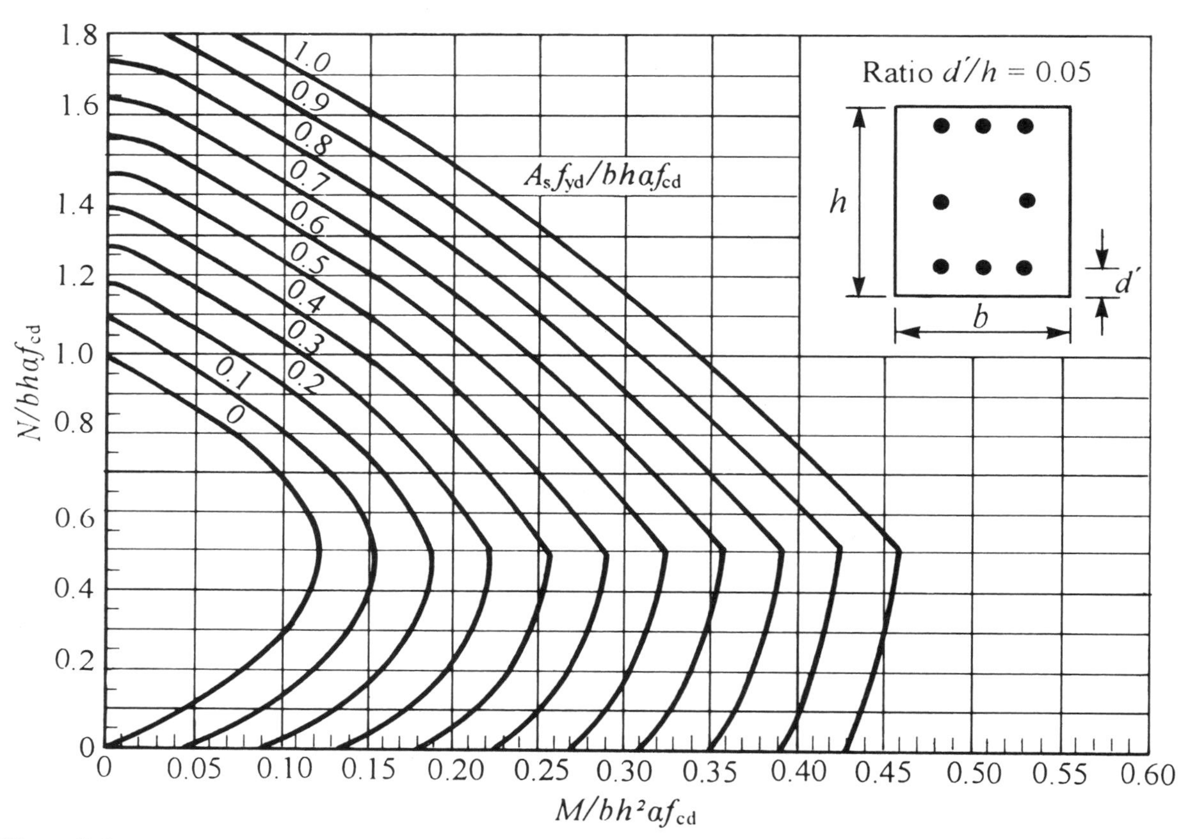

Chart 8.9

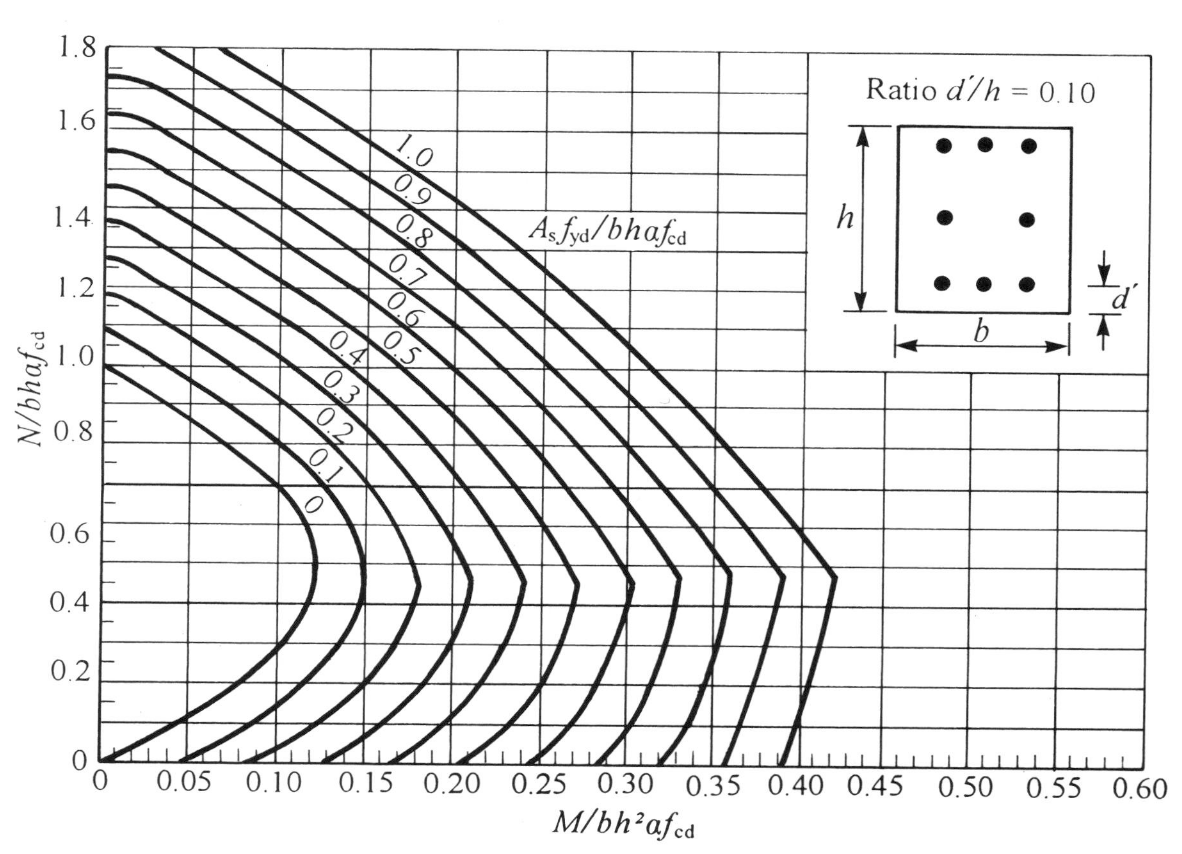

Chart 8.10

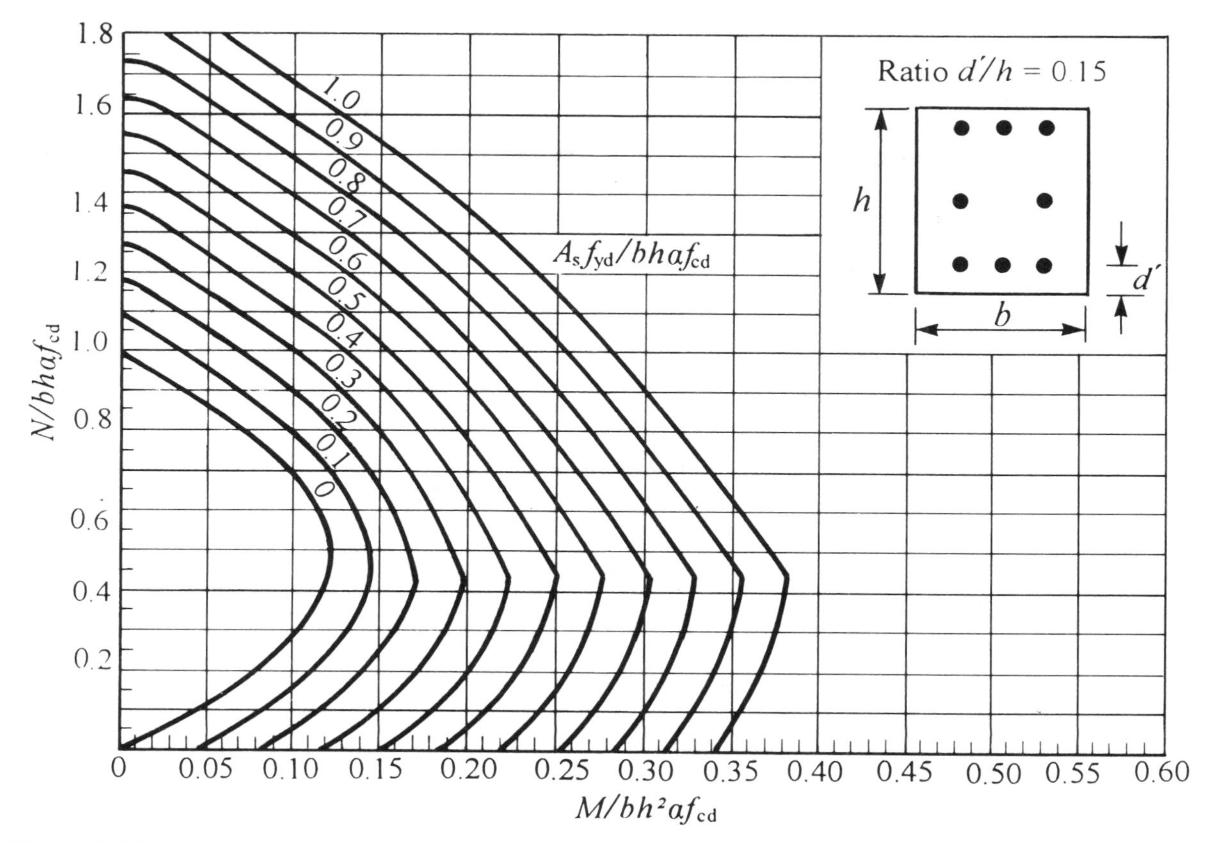

Chart 8.11

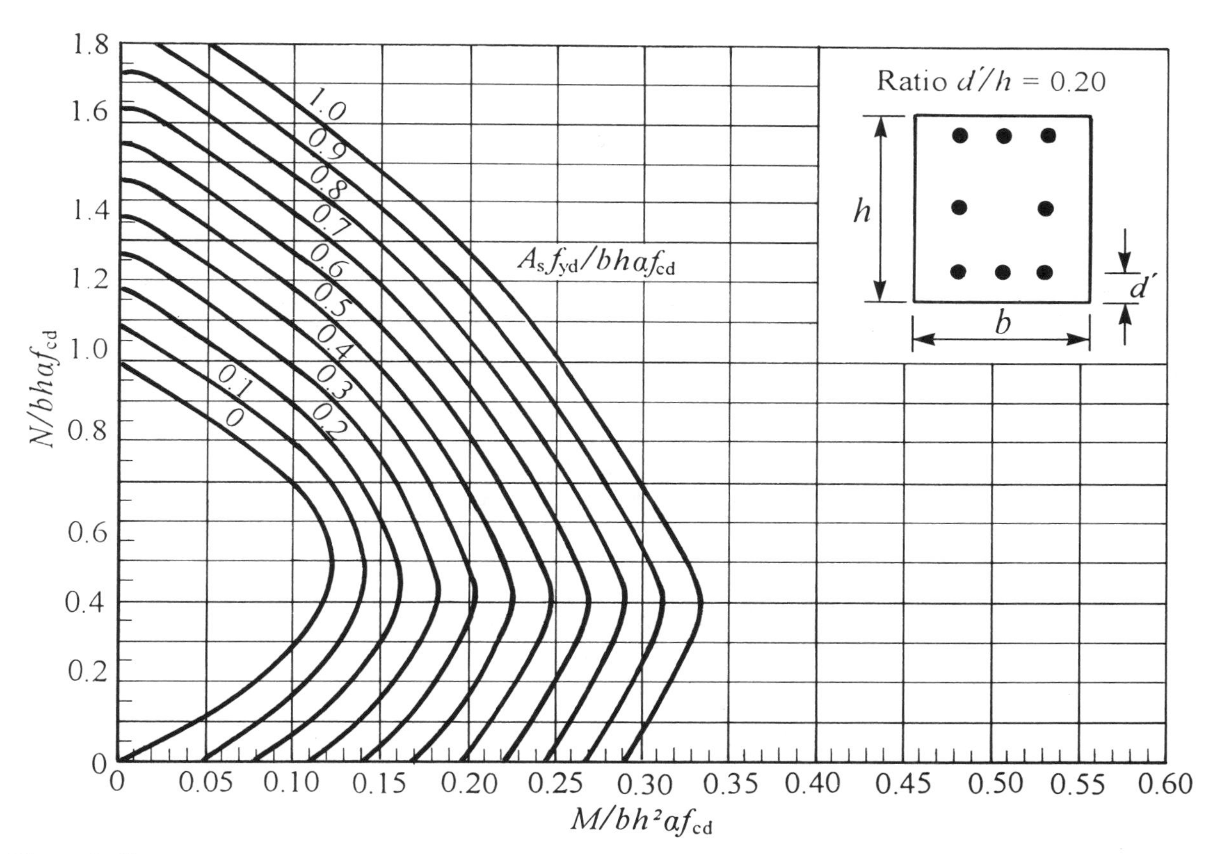

Chart 8.12

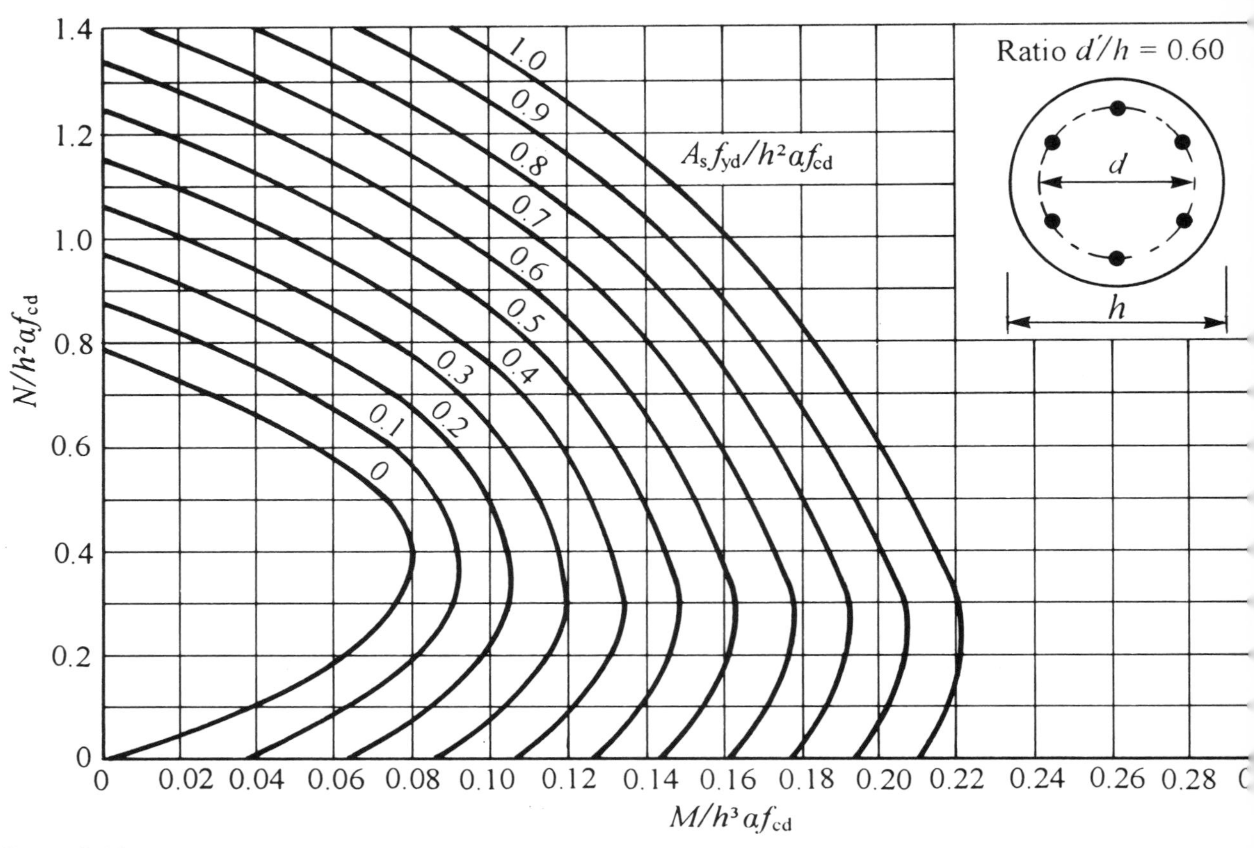

Chart 8.13

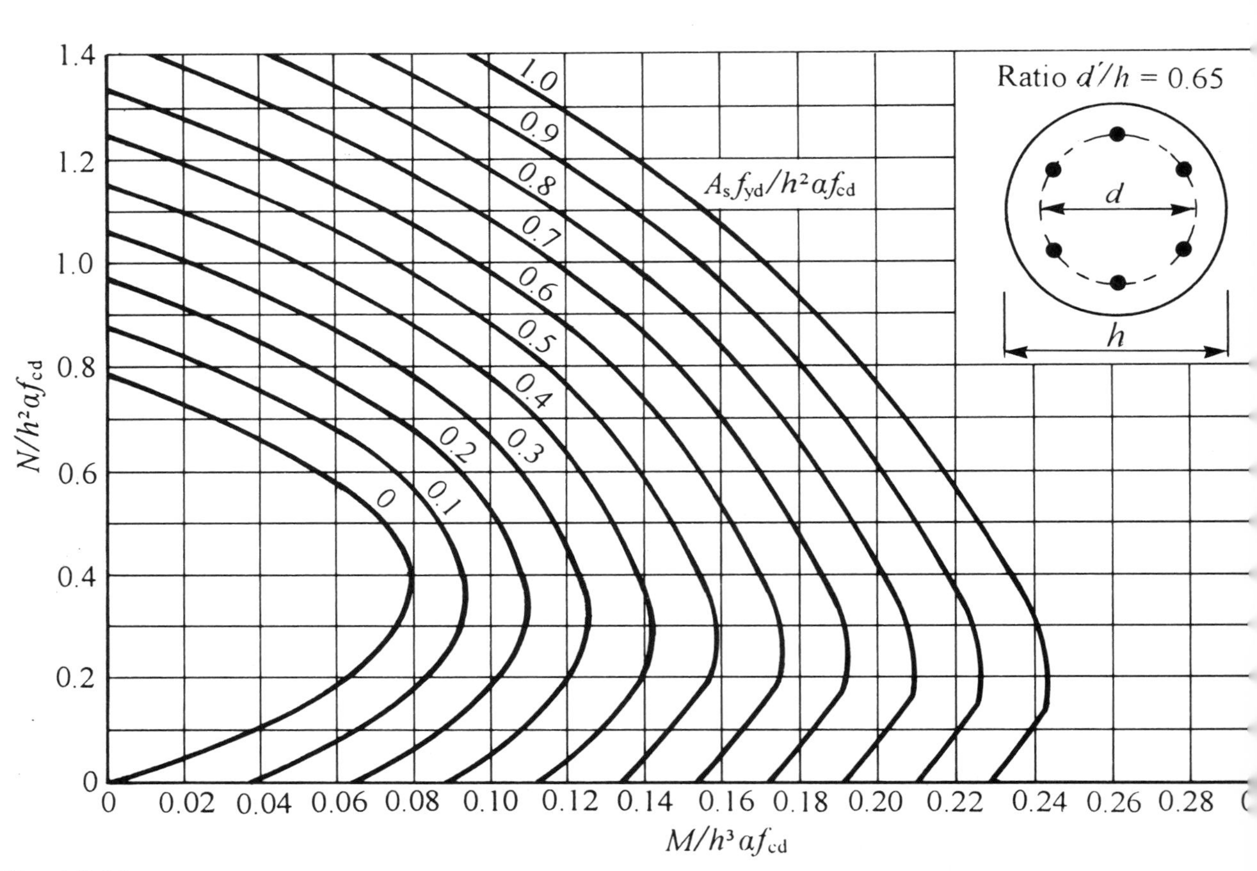

Chart 8.14

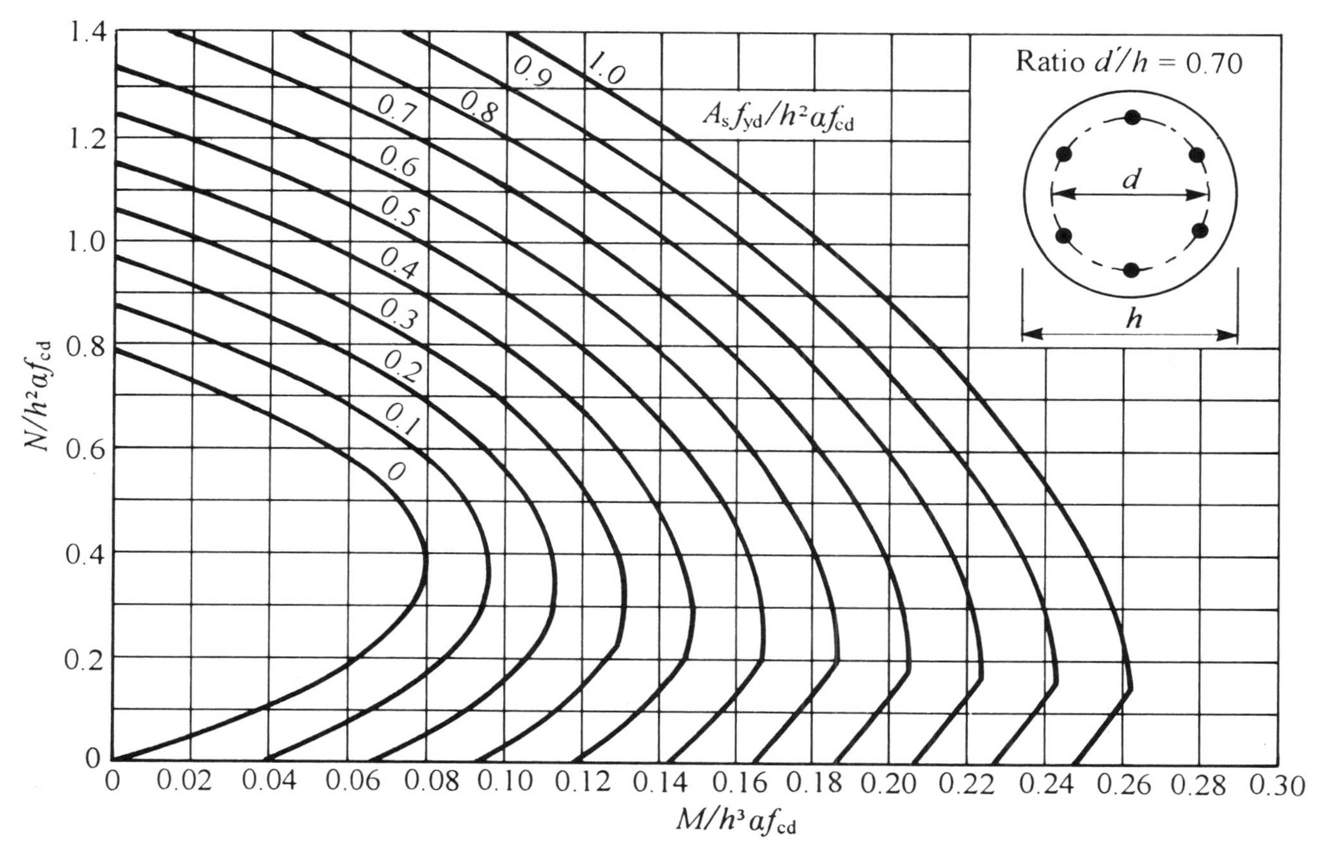

Chart 8.15

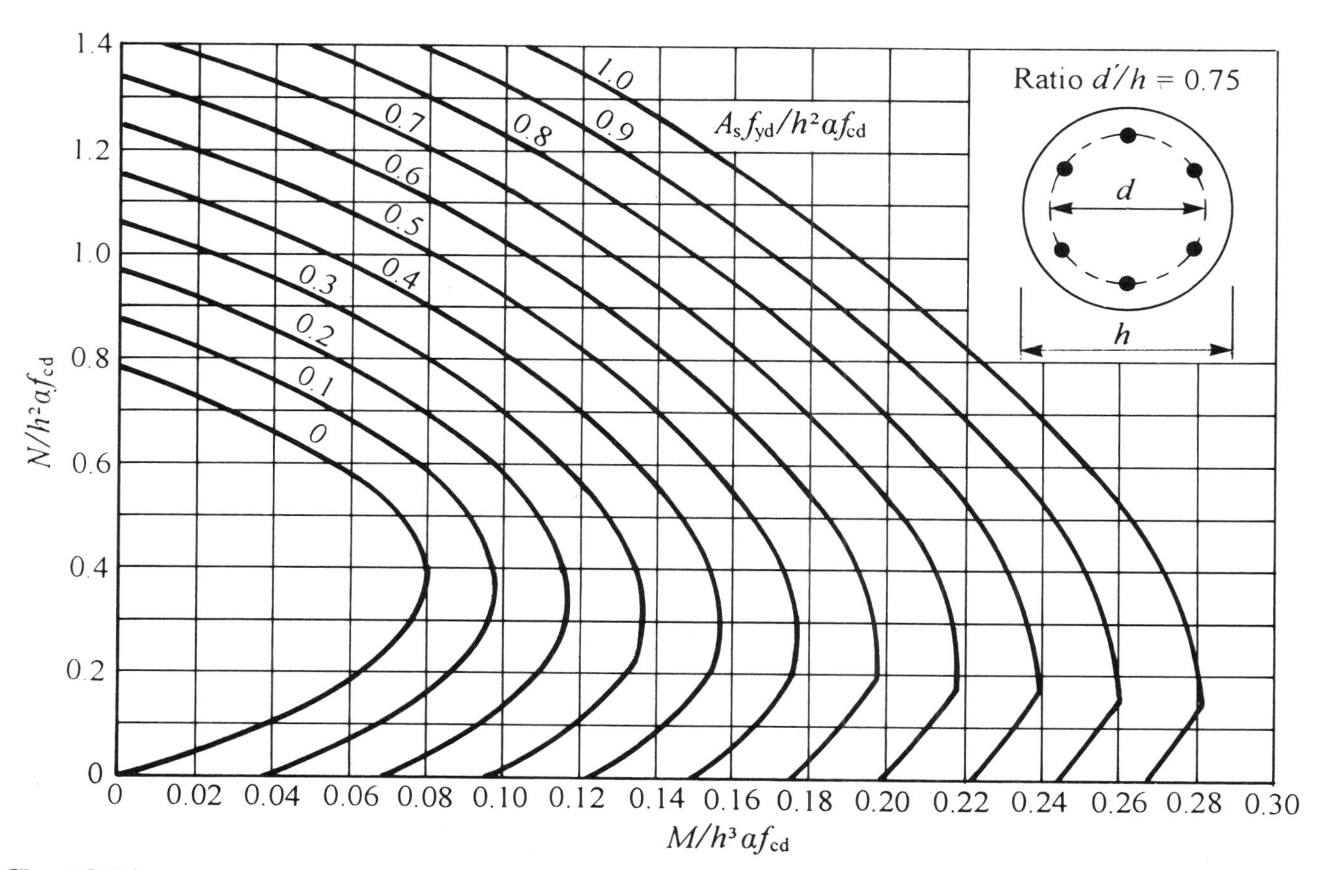

Chart 8.16

Chart 8.17

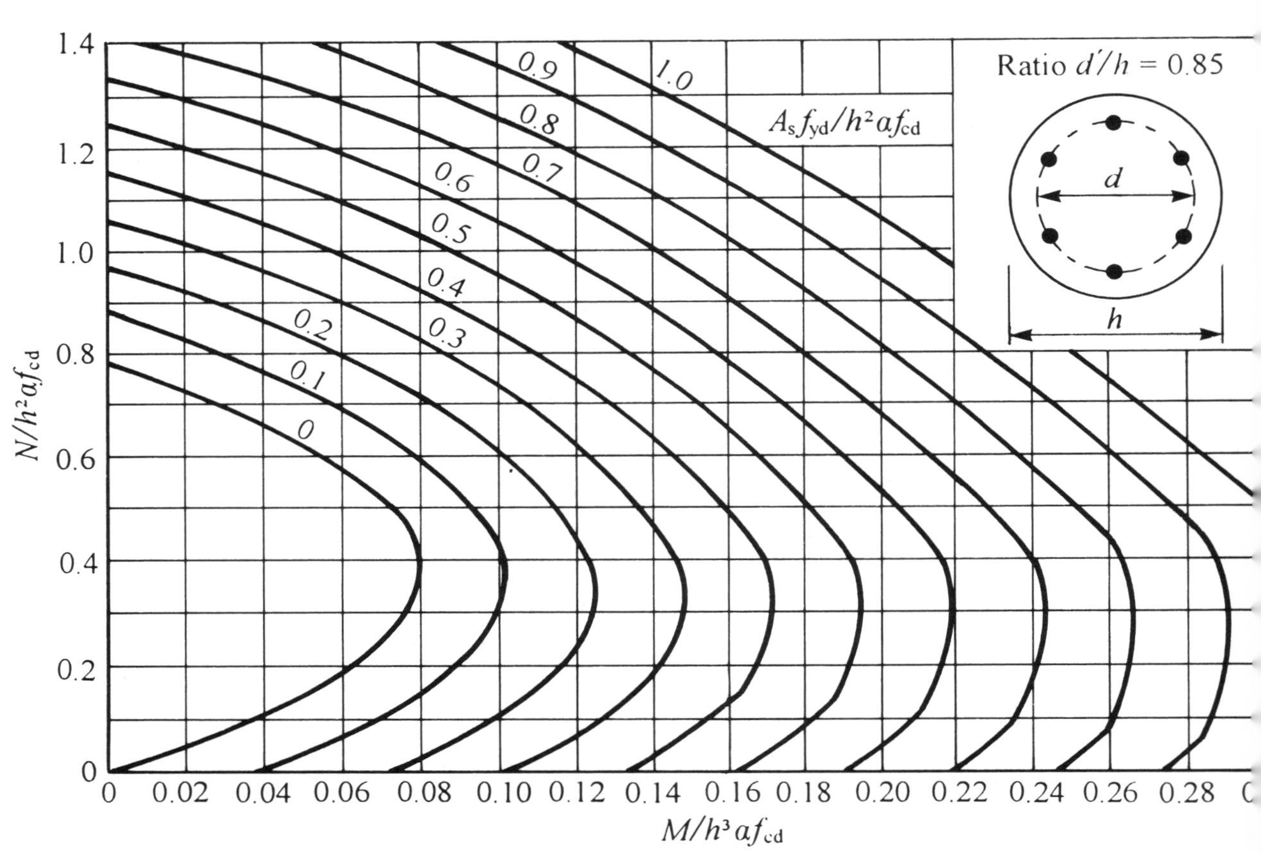

Chart 8.18

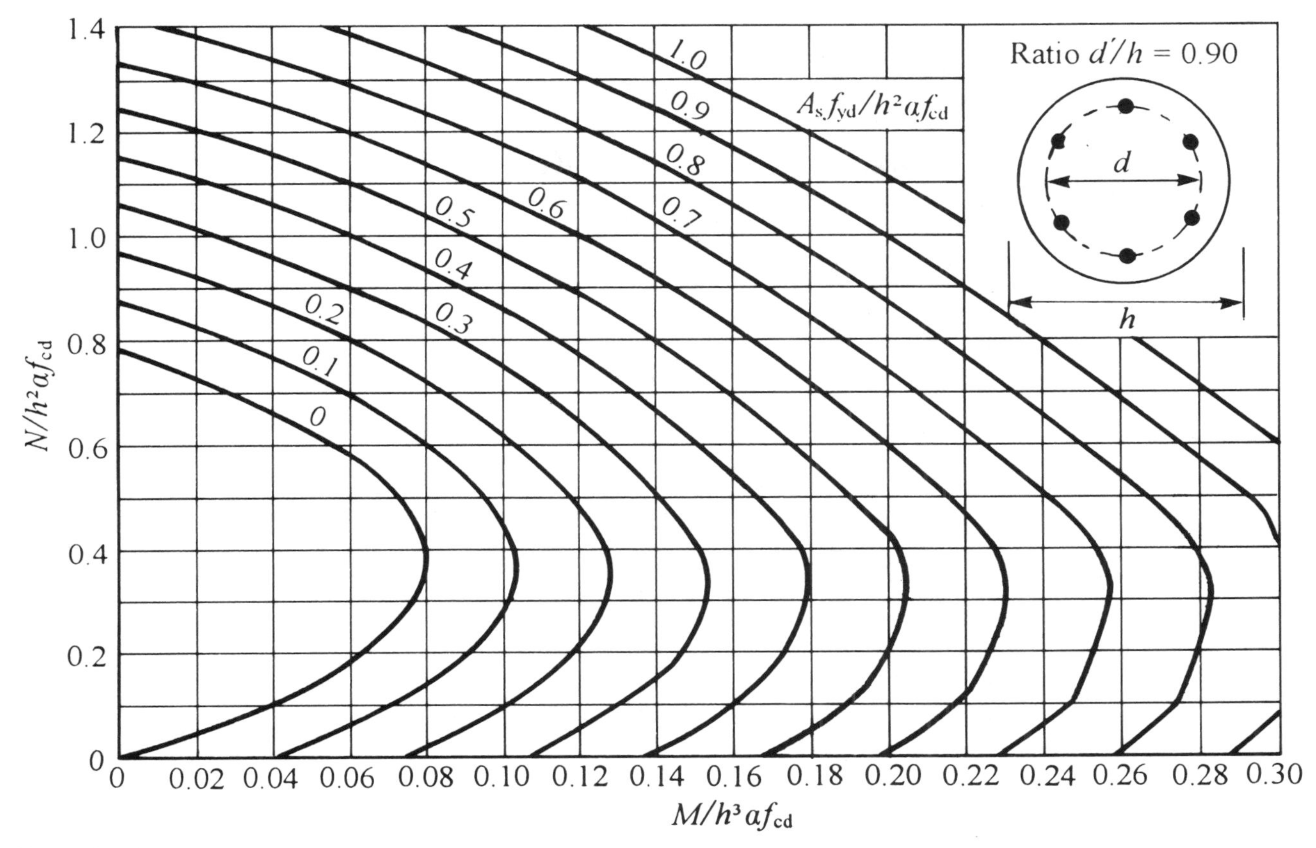

Chart 8.19

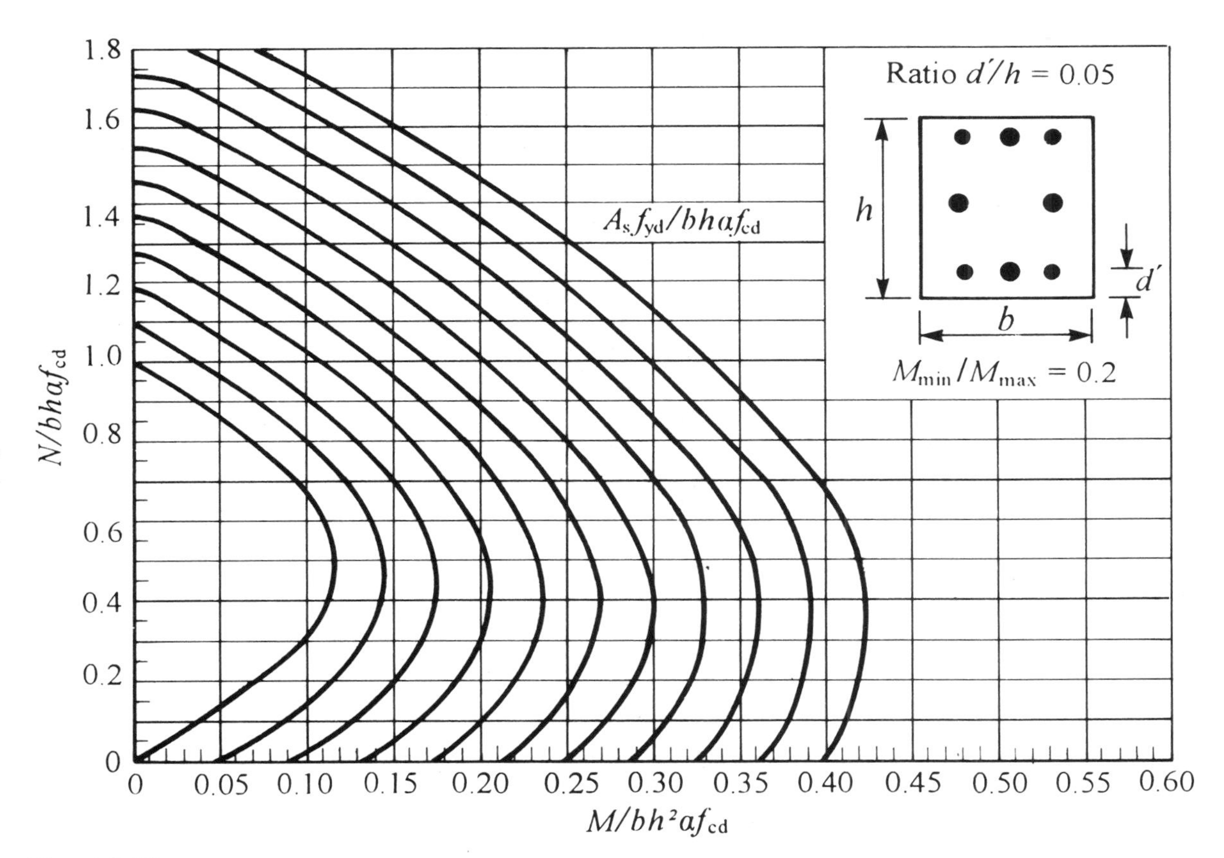

Chart 8.20

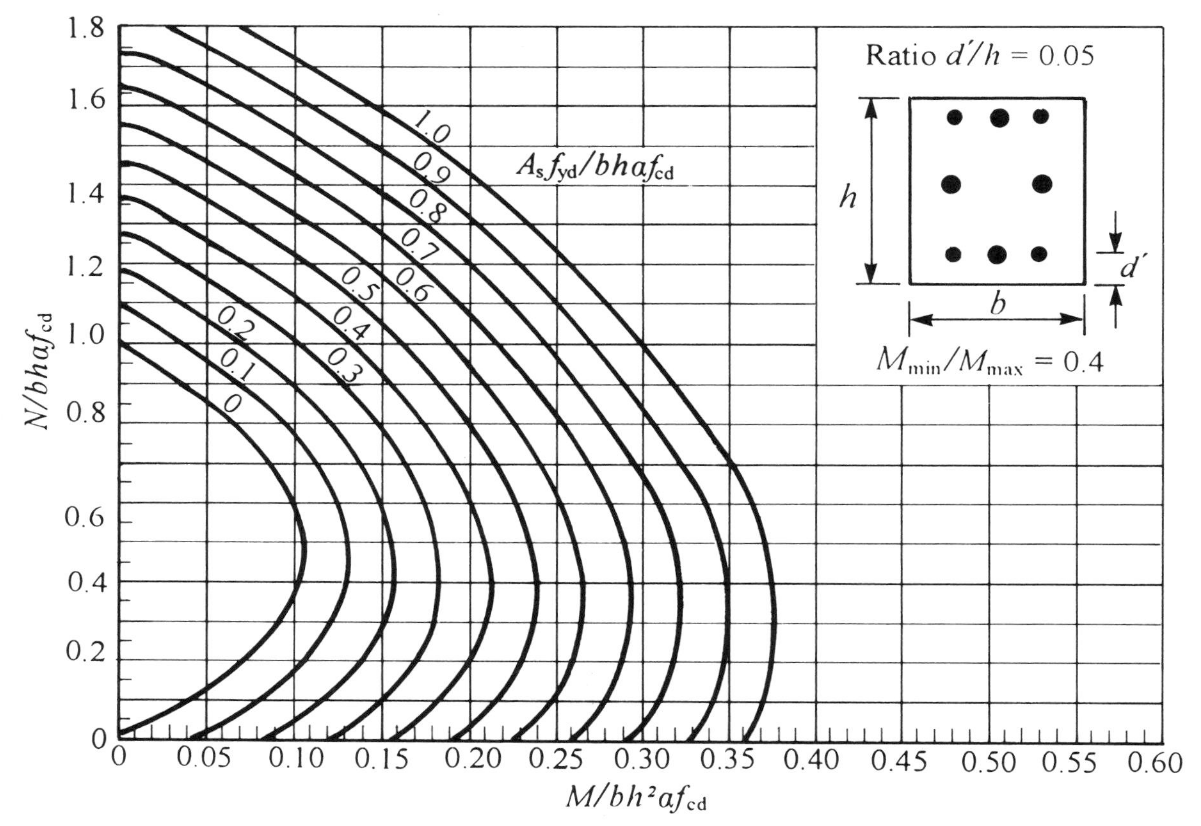

Chart 8.21

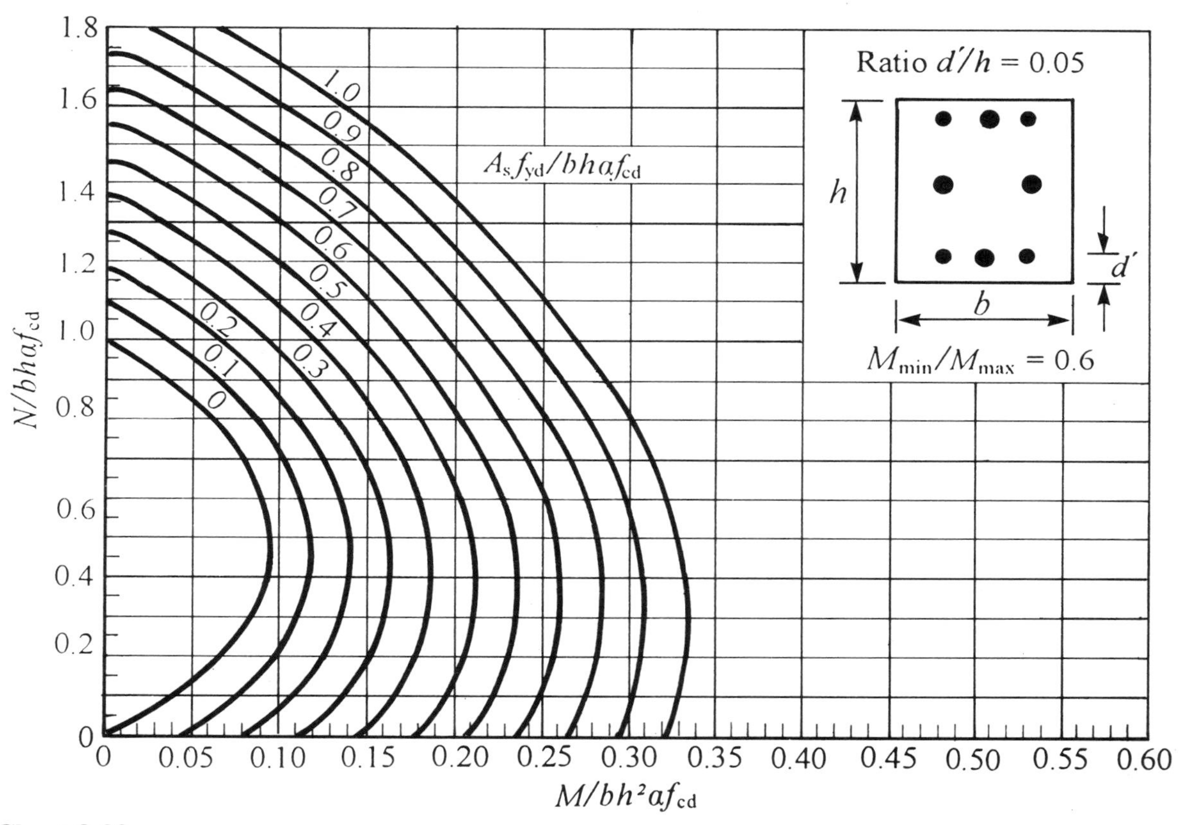

Chart 8.22

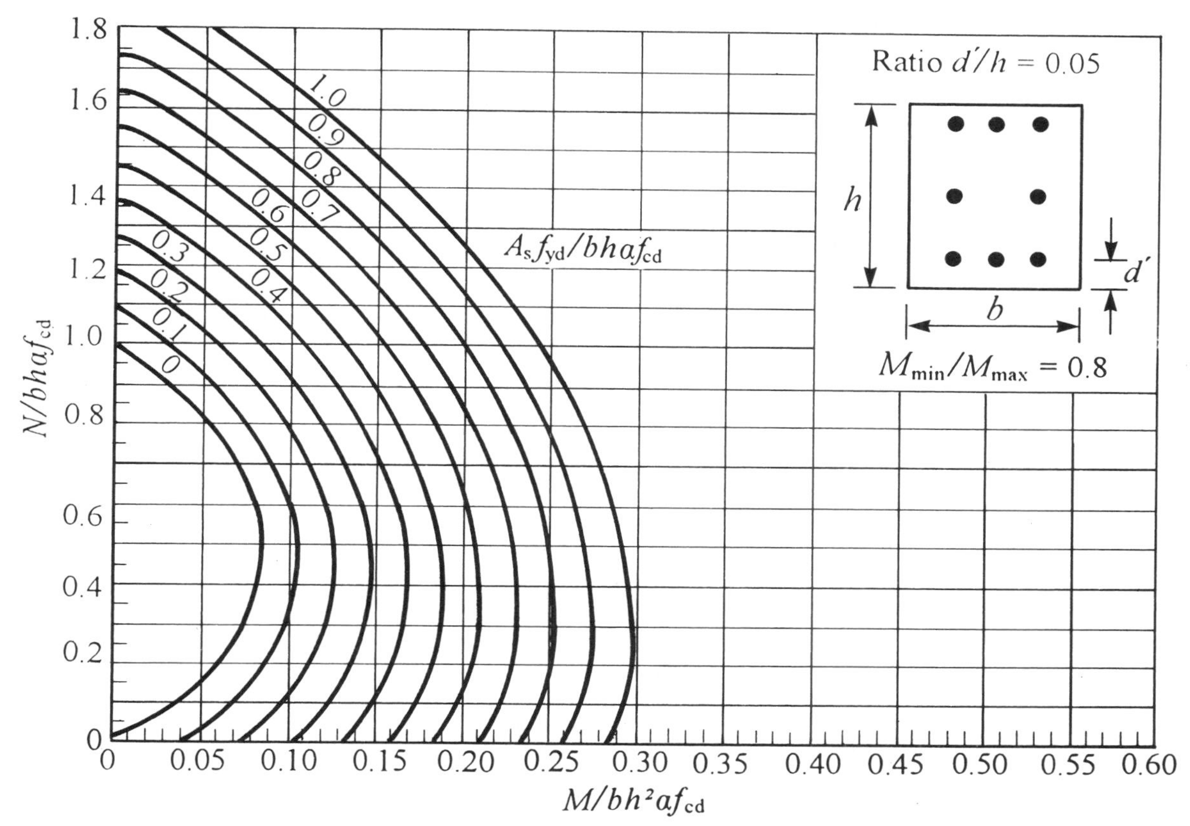

Chart 8.23

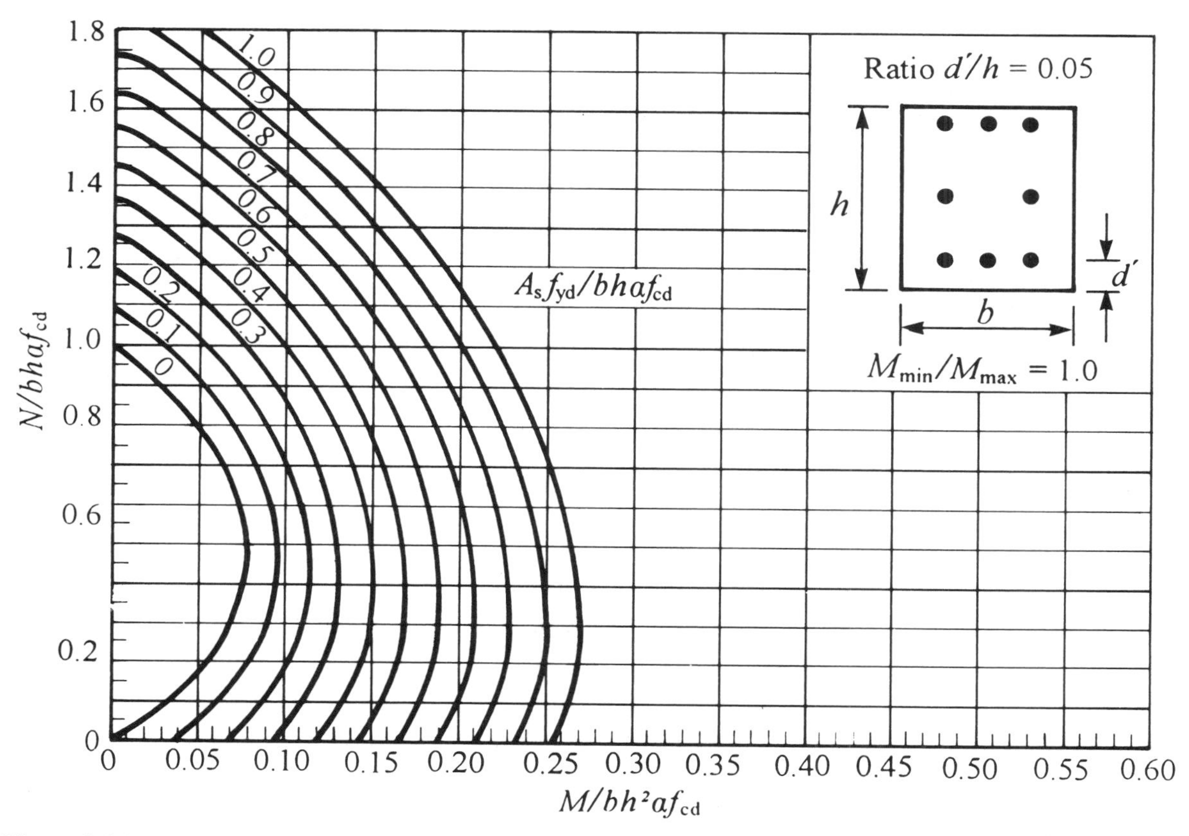

Chart 8.24

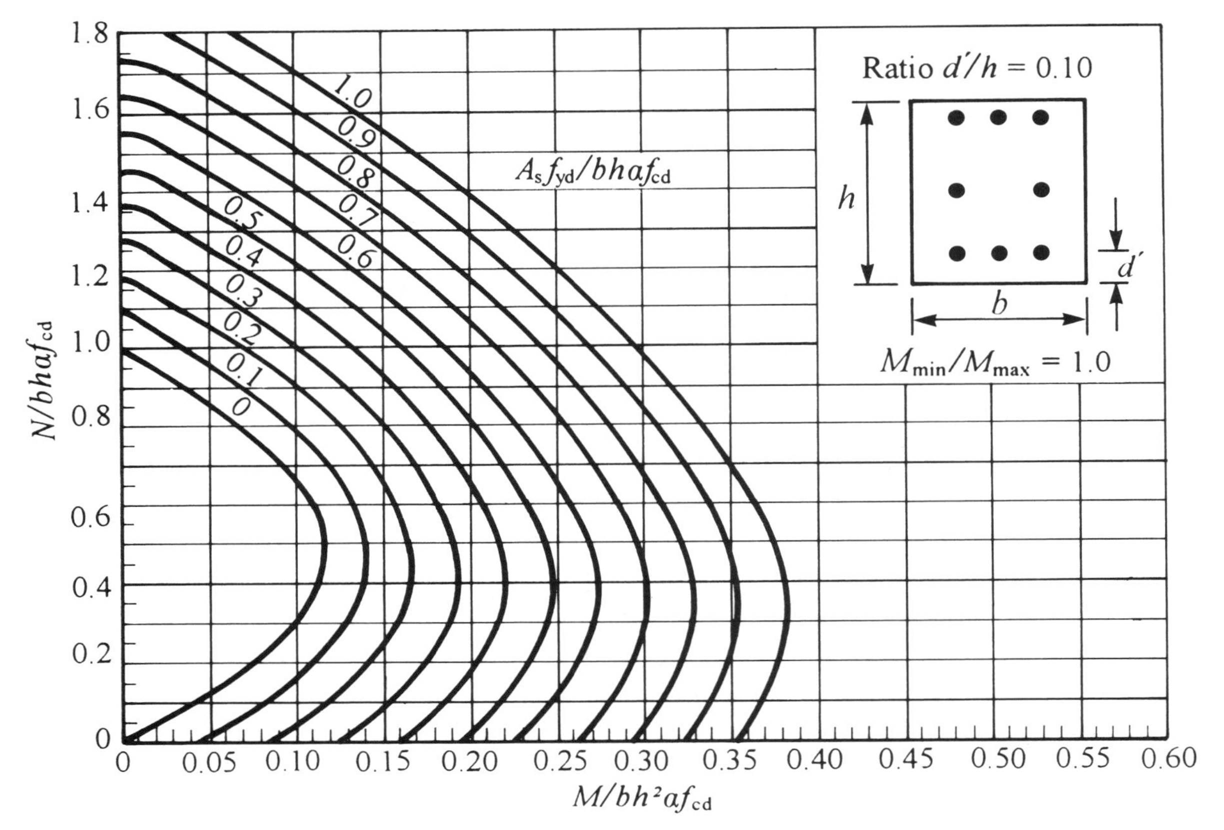

Chart 8.25

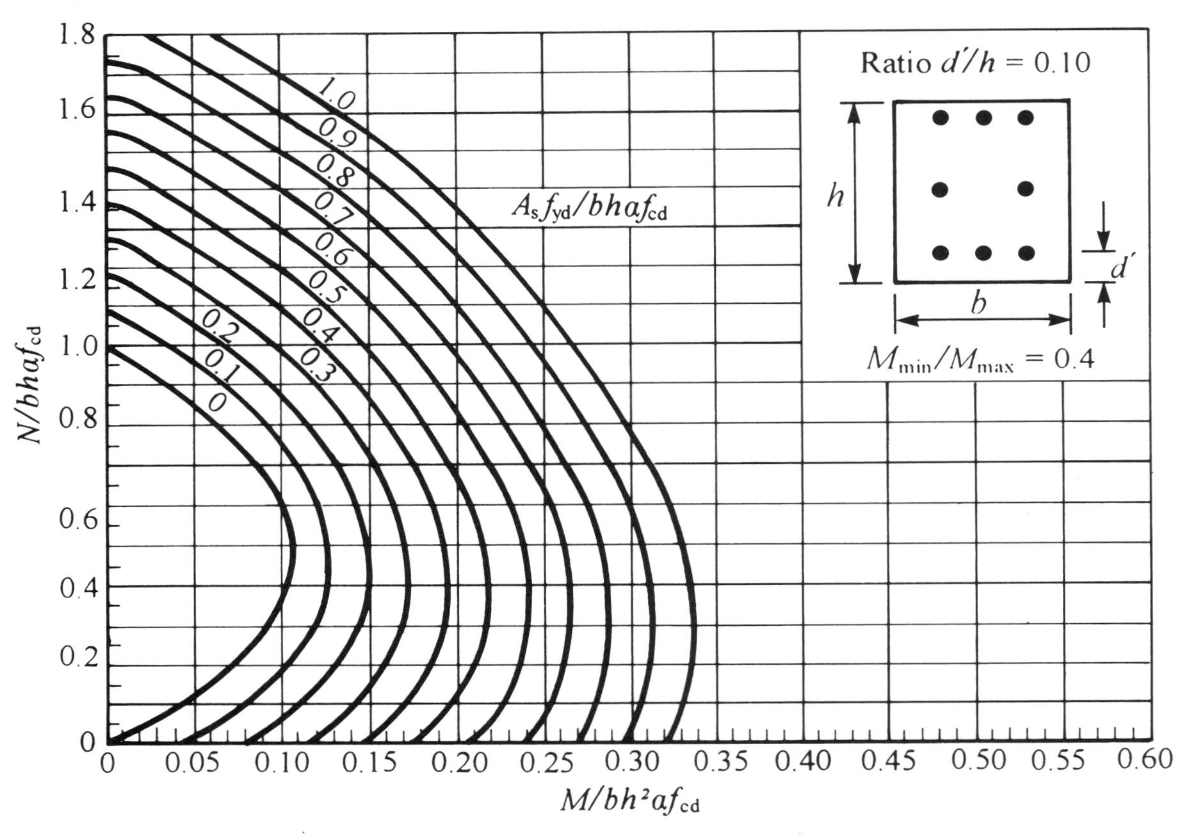

Chart 8.26

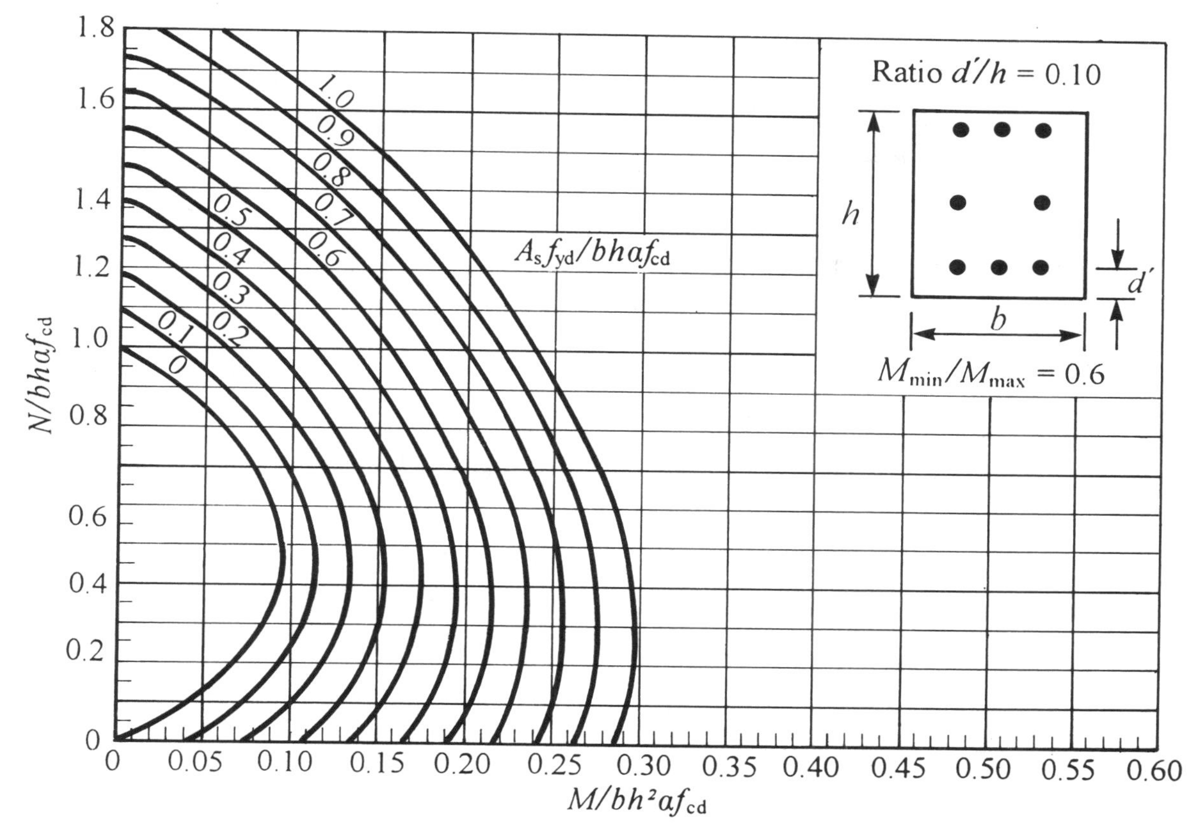

Chart 8.27

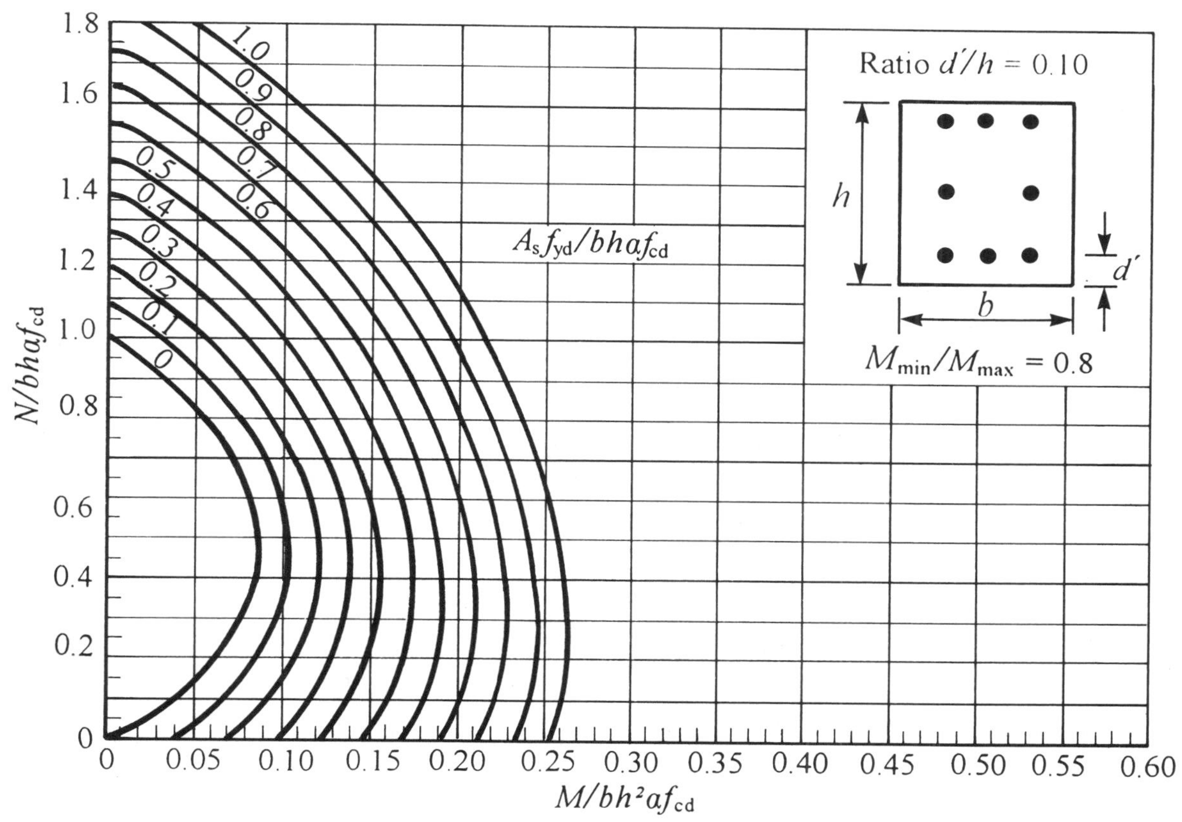

Chart 8.28

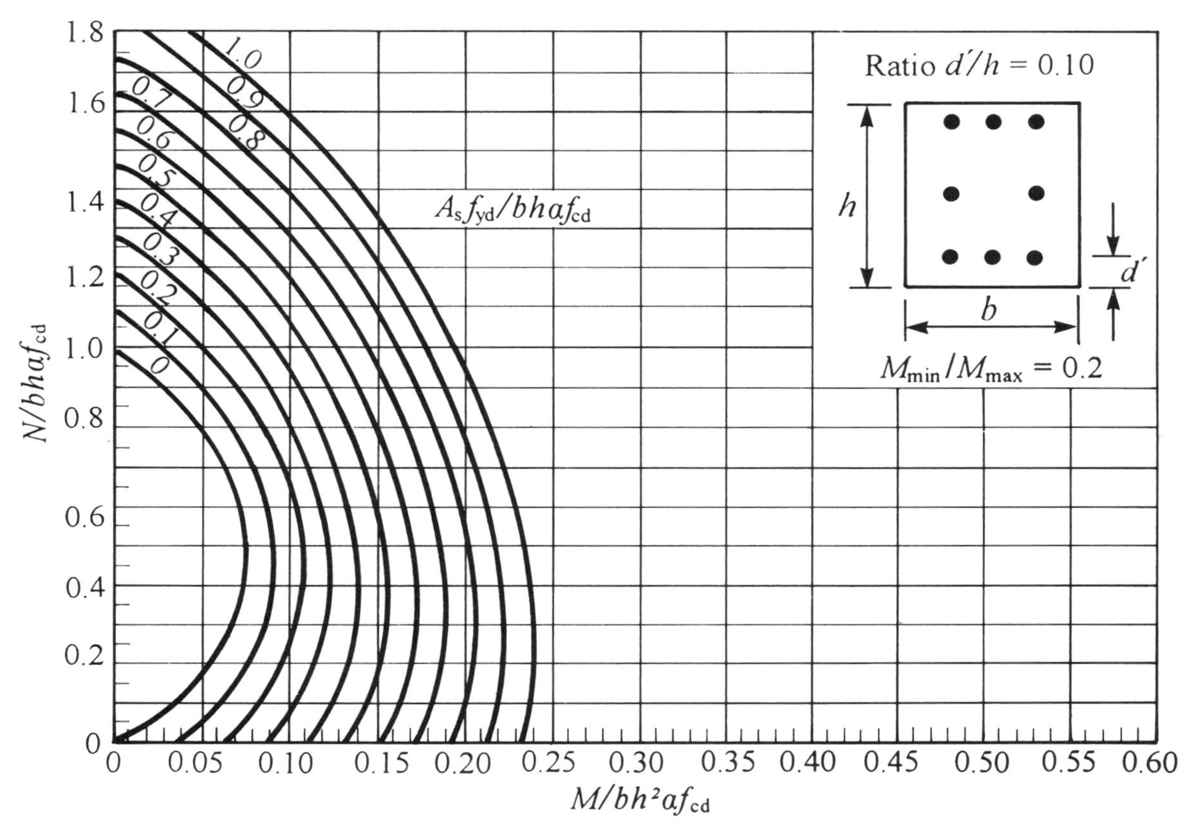

Chart 8.29

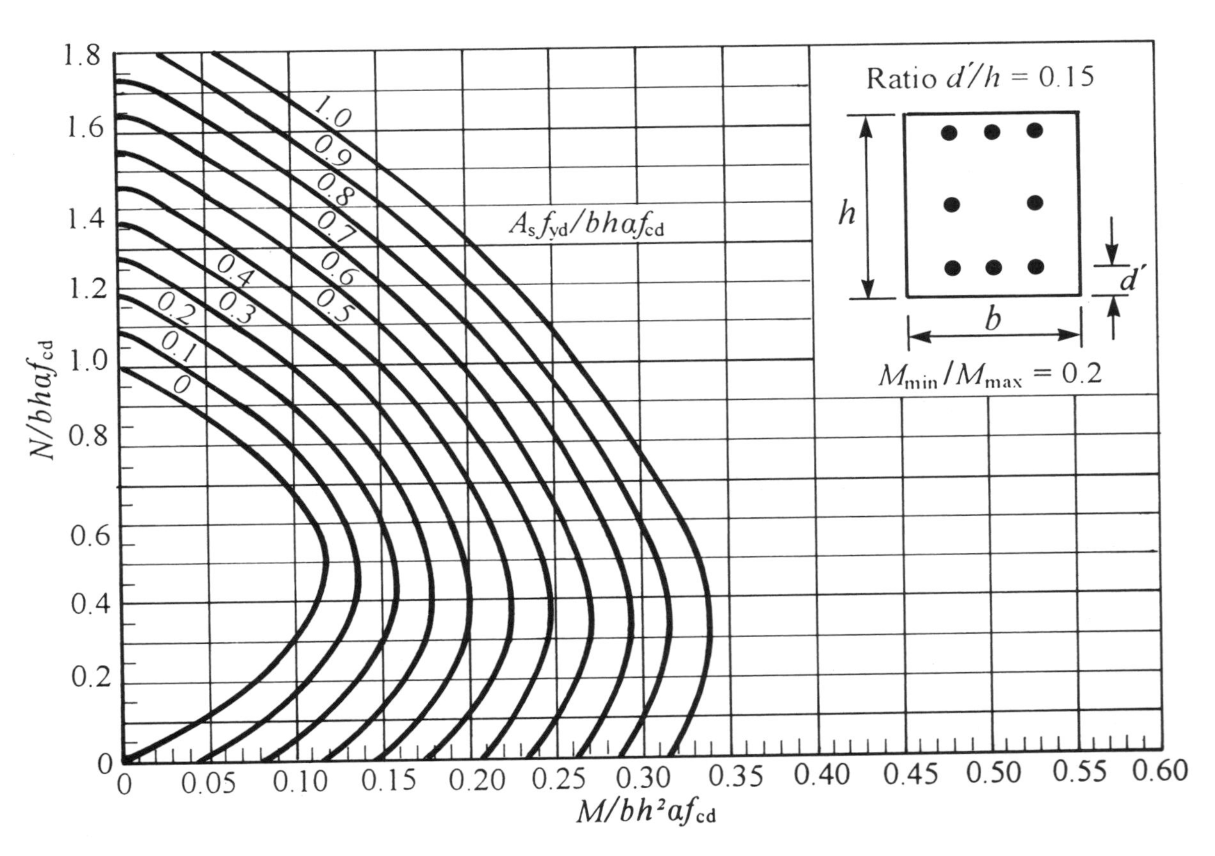

Chart 8.30

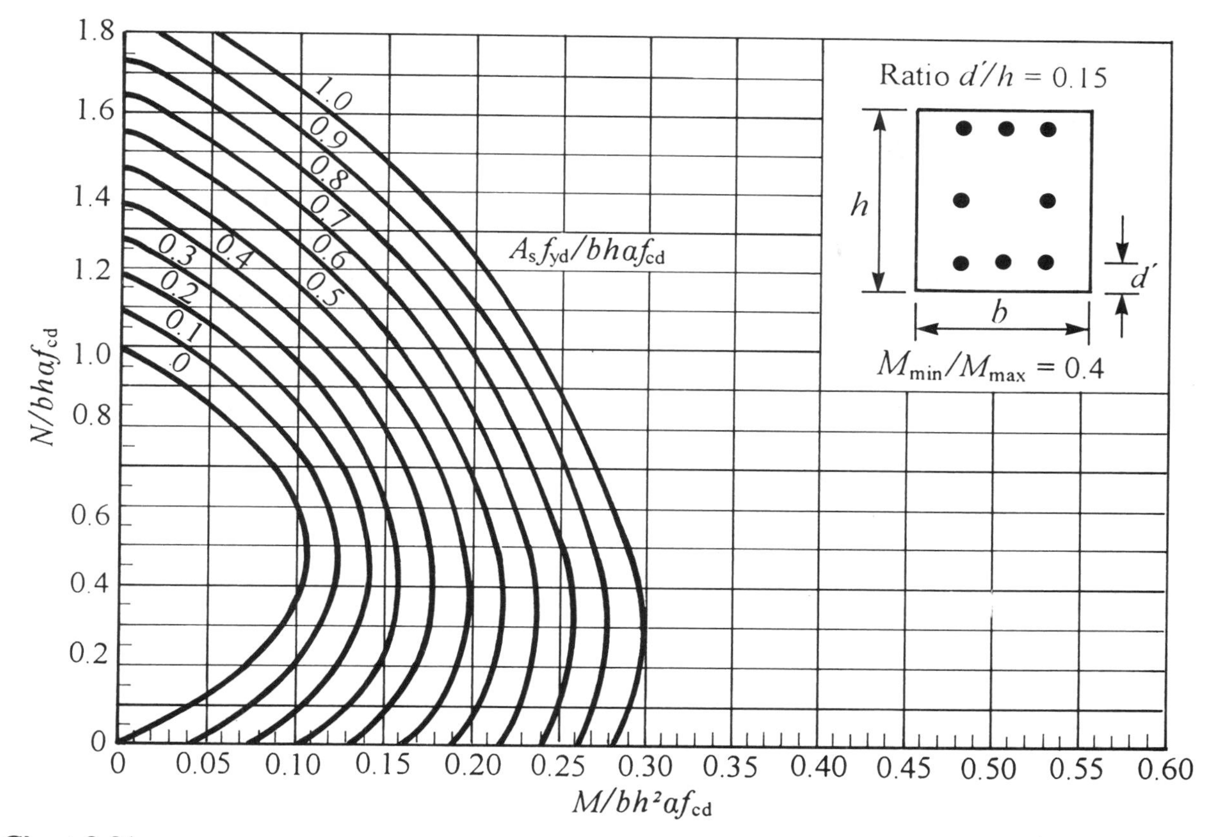

Chart 8.31

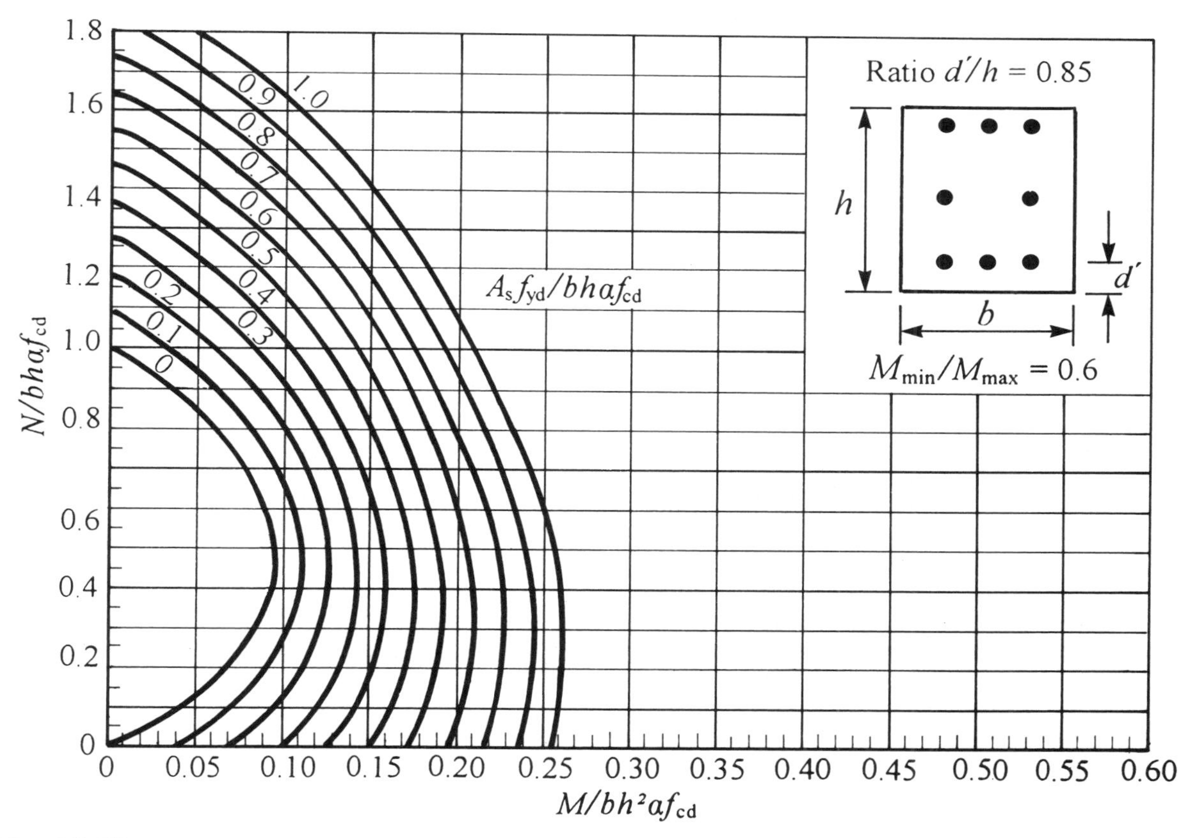

Chart 8.32

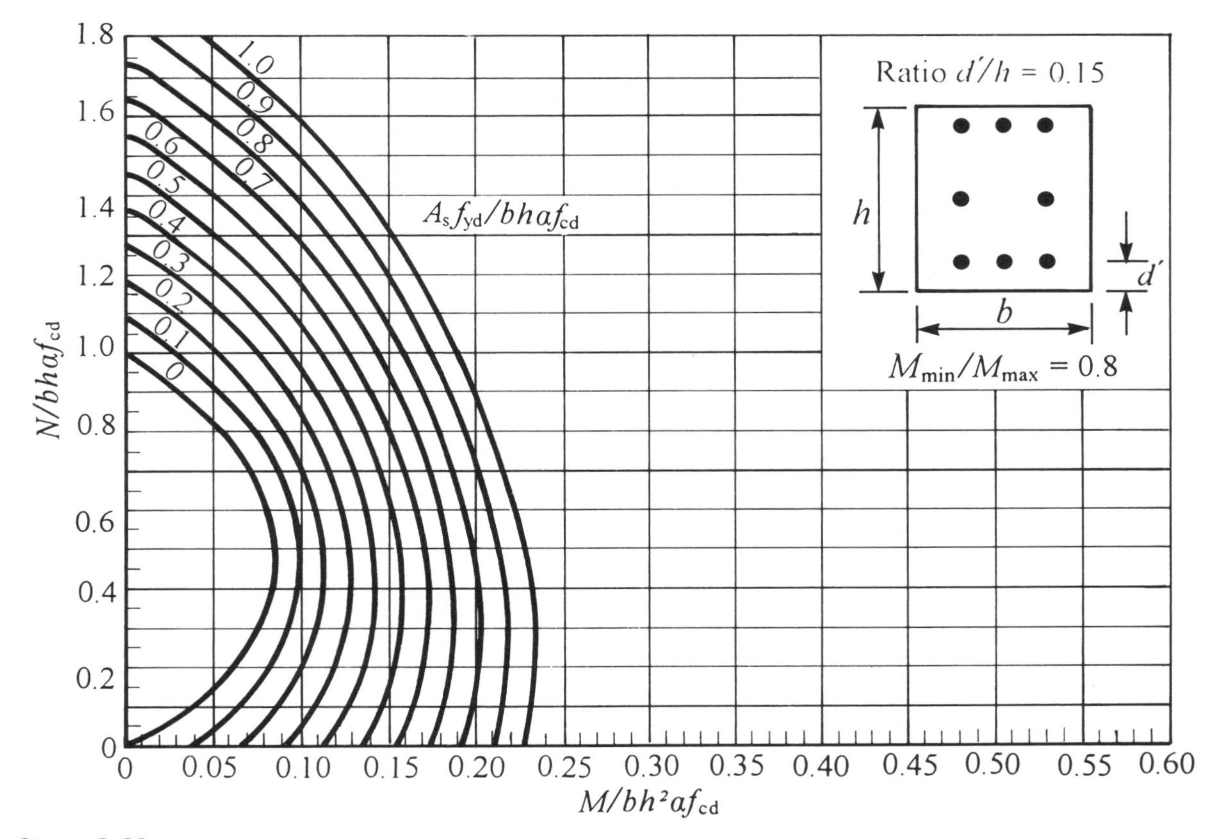

Chart 8.33

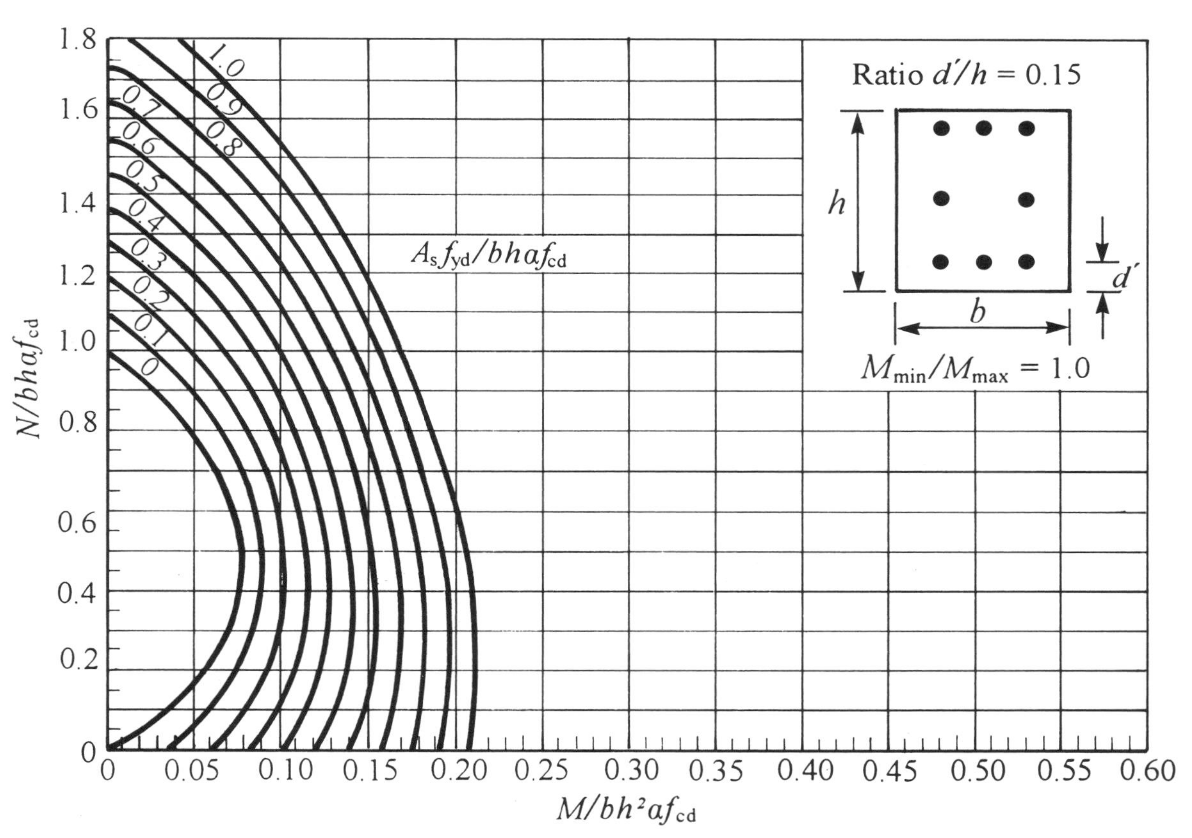

Chart 8.34

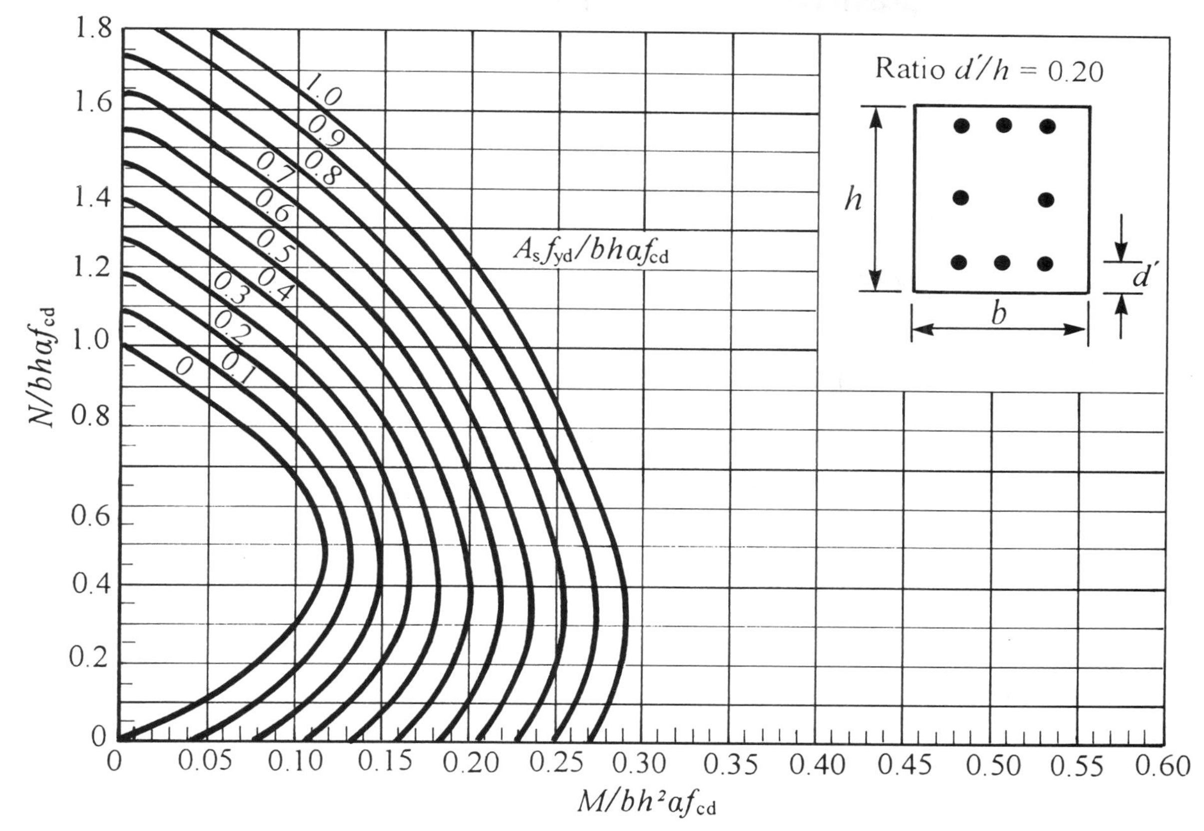

Chart 8.35

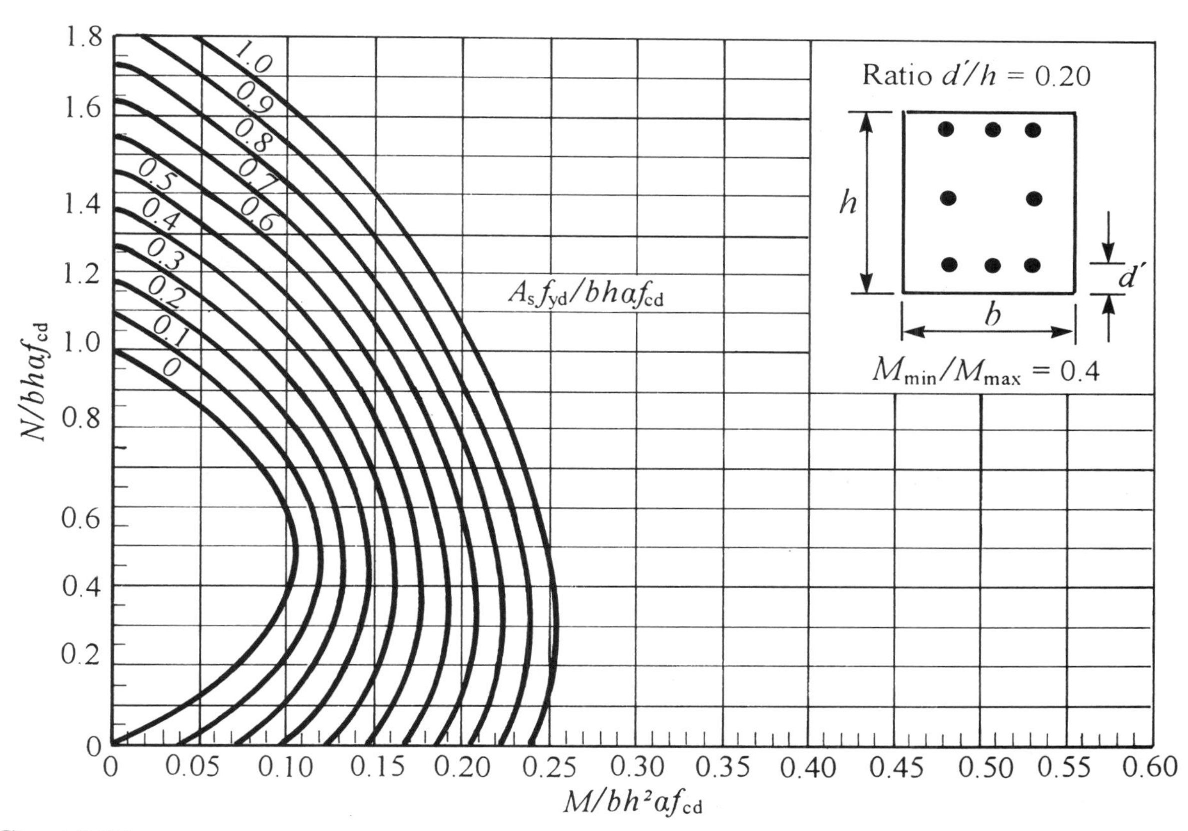

Chart 8.36

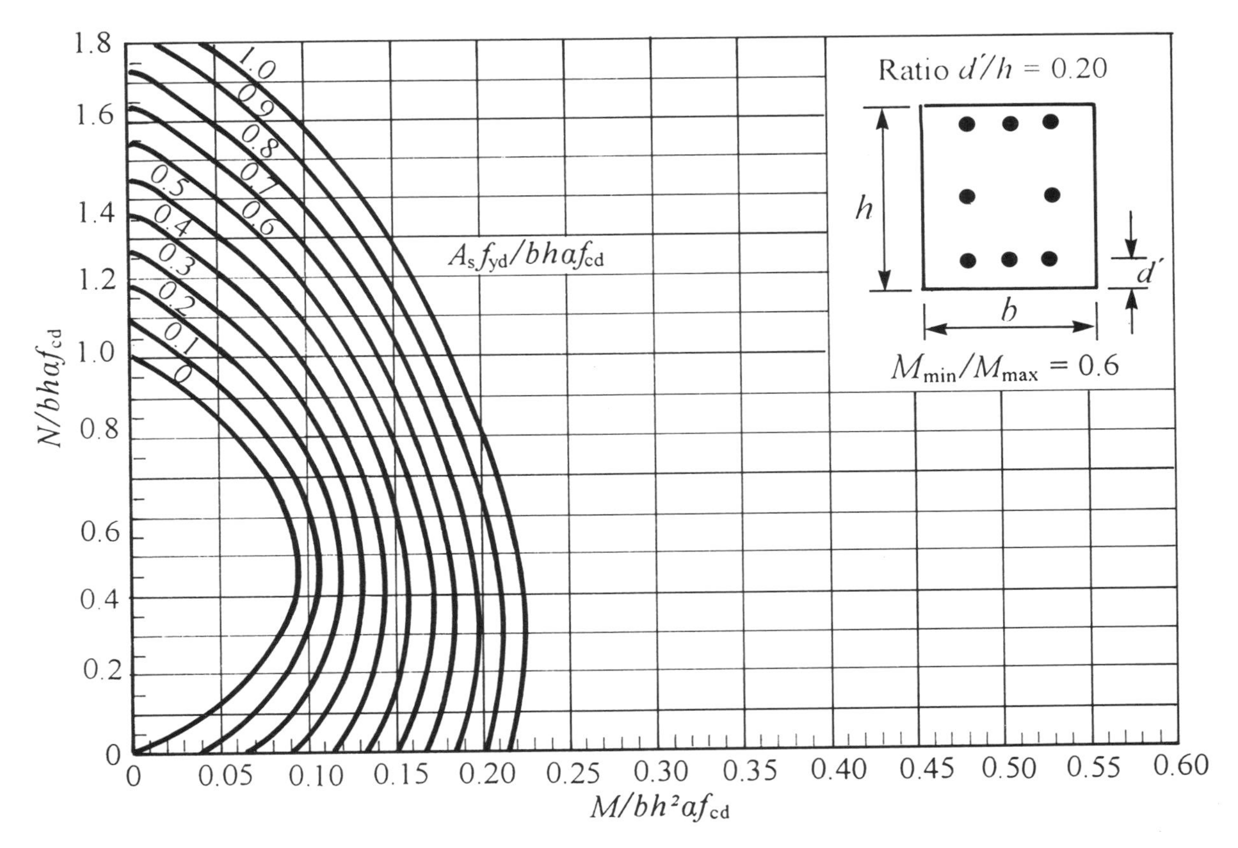

Chart 8.37

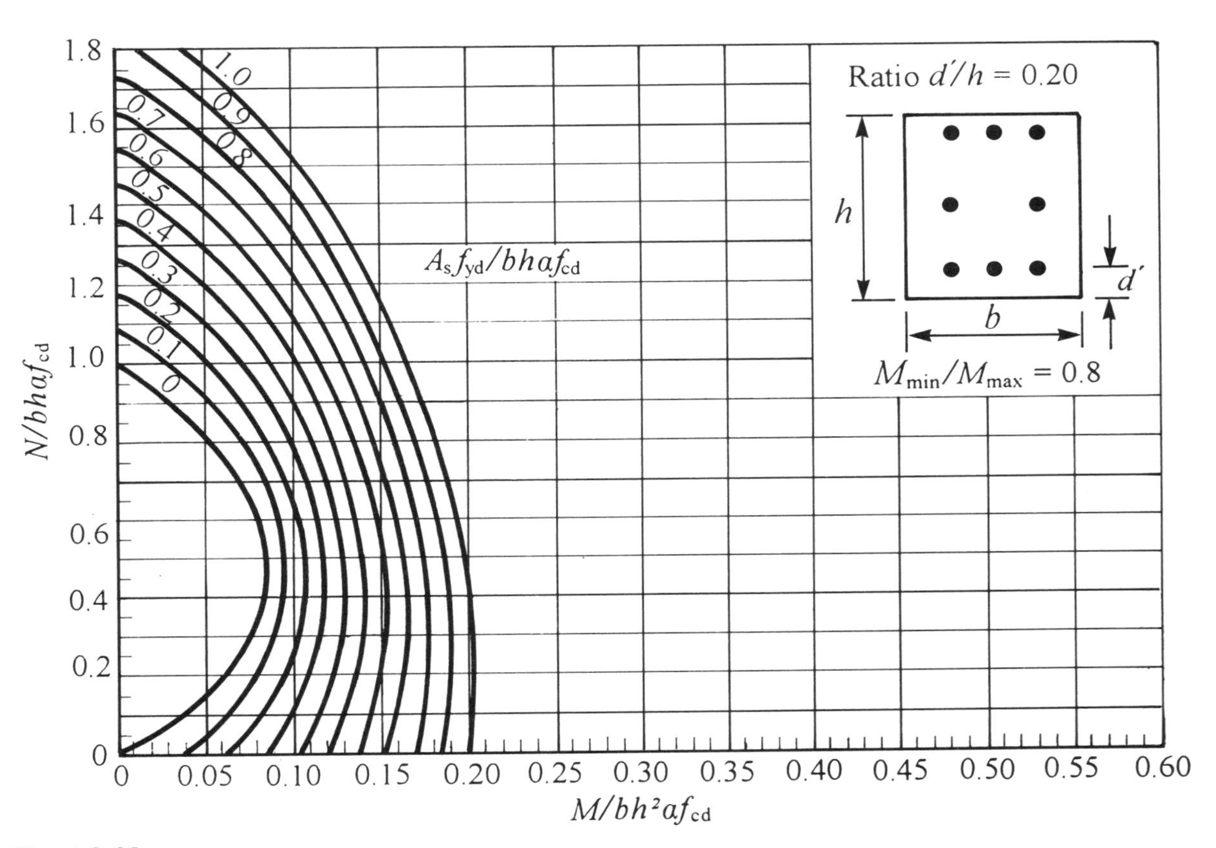

Chart 8.38

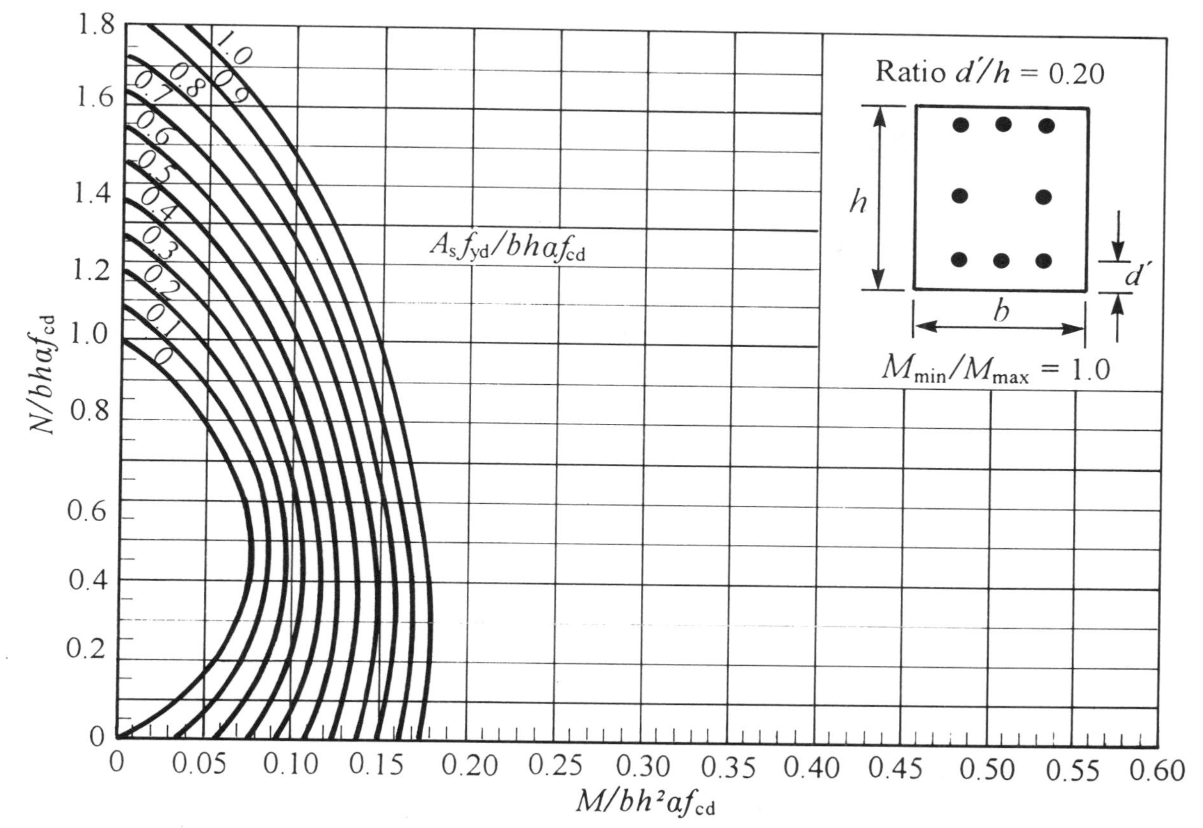

Chart 8.39

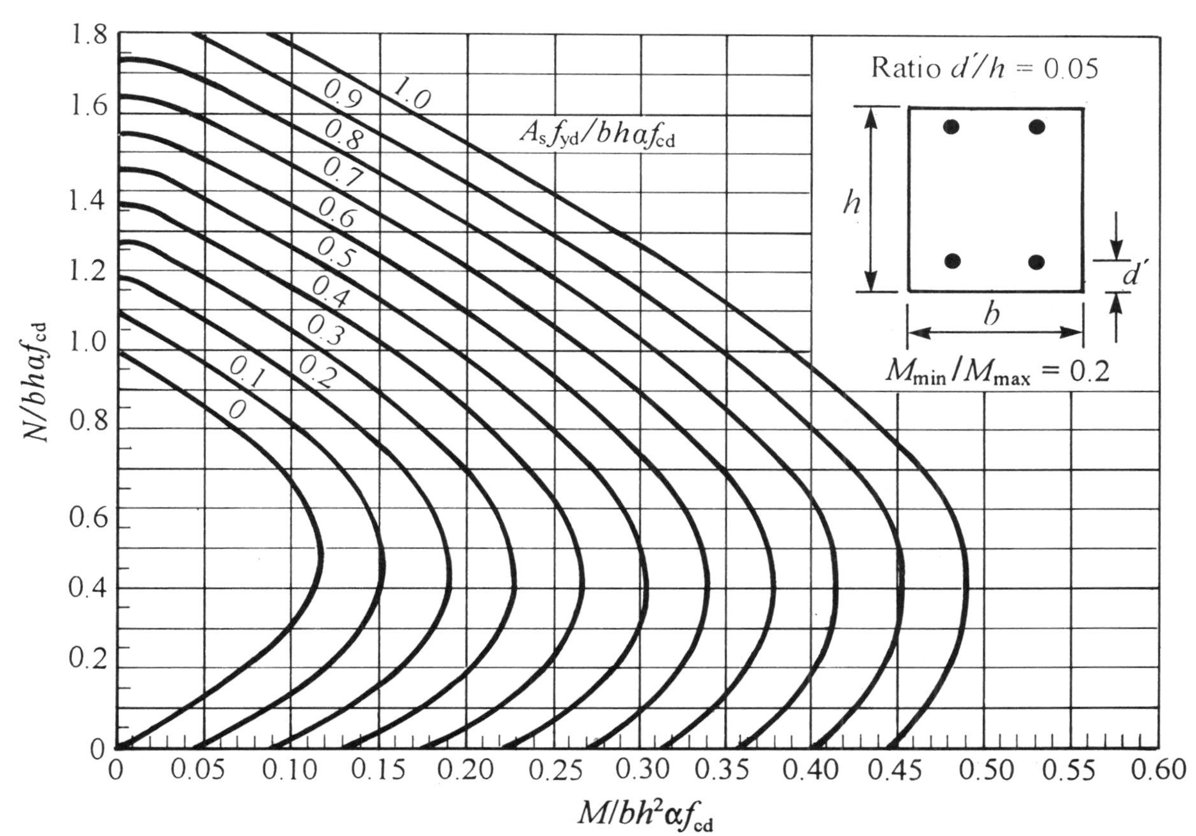

Chart 8.40

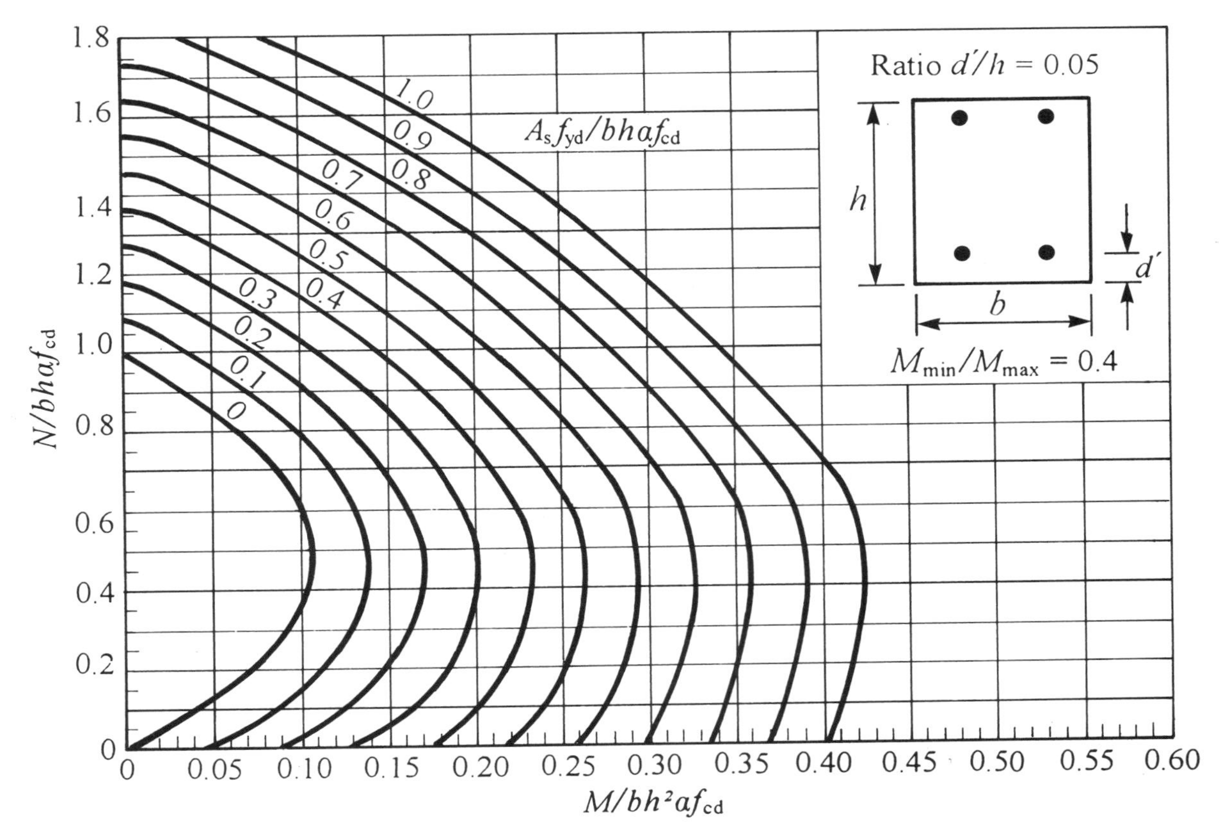

Chart 8.41

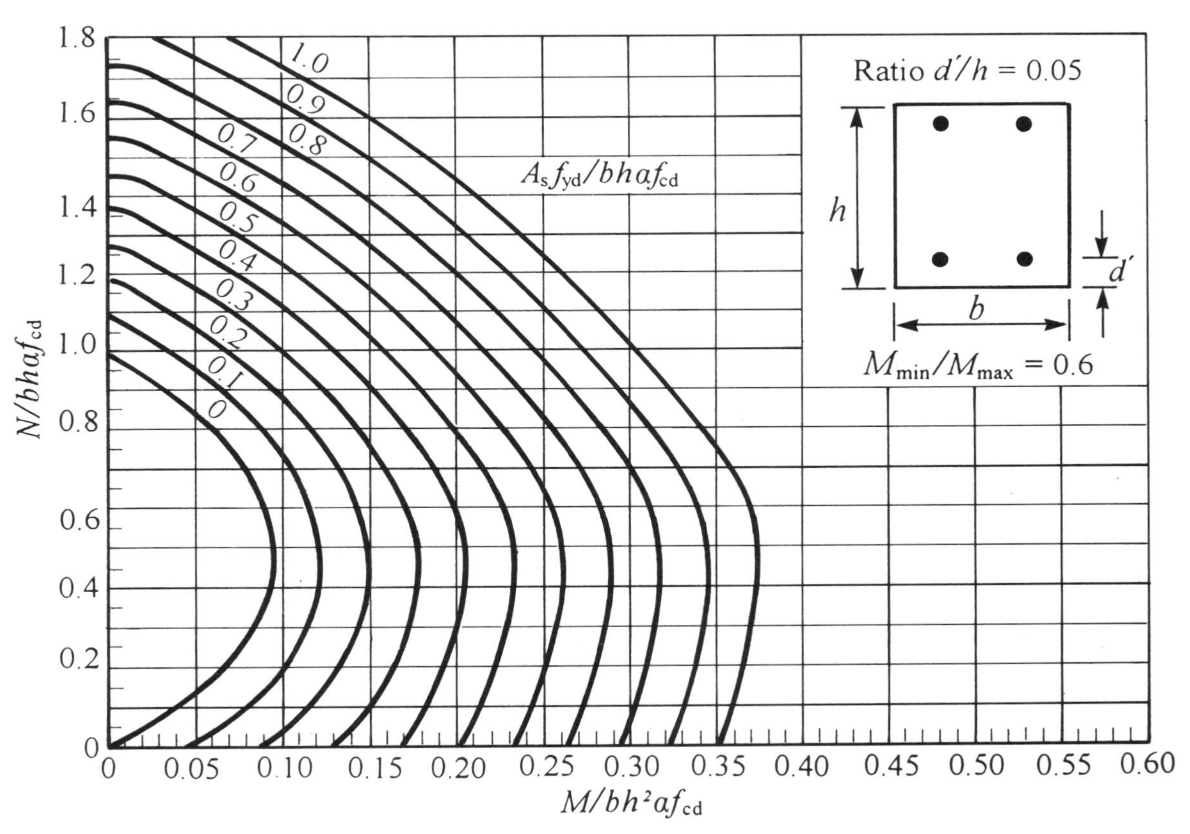

Chart 8.42

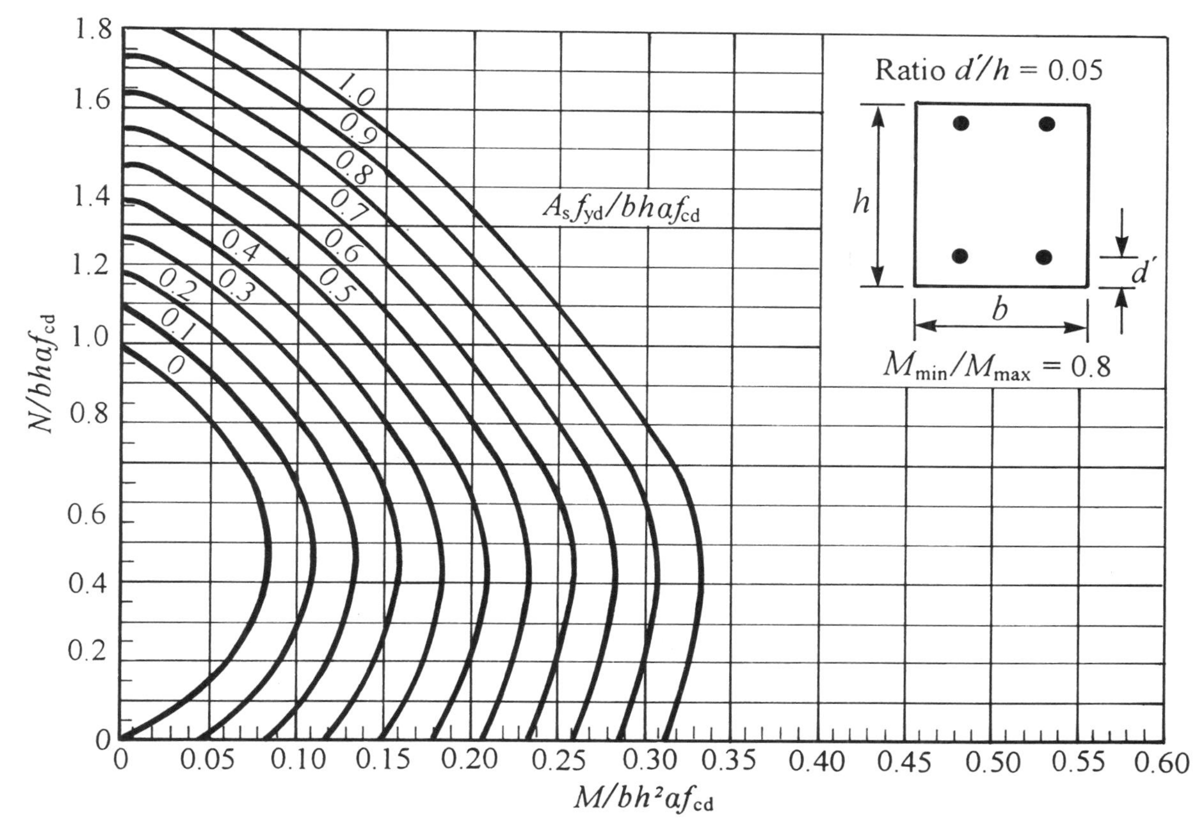

Chart 8.43

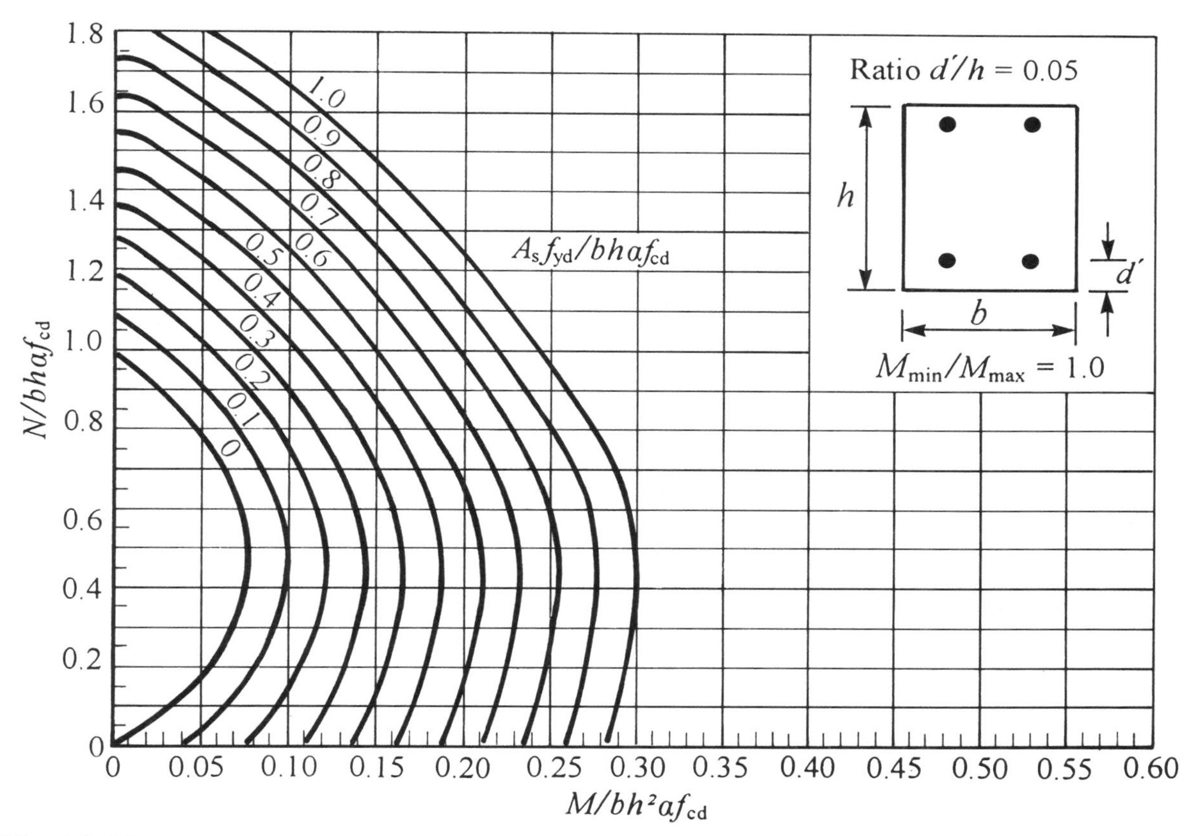

Chart 8.44

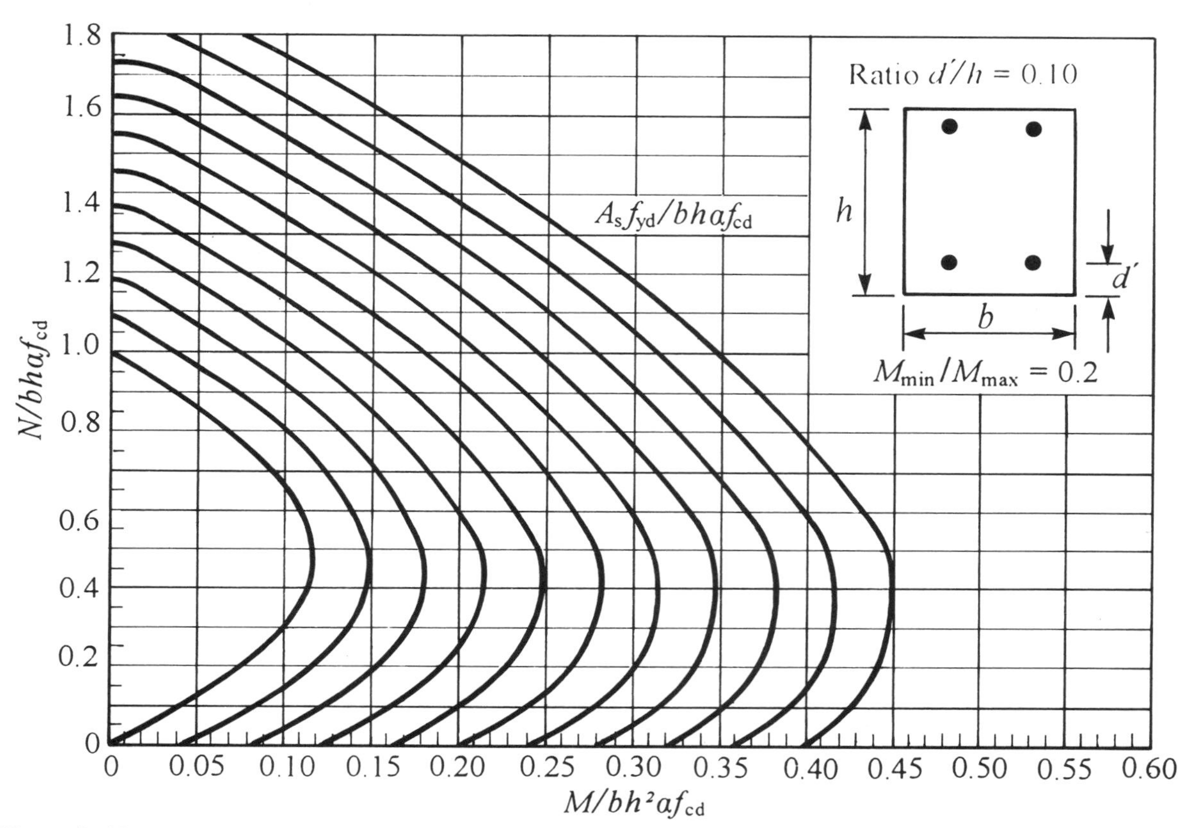

Chart 8.45

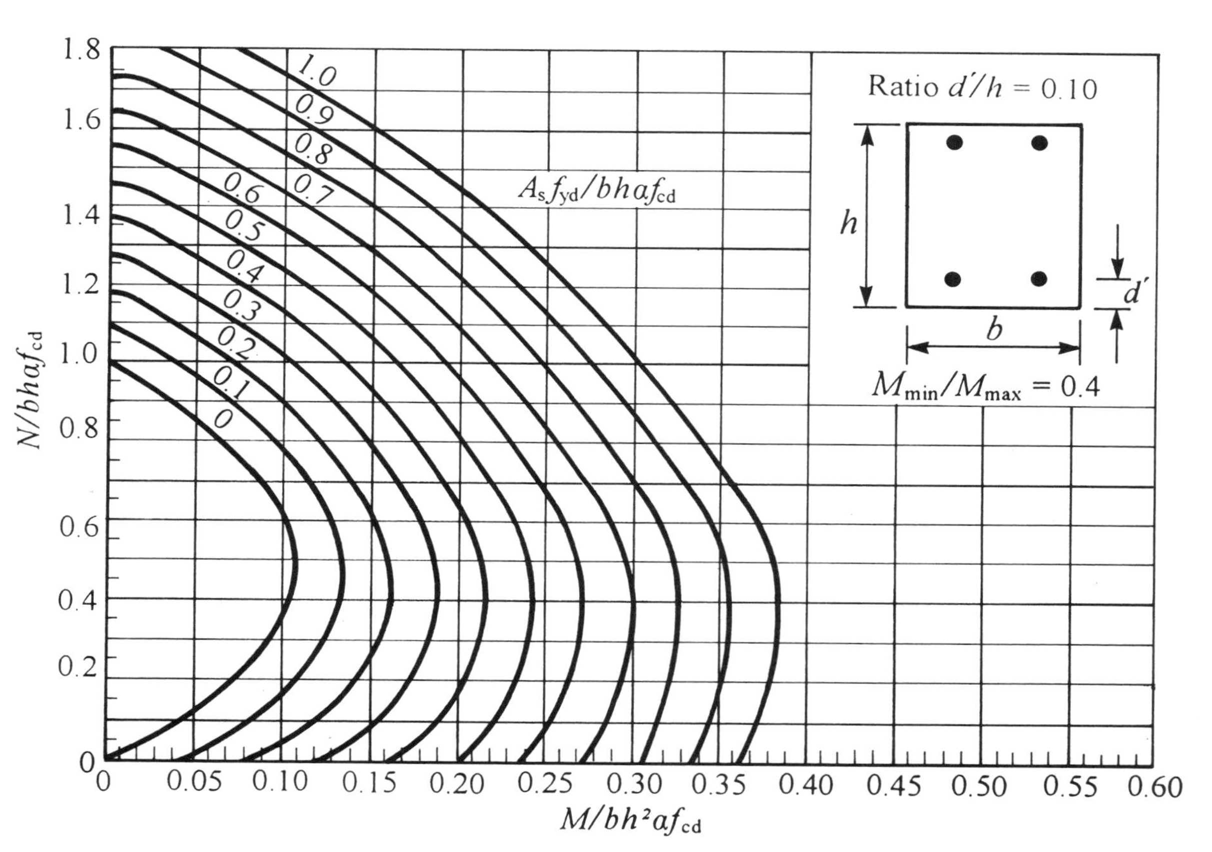

Chart 8.46

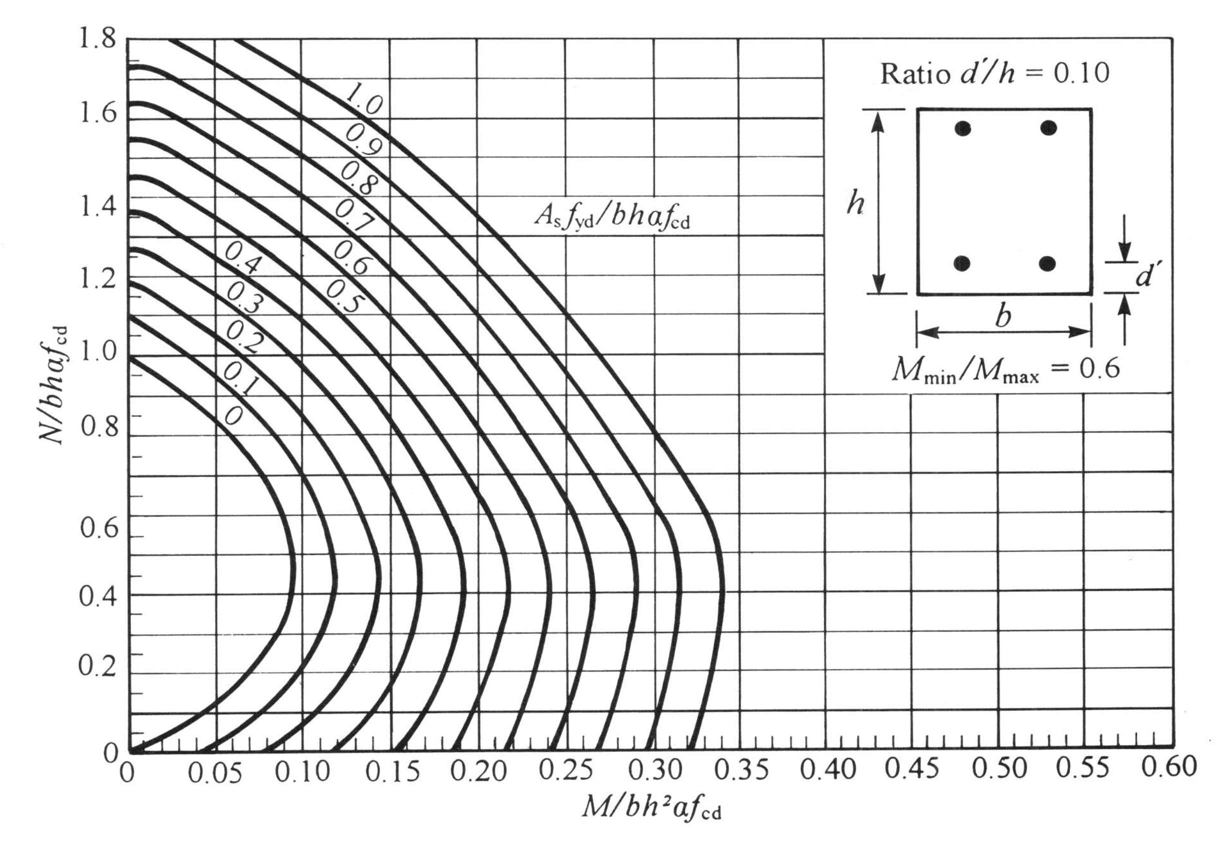

Chart 8.47

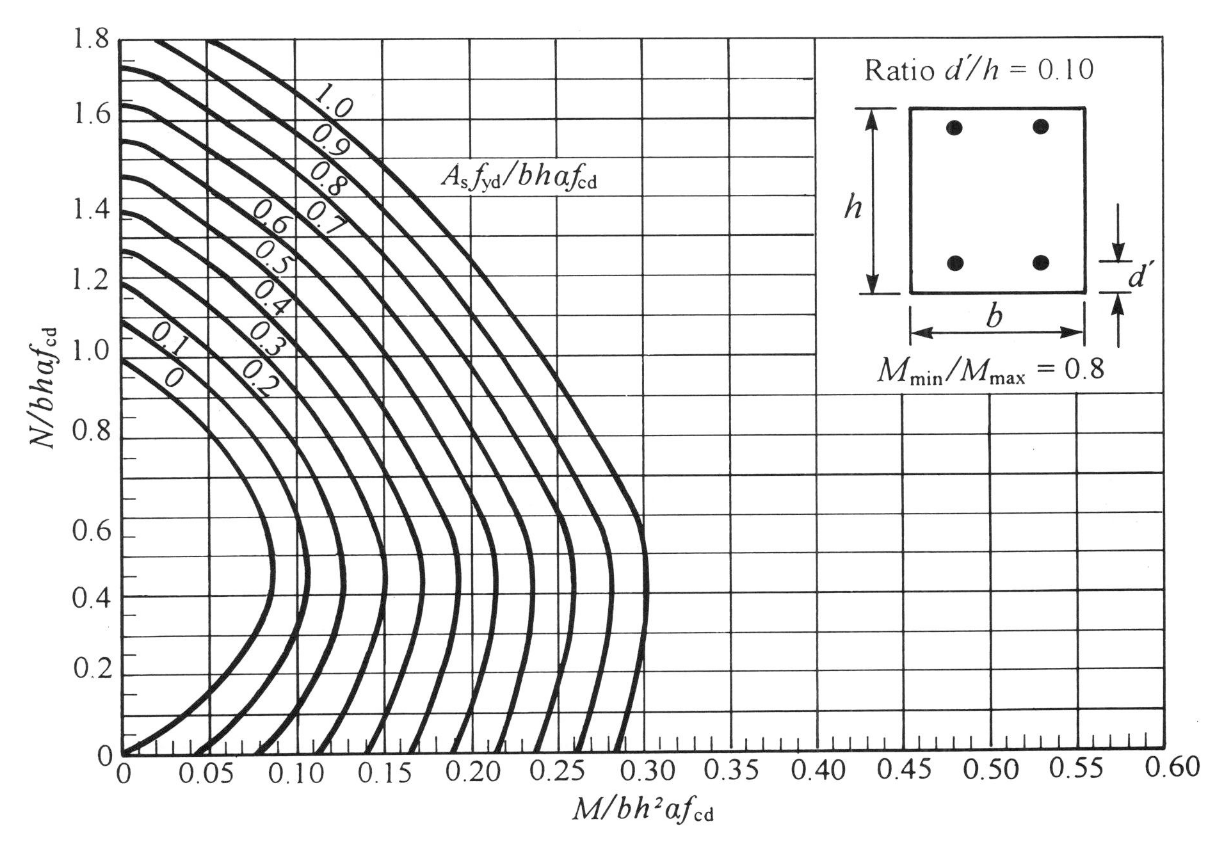

Chart 8.48

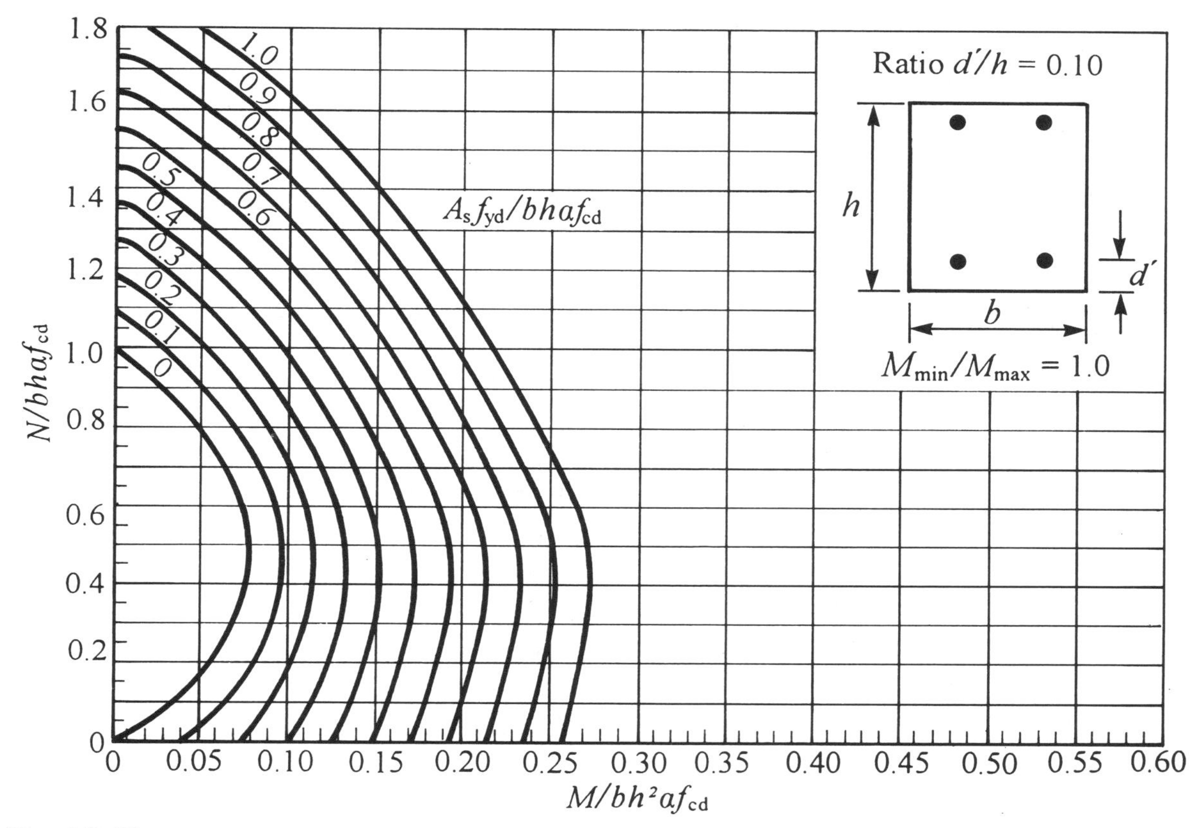

Chart 8.49

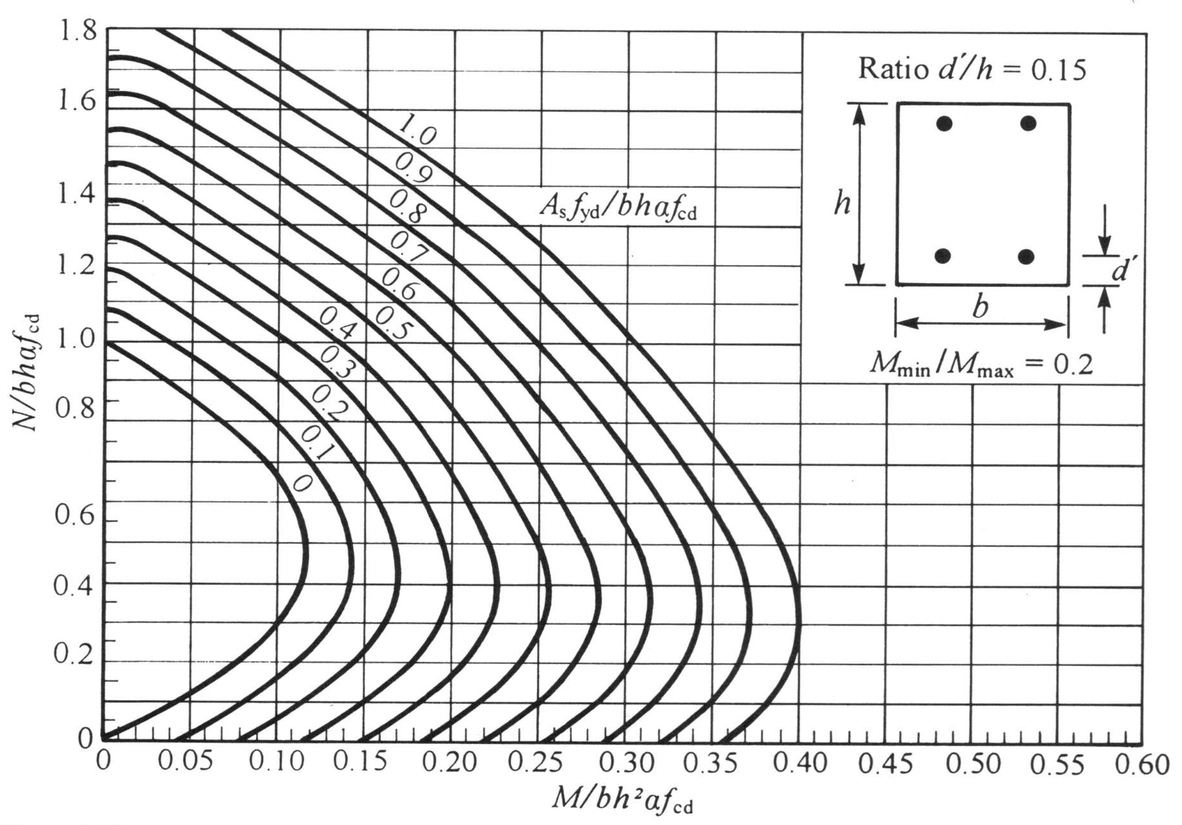

Chart 8.50

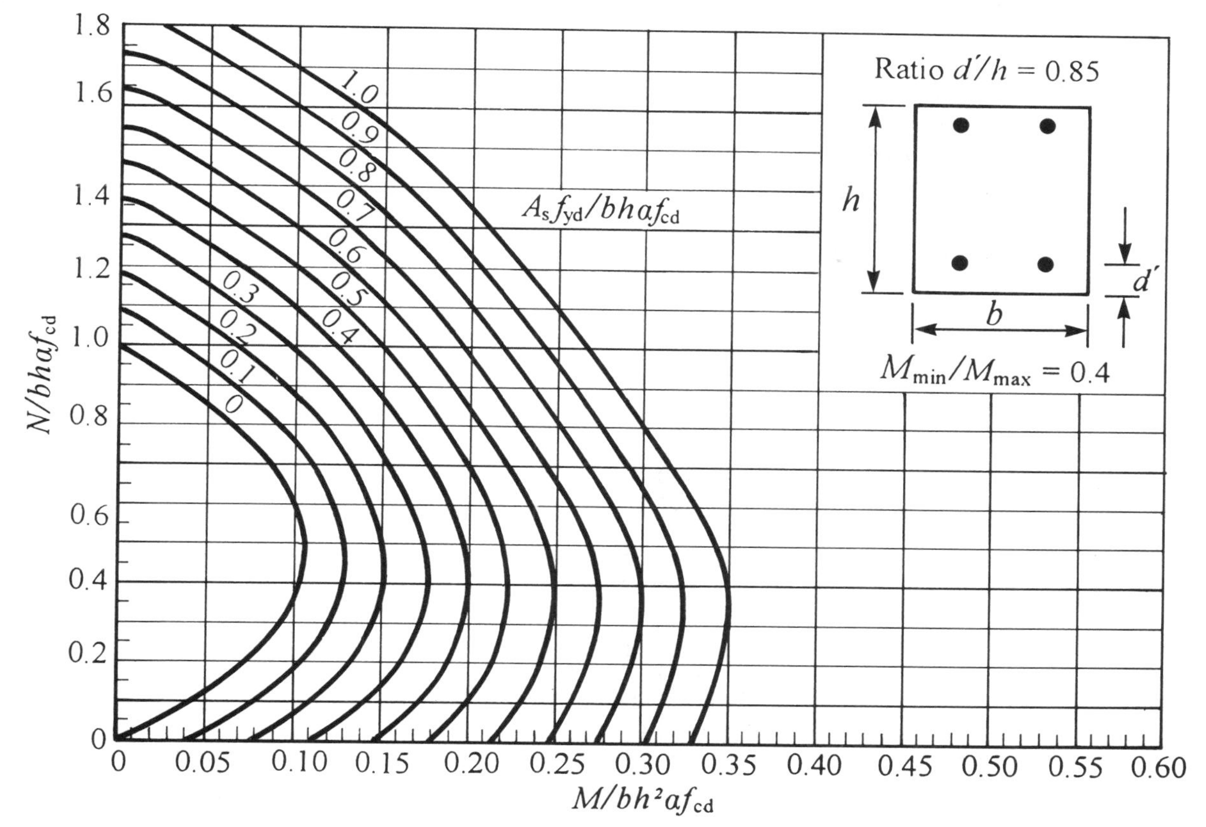

Chart 8.51

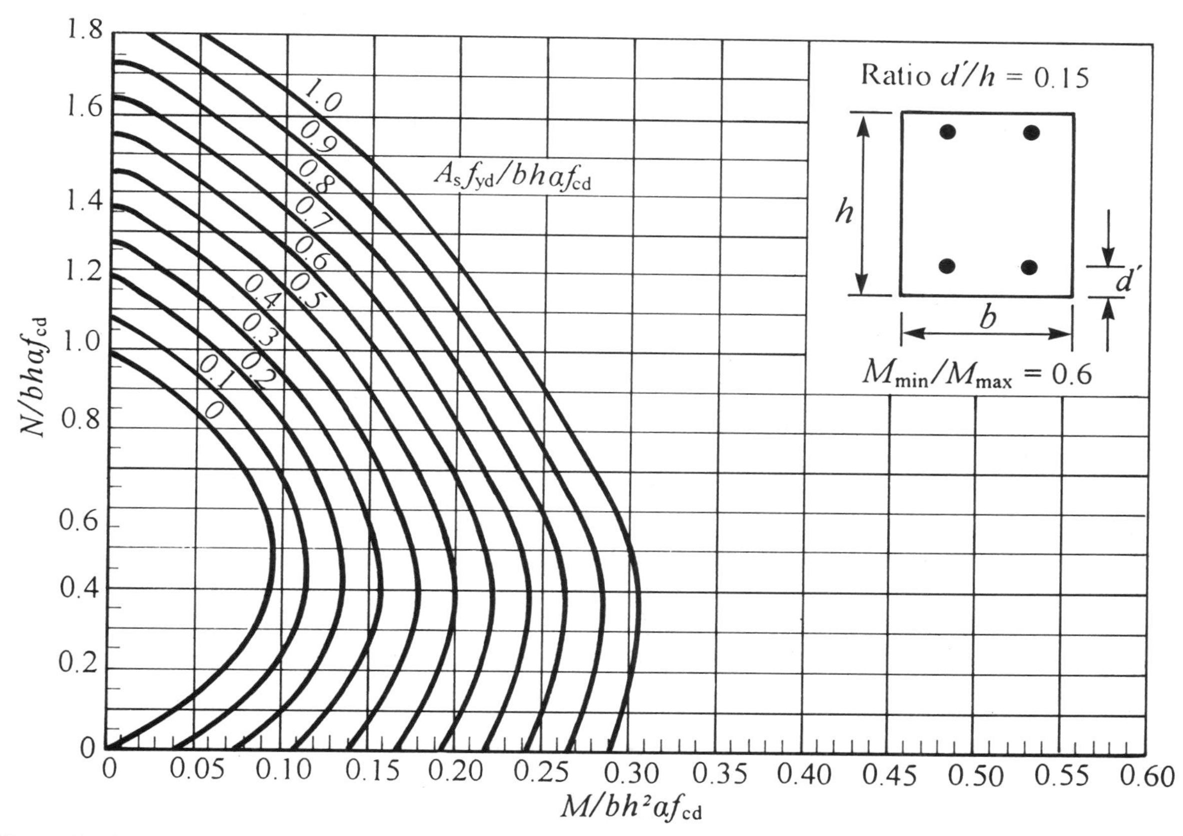

Chart 8.52

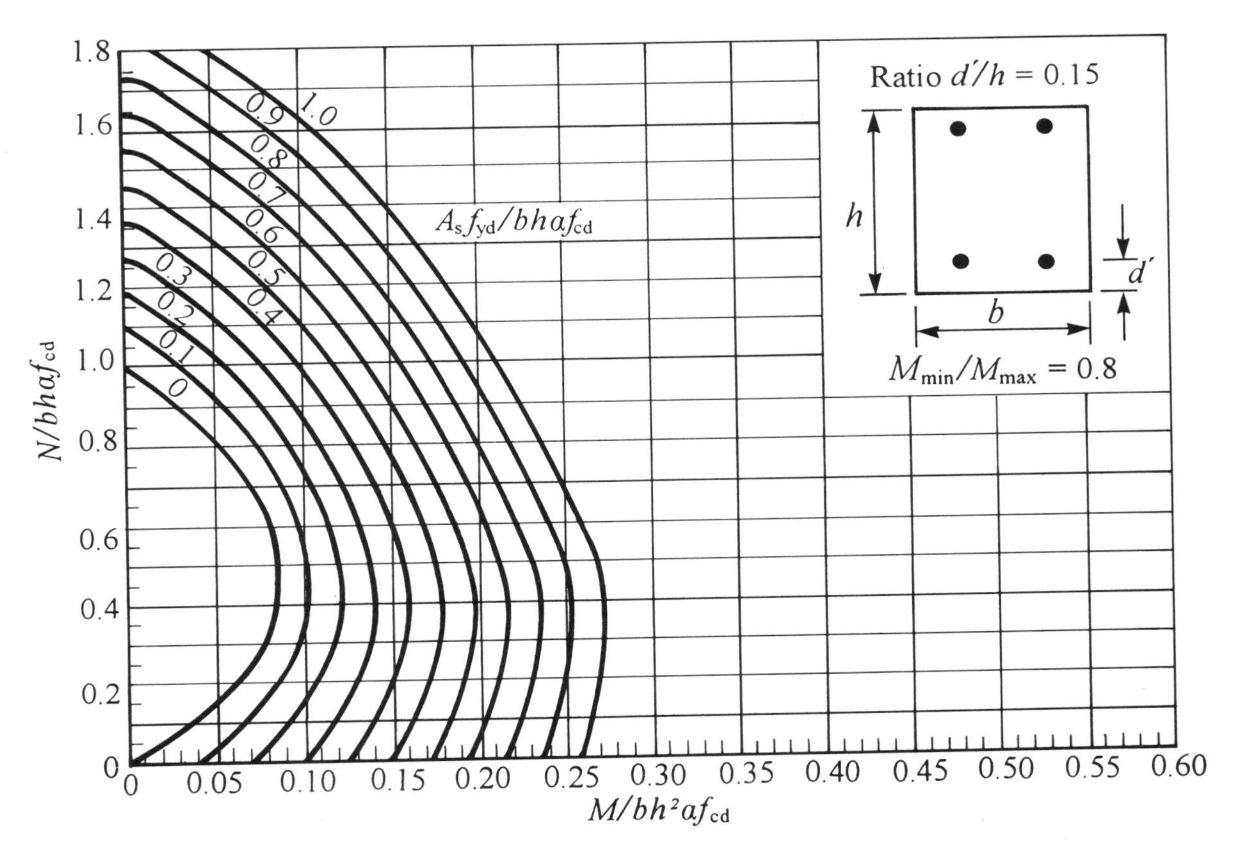

Chart 8.53

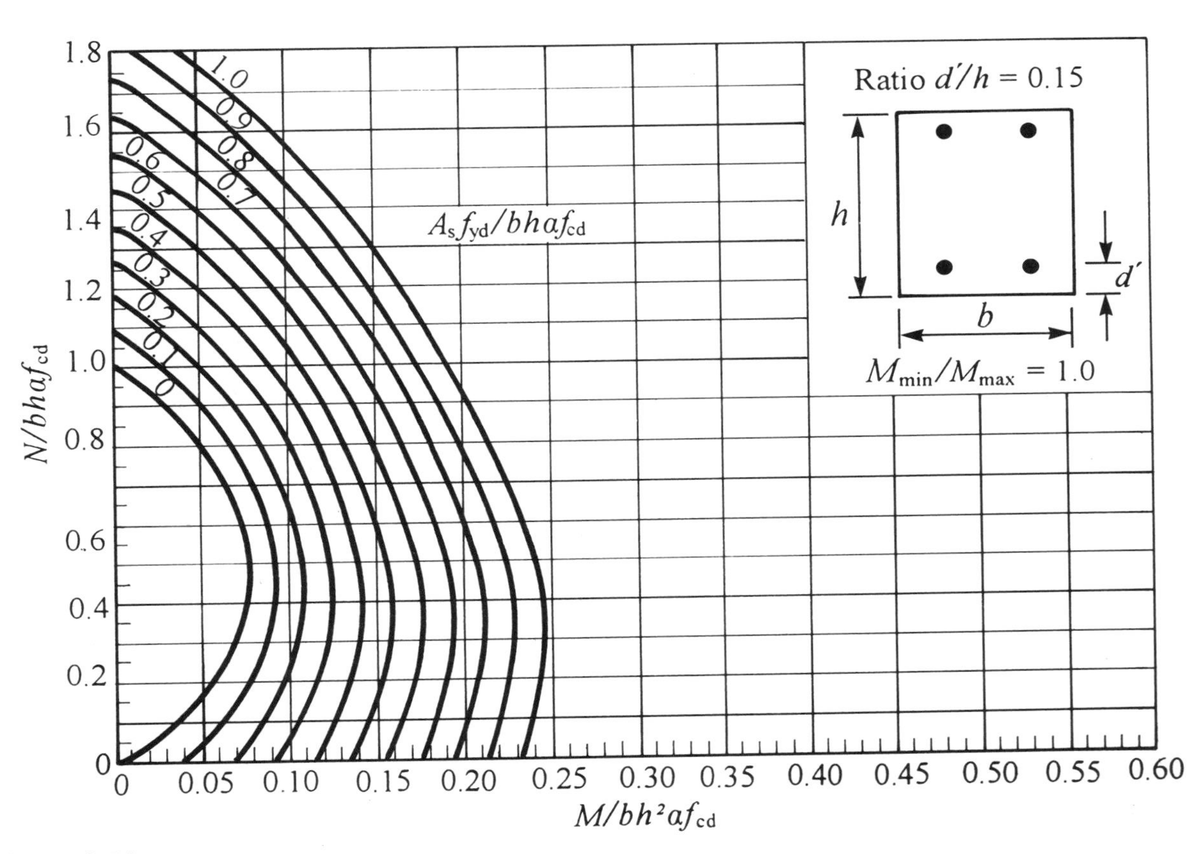

Chart 8.54

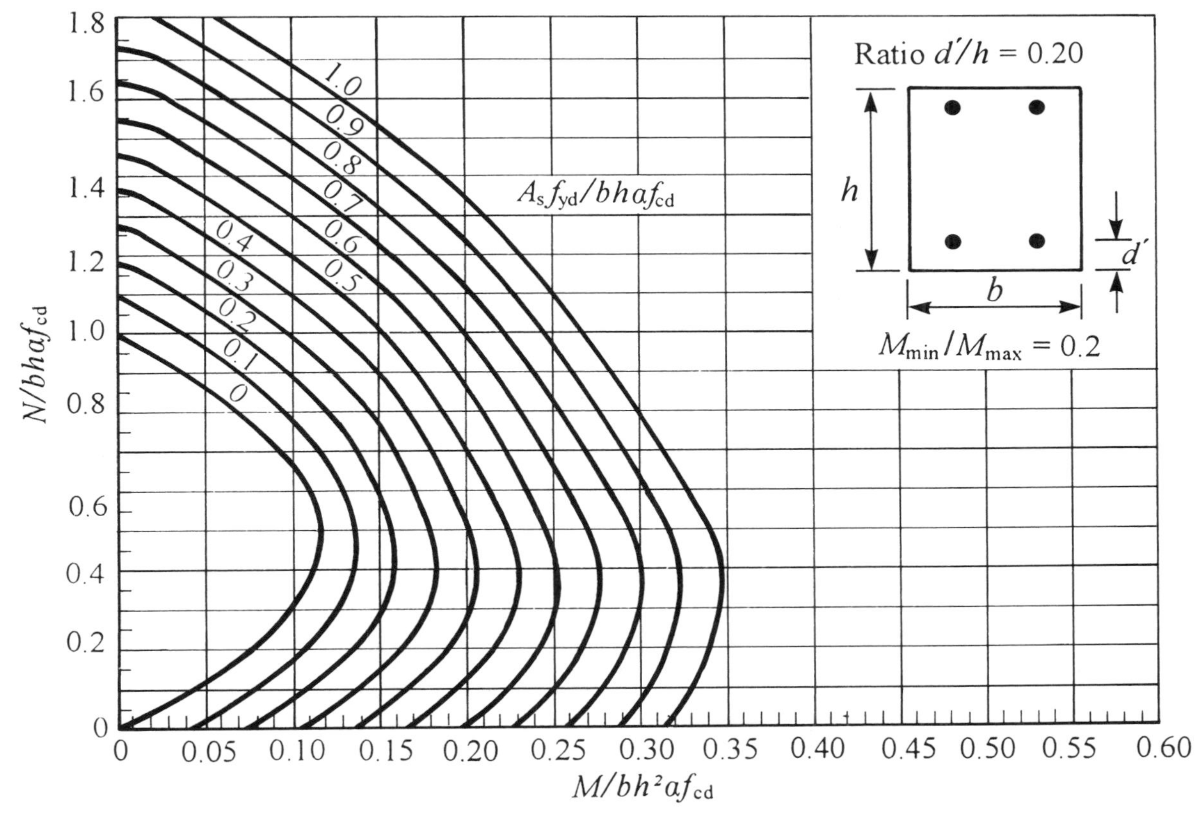

Chart 8.55

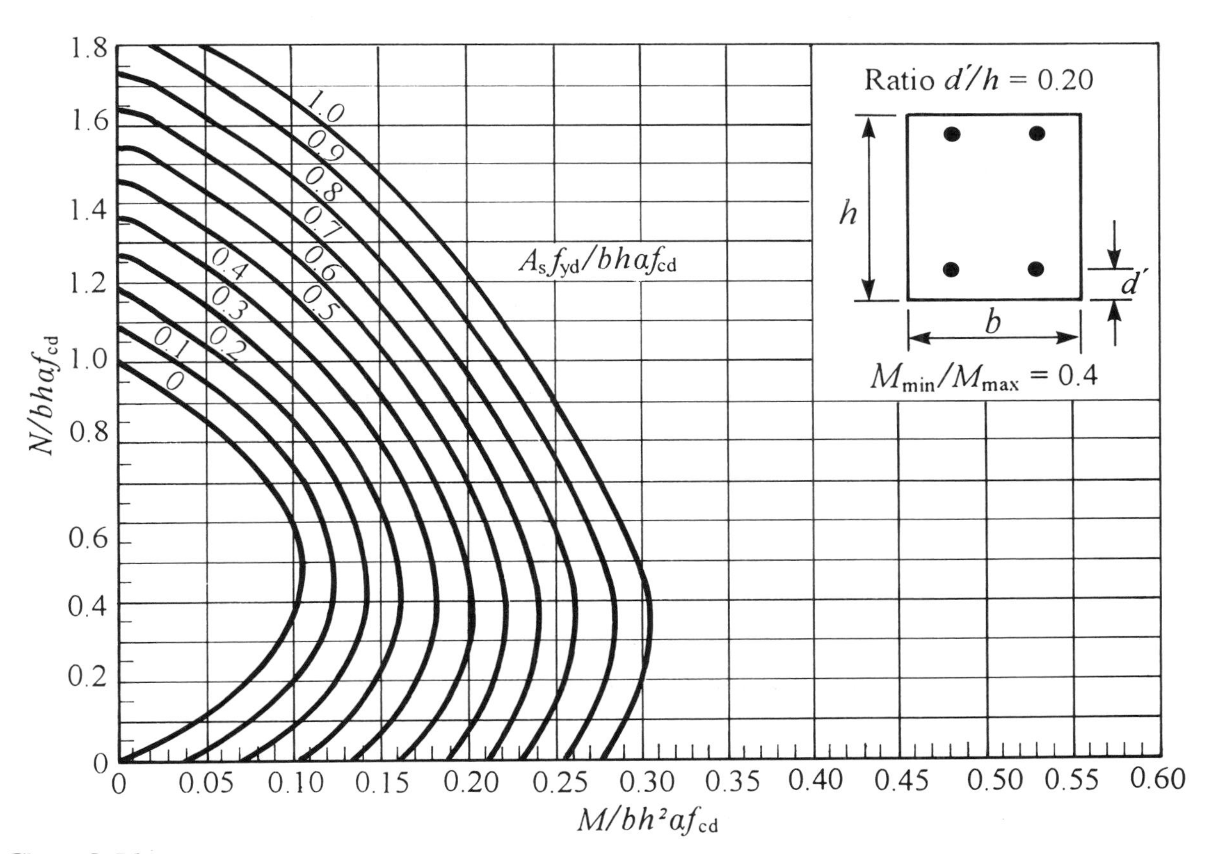

Chart 8.56

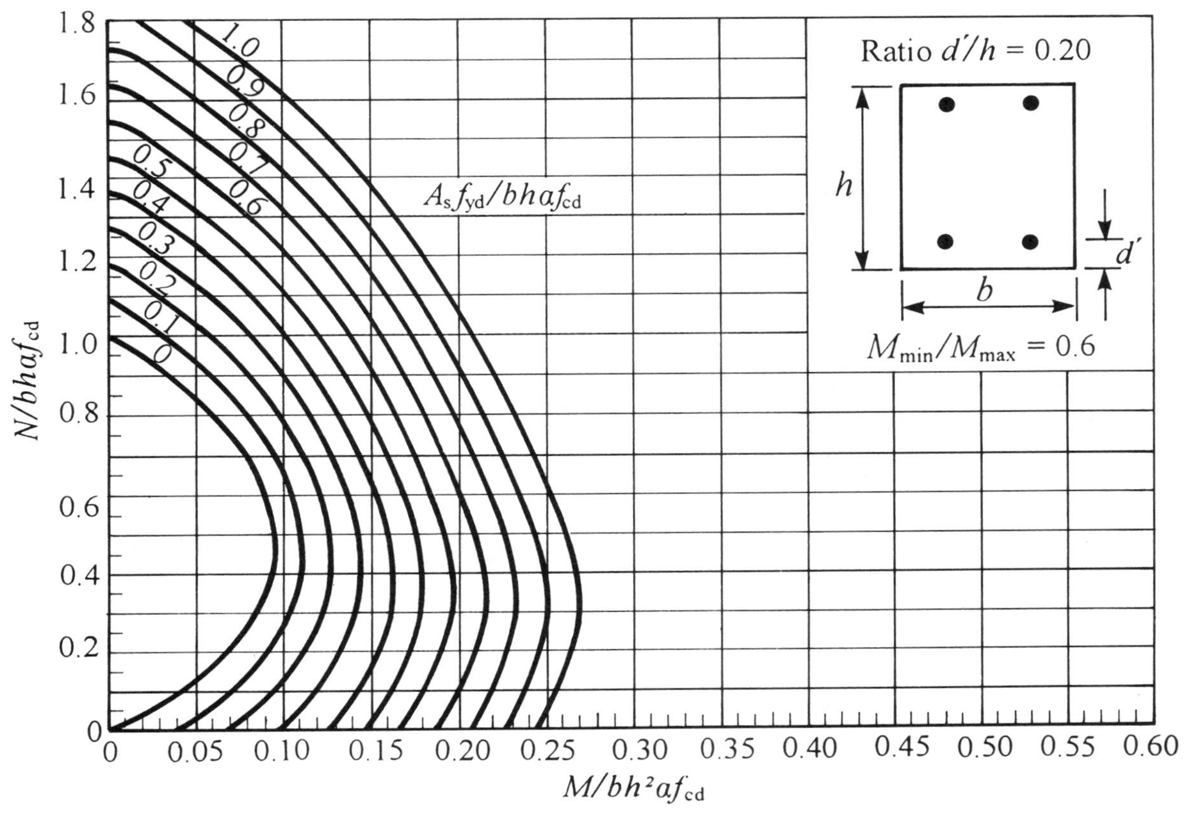

Chart 8.57

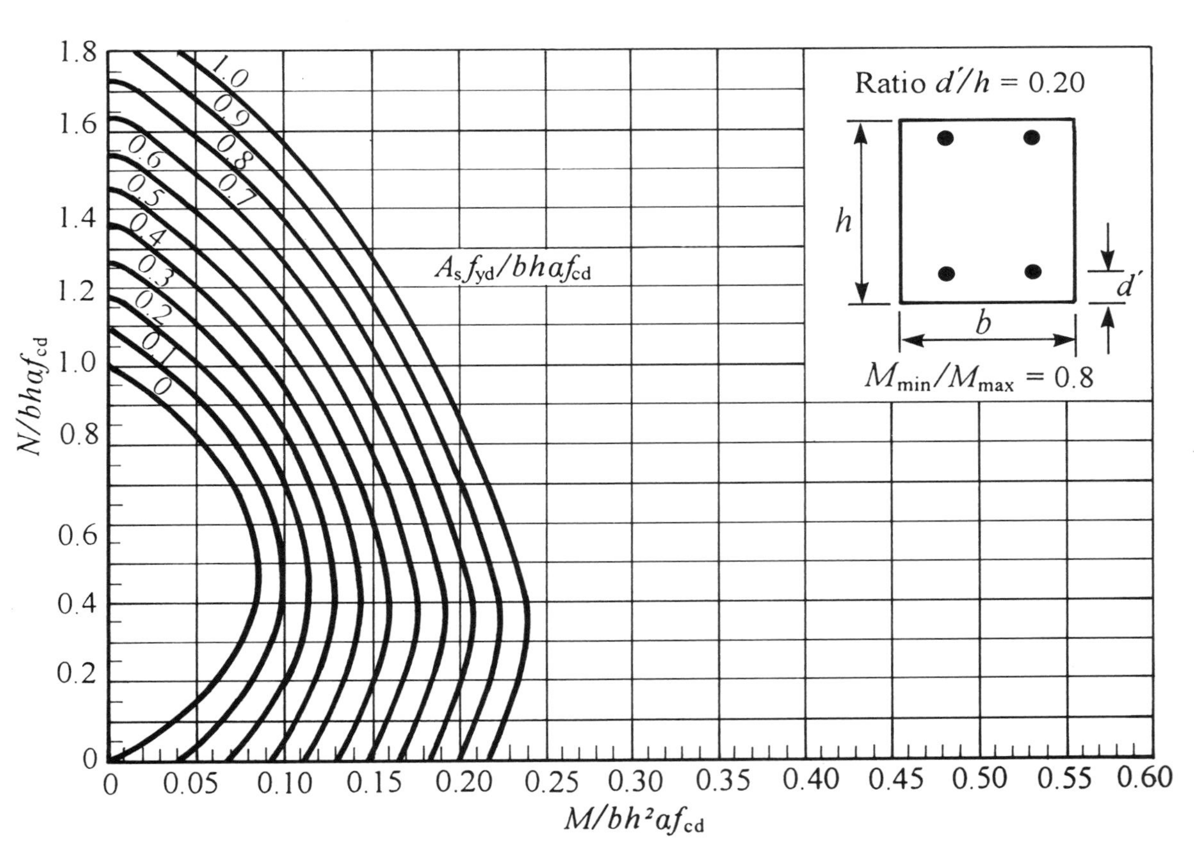

Chart 8.58

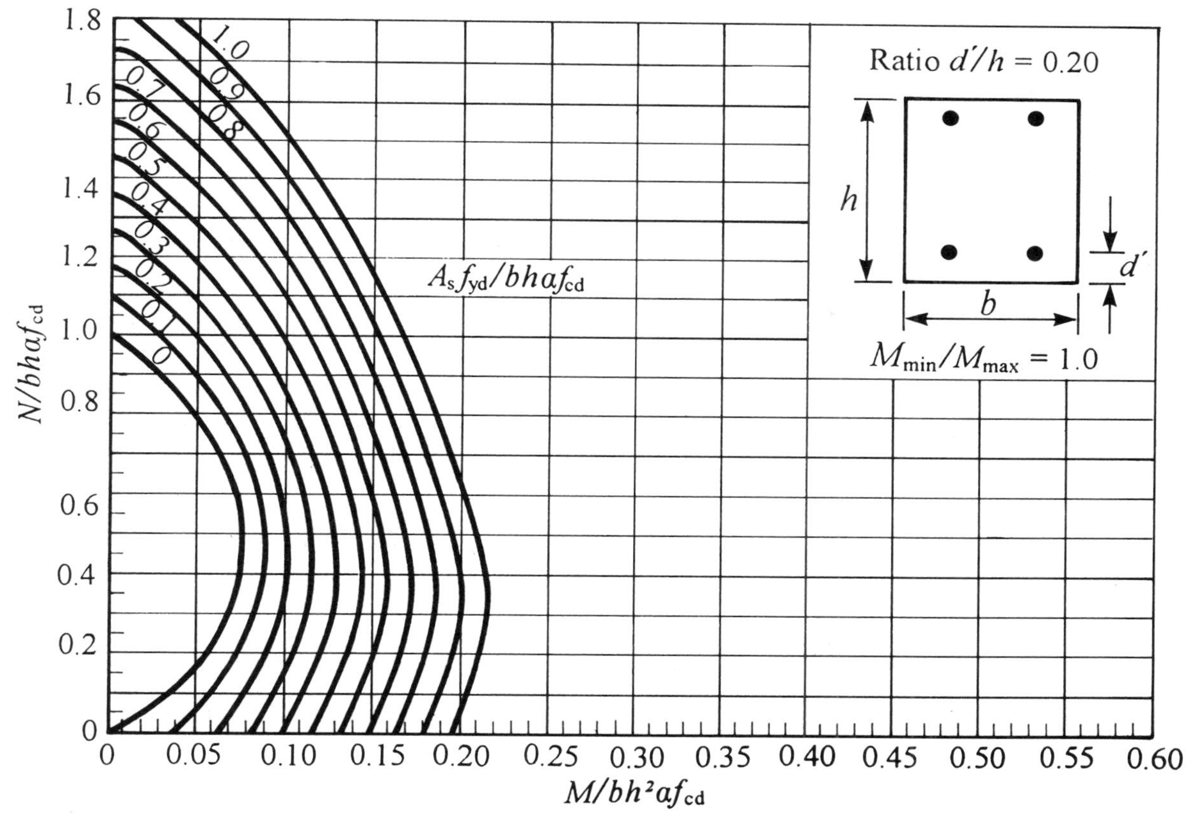

1.8
1.6
1.4
1.2
1.0
0.8
0.6
0.4
0.2
0
$N/bh\alpha f_{cd}$
1.0
0.9
0.8
0.7
0.6
0.5
0.4
0.3
0.2
0.1
0
$A_s f_{yd}/bh\alpha f_{cd}$
Ratio $d'/h = 0.20$
h
d'
b
$M_{min}/M_{max} = 1.0$
0
0.05
0.10
0.15
0.20
0.25
0.30
0.35
0.40
0.45
0.50
0.55
0.60
$M/bh^2\alpha f_{cd}$

Chart 8.59

9 Shear and torsion

9.1 Shear

9.1.1 General

Elements without shear reinforcement 4.3.2.3

Requirement:

$V_{Sd} \leq V_{rd1} \quad \rightarrow \quad \frac{V_{Sd}}{b_w d} \leq \frac{V_{Rd1}}{b_w d}$ 4.3.2.2(2)

V_{Sd} design shear force

V_{Rd1} design shear resistance of the member without shear reinforcement

b_w minimum width of the web

d effective depth

$\frac{V_{Rd1}}{b_w d} = \beta \left(\tau_{Rd}\, k\, (1.2 + 40 \rho_1) + 0.15 \sigma_{cp} \right)$ according to Table 9.1.2 below

Elements with shear reinforcement 4.3.2.4

Standard method 4.3.2.4.3

Requirements:

$V_{sd} \leq V_{Rd3}$ 4.3.2.2(3)

with

$V_{Rd3} = V_{Rd1} + V_{wd} \quad \rightarrow \quad \frac{V_{Sd}}{d} \leq \frac{V_{Rd1}}{d} + \frac{V_{wd}}{d}$ [4.22]

and

$V_{Sd} \leq V_{Rd2} \quad \rightarrow \quad \frac{V_{Sd}}{b_w d} \leq \frac{V_{Rd2}}{b_w d}$ 4.3.2.2(4)

V_{Rd3} design shear resistance of the member with shear reinforcement

V_{wd} contribution of the shear reinforcement

V_{Rd2} maximum design shear force that can be carried without crushing of the notional concrete compressive struts

$$\frac{V_{\mathrm{wd}}}{d} = \frac{A_{\mathrm{sw}}}{s}\, 0.9\, \frac{f_{\mathrm{ywk}}}{\gamma_s}\, (1 + \cot\alpha)\, \sin\alpha$$ according to Table 9.1.5 below

$$\frac{V_{\mathrm{Rd2}}}{b_{\mathrm{w}} d} = \tfrac{1}{2}\, \upsilon\, \frac{f_{\mathrm{ck}}}{\gamma_c}\, 0.9\, (1 + \cot\alpha)$$ according to Table 9.1.3a below

If the effective average stress in the concrete ($\sigma_{\mathrm{cp.eff}}$) is more than 40% of the design value of the compressive cylinder strength of concrete (f_{cd}), V_{Rd2} should be reduced in accordance with the following equation:

$$V_{\mathrm{Rd2.red}} = 1.67\, V_{\mathrm{Rd2}} \left(1 - \frac{\sigma_{\mathrm{cp.eff}}}{f_{\mathrm{ck}}}\, \gamma_c\right) \le V_{\mathrm{Rd2}}$$ according to Table 9.1.4 below

$V_{\mathrm{Rd2.red}}$ reduced maximal design shear force that can be carried without crushing of the notional concrete compressive struts

Variable strut inclination method 4.3.2.4.4 below

Requirements:

$V_{\mathrm{Sd}} \le V_{\mathrm{Rd3}}$ → $\dfrac{V_{\mathrm{Sd}}}{d} \le \dfrac{V_{\mathrm{Rd3}}}{d}$ 4.3.2.2(3)

and

$V_{\mathrm{Sd}} \le V_{\mathrm{Rd2}}$ → $\dfrac{V_{\mathrm{Sd}}}{b_{\mathrm{w}} d} \le \dfrac{V_{\mathrm{Rd2}}}{b_{\mathrm{w}} d}$ 4.3.2.2(4)

$$\frac{V_{\mathrm{wd}}}{d} = \frac{A_{\mathrm{sw}}}{s}\, \frac{z}{d}\, \frac{f_{\mathrm{ywk}}}{\gamma_s}\, (1 + \cot\alpha)\, \sin\alpha\, k_\theta$$ according to Table 9.1.5 below

with $k_\theta = \dfrac{\cot\theta + \cot\alpha}{1 + \cot\alpha}$

$$\frac{V_{\mathrm{Rd2}}}{b_{\mathrm{w}} d} = \frac{\upsilon\, \dfrac{f_{\mathrm{ck}}}{\gamma_c}\, \dfrac{z}{d}\, (\cot\theta + \cot\alpha)}{(1 + \cot^2\theta)}$$ according to Table 9.1.3b below

If the effective average stress in the concrete ($\sigma_{\mathrm{cp.eff}}$) is more than 40% of the design value of the compressive cylinder strength of concrete (f_{cd}), V_{Rd2} should be reduced in accordance with the following equation (4.3.2.2(4)):

$$V_{\mathrm{Rd2.red}} = 1.67\, V_{\mathrm{Rd2}} \left(1 - \frac{\sigma_{\mathrm{cp.eff}}}{f_{\mathrm{ck}}}\, \gamma_c\right) \le V_{\mathrm{Rd2}}$$ according to Table 9.1.4 below

For members with inclined prestressing tendons, V_{Sd} is given by:

$$V_{Sd} = V_{od} - V_{pd}$$ [4.32]

V_{od} design shear force in the section

V_{pd} force component of the inclined prestressed tendons, parallel to V_{od} (see Figure 9.1)

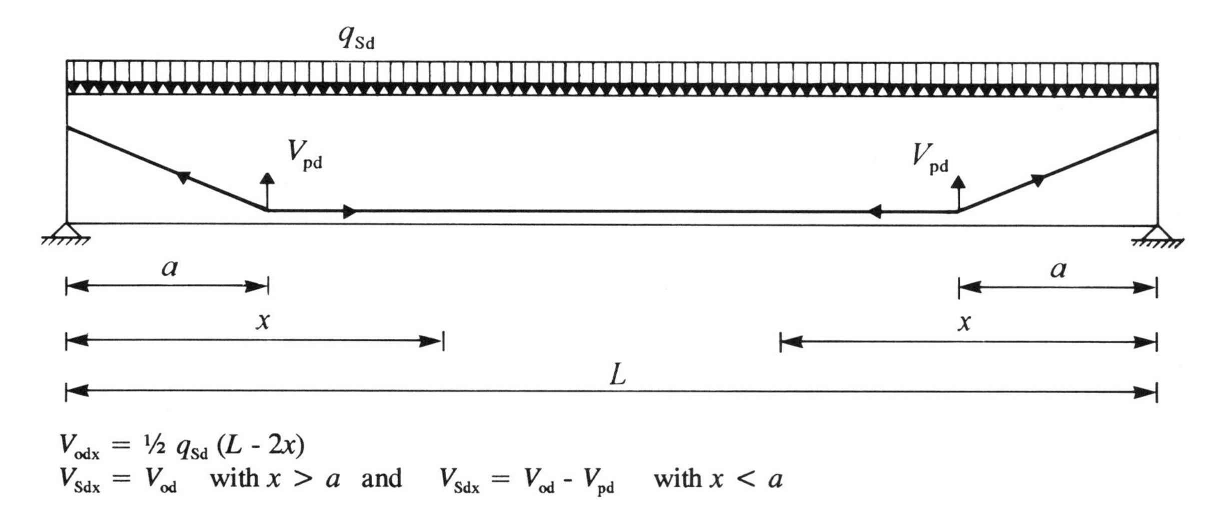

$V_{odx} = ½\, q_{Sd}\,(L - 2x)$
$V_{Sdx} = V_{od}$ with $x > a$ and $V_{Sdx} = V_{od} - V_{pd}$ with $x < a$

Figure 9.1 Force component (V_{pd}) of the inclined prestressed tendons, parallel to the design shear force in the section (V_{od}) .

Apply the detailing requirements according to 4.3.2.4(4)

For the notation for members subjected to shear, see Figure 9.2.

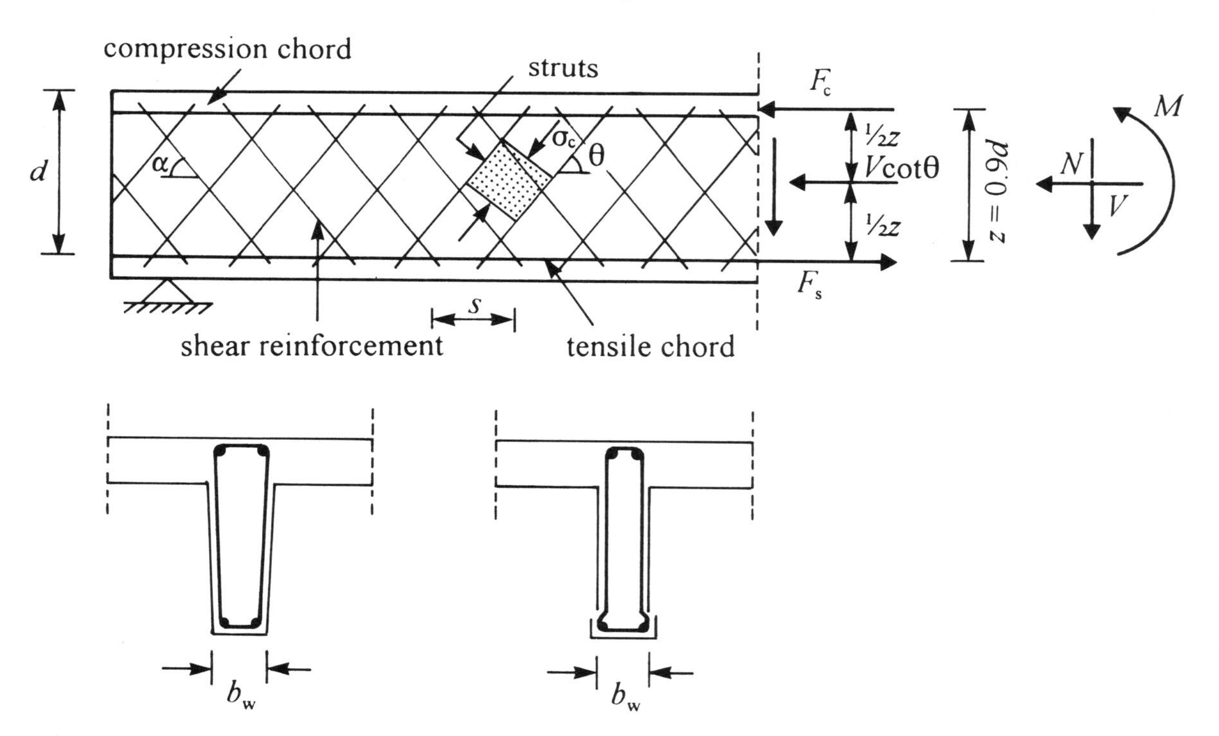

Figure 9.2 Notation for members subjected to shear

9.1.2 $\dfrac{V_{Rd1}}{b_w d}$ **[4.17 and 4.18]**

Table 9.1

$100\rho_l$ (%)	$V_{Rd1}/b_w d$ (with β=1.0; γ=1.5; k=1.0; σ_{cp}=0 N/mm²) in N/mm² per concrete class								
	C12/15	C16/20	C20/25	C25/30	C30/37	C35/45	C40/50	C45/55	C50/60
0.0	0.216	0.264	0.312	0.360	0.408	0.444	0.492	0.528	0.576
0.1	0.223	0.273	0.322	0.372	0.422	0.459	0.508	0.546	0.595
0.2	0.230	0.282	0.333	0.384	0.435	0.474	0.525	0.563	0.614
0.3	0.238	0.290	0.343	0.396	0.449	0.488	0.541	0.581	0.634
0.4	0.245	0.299	0.354	0.408	0.462	0.503	0.558	0.598	0.653
0.5	0.252	0.308	0.364	0.420	0.476	0.518	0.574	0.616	0.672
0.6	0.259	0.317	0.374	0.432	0.490	0.533	0.590	0.634	0.691
0.7	0.266	0.326	0.385	0.444	0.503	0.548	0.607	0.651	0.710
0.8	0.274	0.334	0.395	0.456	0.517	0.562	0.623	0.669	0.730
0.9	0.281	0.343	0.406	0.468	0.530	0.577	0.640	0.686	0.749
1.0	0.288	0.352	0.416	0.480	0.544	0.592	0.656	0.704	0.768
1.1	0.295	0.361	0.426	0.492	0.558	0.607	0.672	0.722	0.787
1.2	0.302	0.370	0.437	0.504	0.571	0.622	0.689	0.739	0.806
1.3	0.310	0.378	0.447	0.516	0.585	0.636	0.705	0.757	0.826
1.4	0.317	0.387	0.458	0.528	0.598	0.651	0.722	0.774	0.845
1.5	0.324	0.396	0.468	0.540	0.612	0.666	0.738	0.792	0.864
1.6	0.331	0.405	0.478	0.552	0.626	0.681	0.754	0.810	0.883
1.7	0.338	0.414	0.489	0.564	0.639	0.696	0.771	0.827	0.902
1.8	0.346	0.422	0.499	0.576	0.653	0.710	0.787	0.845	0.922
1.9	0.353	0.431	0.510	0.588	0.666	0.725	0.804	0.862	0.941
2.0	0.360	0.440	0.520	0.600	0.680	0.740	0.820	0.880	0.960
τ_{Rd} (N/mm²)	0.18	0.22	0.26	0.30	0.34	0.37	0.41	0.44	0.48

$$\frac{V_{Rd1}}{b_w d} = \beta(\tau_{Rd}k(1.2 + 40\rho_l) + 0.15\sigma_{cp})$$ (with β=1.0; γ=1.5; k=1.0; σ_{cp}=0 N/mm²) in N/mm²

- If the distance x of a concentrated load is less than 2.5d from the face of the support, multiply by $\beta = 2.5d/x \leq 5$ to determine the design shear resistance of the member without shear reinforcement for the concentrated load
- If $\gamma_c \neq 1.5$, multiply by $1.5/\gamma_c$
- If $d < 0.6$ m, multiply by $k = 1.6 - d$ (d in metres)
- If $100\rho_l > 2.0$ %, take $100\rho_l = 2.0$ % into account
- If $\sigma_{cp} \neq 0$ N/mm², add $\beta * 0.15\sigma_{cp}$ (compression positive)

9.1.3a Standard method $\dfrac{V_{Rd2}}{b_w d}$ [4.19, 4.20 and 4.25]

Table 9.2

α (degrees)	$\dfrac{V_{Rd2}}{b_w d}$ (with γ_c = 1.5) in N/mm² per concrete class								
	C12/15	C16/20	C20/25	C25/30	C30/37	C35/45	C40/50	C45/55	C50/60
90	2.30	2.98	3.60	4.31	4.95	5.51	6.00	6.75	7.50
85	2.51	3.24	3.92	4.69	5.38	5.99	6.52	7.34	8.16
80	2.71	3.50	4.23	5.07	5.82	6.48	7.06	7.94	8.82
75	2.92	3.77	4.56	5.47	6.28	6.99	7.61	8.56	9.51
70	3.14	4.06	4.91	5.88	6.75	7.52	8.18	9.21	10.23
65	3.38	4.36	5.28	6.32	7.26	8.08	8.80	9.90	11.00
60	3.63	4.69	5.68	6.80	7.81	8.70	9.46	10.65	11.83
55	3.92	5.06	6.12	7.33	8.42	9.37	10.20	11.48	12.75
50	4.24	5.47	6.62	7.93	9.10	10.14	11.03	12.41	13.79
45	4.61	5.95	7.20	8.63	9.90	11.03	12.00	13.50	15.00

$\dfrac{V_{Rd2}}{b_w d} = \frac{1}{2}\,\upsilon\,\dfrac{f_{ck}}{\gamma_c}\,0.9\,(1 + \cot\alpha)$ (with $\upsilon = 0.7 - \dfrac{f_{ck}}{200} \not< 0.5$ (f_{ck} in N/mm²) and γ_c = 1.5) in N/mm²

For sections without designed shear reinforcement, α = 90° should be taken [4.19].

- If $\gamma_c \neq 1.5$, multiply by $1.5/\gamma_c$

9.1.3b Variable strut inclination method $\dfrac{V_{Rd2}}{b_w d}$ [4.26 and 4.28]

Table 9.3

α (degrees)	θ (degrees)	$\frac{V_{Rd2}}{b_w d}$ (with γ_c = 1.5 and $\frac{z}{d}$ = 0.9) in N/mm² per concrete class								
		C12/15	C16/20	C20/25	C25/30	C30/37	C35/45	C40/50	C45/55	C50/60
90	68	1.60	2.07	2.50	3.00	3.44	3.83	4.17	4.69	5.21
	60	2.00	2.58	3.12	3.73	4.29	4.77	5.20	5.85	6.50
	45	2.30	2.98	3.60	4.31	4.95	5.51	6.00	6.75	7.50
	30	2.00	2.58	3.12	3.73	4.29	4.77	5.20	5.85	6.50
	22	1.60	2.07	2.50	3.00	3.44	3.83	4.17	4.69	5.21
75	68	2.66	3.44	4.16	4.98	5.72	6.37	6.93	7.80	8.67
	60	2.92	3.77	4.56	5.47	6.28	6.99	7.61	8.56	9.51
	45	2.92	3.77	4.56	5.47	6.28	6.99	7.61	8.56	9.51
	30	2.30	2.98	3.60	4.31	4.95	5.51	6.00	6.75	7.50
	22	1.77	2.29	2.77	3.32	3.81	4.24	4.62	5.20	5.77
60	68	3.89	5.02	6.07	7.28	8.35	9.30	10.12	11.39	12.65
	60	3.99	5.15	6.24	7.47	8.57	9.55	10.39	11.69	12.99
	45	3.63	4.69	5.68	6.80	7.81	8.70	9.46	10.65	11.83
	30	2.66	3.44	4.16	4.98	5.72	6.37	6.93	7.79	8.66
	22	1.97	2.55	3.08	3.69	4.24	4.72	5.14	5.78	6.43
45	68	5.56	7.18	8.69	10.41	11.95	13.31	14.48	16.29	18.10
	60	5.45	7.04	8.52	10.20	11.71	13.04	14.20	15.97	17.75
	45	4.61	5.95	7.20	8.63	9.90	11.03	12.00	13.50	15.00
	30	3.15	4.07	4.92	5.89	6.76	7.53	8.20	9.22	10.25
	22	2.25	2.90	3.51	4.21	4.83	5.38	5.85	6.58	7.31

$$\frac{V_{Rd2}}{b_w d} = \frac{\upsilon \dfrac{f_{ck}}{\gamma_c} \dfrac{z}{d} (\cot\theta + \cot\alpha)}{(1 + \cot^2\theta)}$$

(with $\upsilon = 0.7 - \dfrac{f_{ck}}{200} \nless 0.5$ (f_{ck} in N/mm²); γ_c = 1.5 and $\dfrac{z}{d}$ = 0.9) in N/mm²

- If $\gamma_c \neq 1.5$, multiply by $1.5/\gamma_c$
- If $z \neq 0.9d$, multiply by $z/(0.9d)$

9.1.4 $\dfrac{V_{Rd2.red}}{V_{Rd2}}$ **[4.15 and 4.16]**

Table 9.4

$$\frac{V_{Rd2.red}}{V_{Rd2}} = 1.67\left(1 - \frac{\sigma_{cp.eff}}{f_{ck}}\gamma_c\right) \leq 1 \qquad [4.15]$$

$$\sigma_{cp.eff} = \frac{N_{Sd}}{A_c} - \frac{A_{s2}f_{yk}}{A_c\gamma_s} \qquad [4.16]$$

- If f_{yk}/γ_s > 400 N/mm², take f_{yk}/γ_s = 400 N/mm² into account

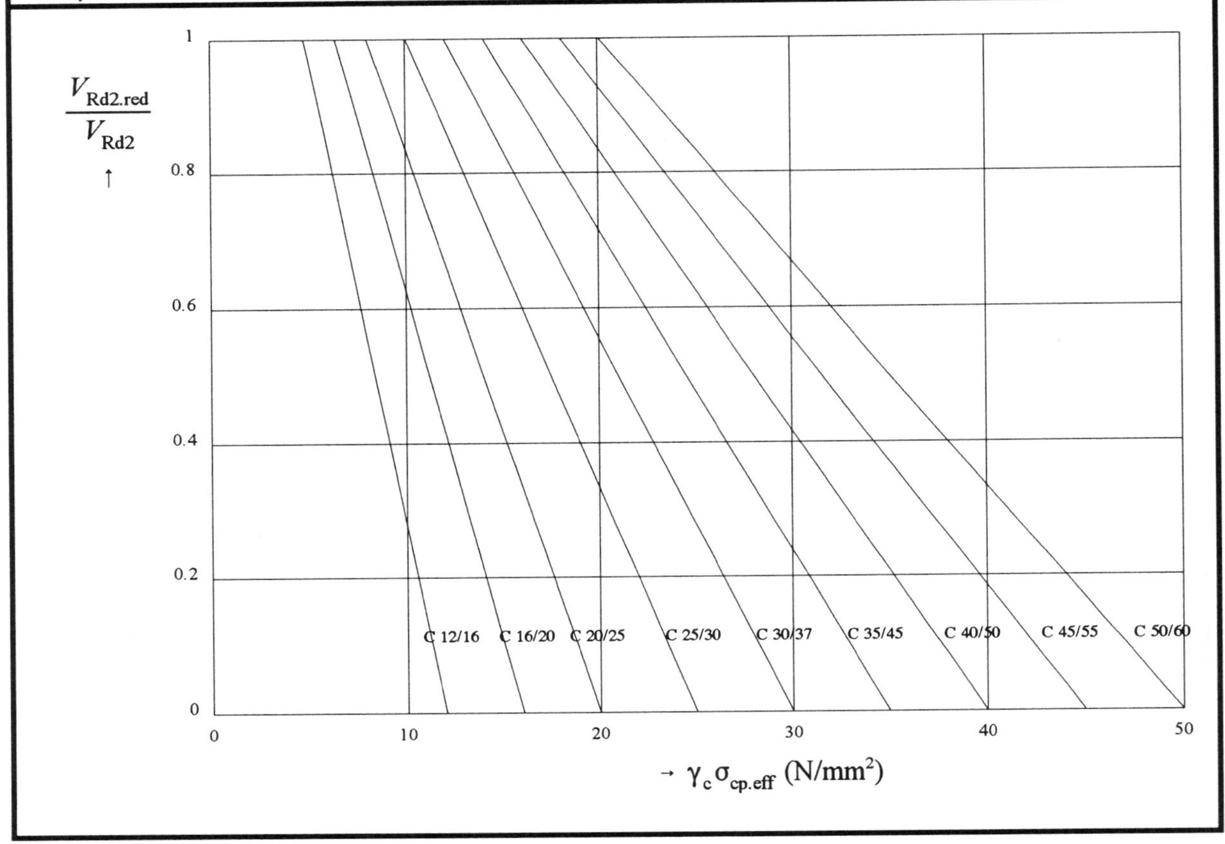

9.1.5 $\frac{V_{wd}}{d}$ and $\frac{V_{Rd3}}{d}$ [4.24 and 4.29]

Table 9.5

Stirrups with two legs		$\frac{A_{sw}}{s}$ (mm²/m)	$\frac{V_{wd}}{d}$ in kN/m for α =				Stirrups with two legs		$\frac{A_{sw}}{s}$ (mm²/m)	$\frac{V_{wd}}{d}$ in kN/m for α =			
			90°	75°	60°	45°				90°	75°	60°	45°
ø 5	100	393	154	188	210	217	ø 10	100	1571	615	753	840	870
	150	262	103	126	140	145		150	1047	410	502	560	580
	200	196	77	94	105	109		200	785	307	377	420	435
	250	157	61	75	84	87		250	628	246	301	336	348
	300	131	51	63	70	72		300	524	205	251	280	290
ø 6	100	565	221	271	302	313	ø 12	100	2262	886	1085	1210	1252
	150	377	148	181	202	209		150	1508	590	723	806	835
	200	283	111	136	151	157		200	1131	443	542	605	626
	250	226	88	108	121	125		250	905	354	434	484	501
	300	188	74	90	101	104		300	754	295	362	403	417
ø 8	100	1005	393	482	538	557	ø 16	100	4021	1,574	1928	2151	2226
	150	670	262	321	358	371		150	2681	1,050	1285	1434	1484
	200	503	197	241	269	278		200	2011	787	964	1075	1113
	250	402	157	193	215	223		250	1608	630	771	860	891
	300	335	131	161	179	186		300	1340	525	643	717	742

Standard method

- If $\theta = 45°$

$$\frac{V_{wd}}{d} = \frac{A_{sw}}{s}\, 0.9\, \frac{f_{ywk}}{\gamma_s}\,(1 + \cot\alpha)\sin\alpha \qquad (\text{with } \gamma_s = 1.15 \text{ and } f_{ywk} = 500 \text{ N/mm}^2) \text{ in N/mm}^2$$

- If $\gamma_s \neq 1.15$ multiply by $1.15/\gamma_s$
- If $f_{ywk} \neq 500$ N/mm², multiply by $f_{ywk}/500$

Variable strut inclination method

- If $\theta \neq 45°$

$$\frac{V_{wd}}{d} = \frac{A_{sw}}{s}\,\frac{z}{d}\,\frac{f_{ywk}}{\gamma_s}\,(1 + \cot\alpha)\sin\alpha\, k_\theta \qquad \left(\text{with } \frac{z}{d} = 0.9\right)$$

k_θ according to:

θ	α	k_θ	θ	α	k_θ	θ	α	k_θ	θ	α	k_θ
68	90	0.404	60	90	0.577	30	90	1.732	22	90	2.475
	75	0.530		75	0.667		75	1.577		75	2.163
	60	0.622		60	0.732		60	1.464		60	1.935
	45	0.702		45	0.789		45	1.366		45	1.738

$$k_\theta = \frac{\cot\theta + \cot\alpha}{1 + \cot\alpha}$$

- If $z \neq 0.9d$, multiply by $z/(0.9d)$

The upper part of the Table represents the values V_{wd}/d according to equation 4.24

The values in the upper part multiplied by k_θ represent V_{Rd3}/d according to equation 4.29

9.2 Torsion

9.2.1 General

Requirements

$T_{Sd} \leq T_{Rd1}$ $\rightarrow$ $\frac{T_{Sd}}{h^3} \leq \frac{T_{Rd1}}{h^3}$ [4.38]

and

$T_{Sd} \leq T_{Rd2}$ $\rightarrow$ $\frac{T_{Sd}}{h^2} \leq \frac{T_{Rd2}}{h^2}$ [4.39]

or

$\rightarrow$ $\frac{T_{Sd}}{h^3} \leq \frac{T_{Rd2}}{h^3}$

T_{Sd} design torsional moment

T_{Rd1} maximum torsional moment that can be resisted by the compressive struts in the concrete

T_{Rd2} maximum torsional moment that can be resisted by the torsion reinforcement

$$\frac{T_{Rd1}}{h^3} = \upsilon\, f_{cd}\, \frac{t}{h} \left(\frac{b}{h} - \frac{t}{h}\right)\left(1 - \frac{t}{h}\right) k_{\theta,1}$$ according to Table 9.2.2 below

with $k_{\theta,1} = \dfrac{2}{\cot\theta + \tan\theta}$

$$\frac{T_{Rd2}}{h^2} = 2\left(\frac{b}{h} - \frac{t}{h}\right)\left(1 - \frac{t}{h}\right) \frac{f_{ywk}}{\gamma_s} \frac{A_{sw}}{s} k_{\theta,2}$$ according to Table 9.2.3 below

with $k_{\theta,2} = \cot\theta$

$$\frac{T_{Rd2}}{h^3} = \frac{A_{sl}}{bh} \frac{f_{ylk}}{\gamma_s} \frac{\frac{b}{h}\left(\frac{b}{h} - \frac{t}{h}\right)\left(1 - \frac{t}{h}\right)}{\frac{b}{h} - 2\frac{t}{h} + 1} k_{\theta,3}$$ according to Table 9.2.4 below

with $k_{\theta,3} = \dfrac{1}{\cot\theta}$

For the notations used in relation to torsion, see Figure 9.3.

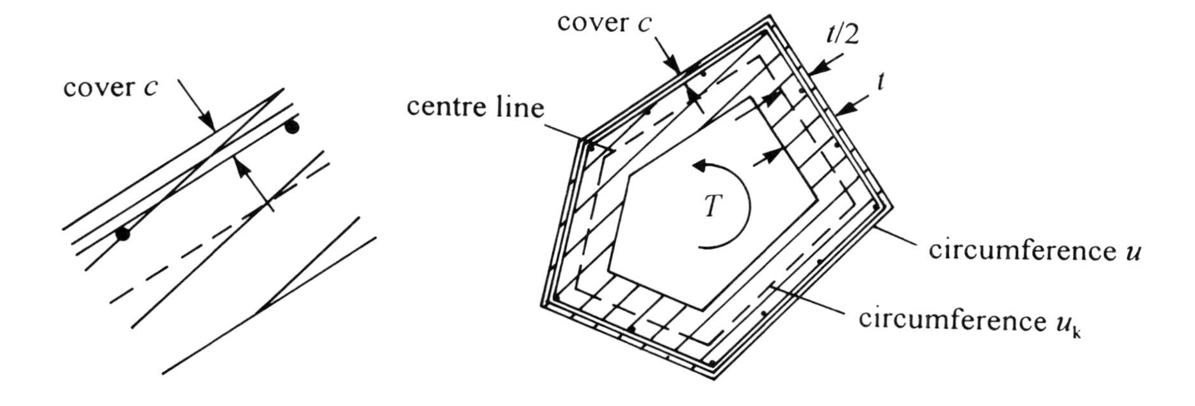

Figure 9.3 Notation used in relation to torsion

9.2.2 $\frac{T_{Rd1}}{h^3}$ **[4.40]**

Table 9.6

$\frac{b}{h}$	$\frac{t}{h}$	$\frac{T_{Rd1}}{h^3}$ in N/mm² per concrete class								
		C12/15	C16/20	C20/25	C25/30	C30/37	C35/45	C40/50	C45/55	C50/60
1.0	0.250	0.504	0.651	0.788	0.943	1.083	1.206	1.313	1.477	1.641
	0.200	0.459	0.593	0.717	0.859	0.986	1.098	1.195	1.344	1.493
	0.150	0.388	0.502	0.607	0.727	0.834	0.929	1.012	1.138	1.264
	0.100	0.290	0.375	0.454	0.543	0.624	0.695	0.756	0.851	0.945
	0.050	0.162	0.209	0.253	0.303	0.347	0.387	0.421	0.474	0.526
0.8	0.222	0.358	0.462	0.559	0.670	0.769	0.856	0.932	1.049	1.165
	0.200	0.344	0.444	0.538	0.644	0.739	0.823	0.896	1.008	1.120
	0.150	0.297	0.384	0.464	0.556	0.638	0.711	0.774	0.870	0.967
	0.100	0.226	0.292	0.353	0.423	0.485	0.540	0.588	0.662	0.735
	0.050	0.128	0.165	0.200	0.239	0.274	0.305	0.333	0.374	0.416
0.6	0.188	0.225	0.291	0.352	0.422	0.484	0.539	0.587	0.660	0.733
	0.150	0.206	0.266	0.321	0.385	0.442	0.492	0.536	0.602	0.669
	0.100	0.161	0.208	0.252	0.302	0.347	0.386	0.420	0.473	0.525
	0.050	0.094	0.121	0.146	0.175	0.201	0.224	0.244	0.274	0.305
0.4	0.143	0.113	0.146	0.176	0.211	0.242	0.270	0.294	0.331	0.367
	0.100	0.097	0.125	0.151	0.181	0.208	0.232	0.252	0.284	0.315
	0.050	0.060	0.077	0.093	0.112	0.128	0.143	0.155	0.175	0.194
0.2	0.083	0.032	0.041	0.050	0.060	0.069	0.076	0.083	0.094	0.104
	0.050	0.026	0.033	0.040	0.048	0.055	0.061	0.067	0.075	0.083

$$\frac{T_{Rd1}}{h^3} = \upsilon f_{cd} \frac{t}{h}\left(\frac{b}{h} - \frac{t}{h}\right)\left(1 - \frac{t}{h}\right) k_{\theta,1} \quad \text{with } \upsilon = 0.7\left(0.7 - \frac{f_{ck}}{200}\right) \nless 0.35\ (f_{ck} \text{ in N/mm}^2)\ ;\ \gamma_c = 1.5;\ \theta = 45°$$

and so $k_{\theta,1} = 1.0$) in N/mm²

For t/h any value meeting the requirements $t \leq A/u$, $t \ngtr$ the actual wall thickness of a hollow section, and $t \nless 2c$ may be chosen. The maximum value $T = A/u$ is given as the maximum value in the Table for each value of b/h.

- If $\gamma_c \neq 1.5$, multiply by $1.5/\gamma_c$
- If $\theta \neq 45°$, multiply by $k_{\theta,1}$ according to:

θ (degrees)	68 22	65 25	60 30	55 35	50 40	45 45
$k_{\theta,1}$	0.69	0.77	0.87	0.94	0.98	1.00

$k_{\theta,1} = 2/(\cot\theta + \tan\theta)$

9.2.3a $\frac{T_{Rd2}}{h^2}$ **[4.43]**

Table 9.7

$\frac{b}{h}$	$\frac{t}{h}$	$\frac{T_{Rd2}}{h^2}$ in kN/m per configuration of stirrups with one leg									
		Stirrups Φ 5 - s					Stirrups Φ 6 - s				
		100	150	200	250	300	100	150	200	250	300
		A_{sw}/s (mm²/m)					A_{sw}/s (mm²/m)				
		196	131	98	79	65	283	188	141	113	94
1.0	0.250	96.1	64.1	48.0	38.4	32.0	138.4	92.2	69.2	55.3	46.1
	0.200	109.3	72.9	54.7	43.7	36.4	157.4	105.0	78.7	63.0	52.5
	0.150	123.4	82.3	61.7	49.4	41.1	177.7	118.5	88.9	71.1	59.2
	0.100	138.4	92.2	69.2	55.3	46.1	199.2	132.8	99.6	79.7	66.4
	0.050	154.2	102.8	77.1	61.7	51.4	222.0	148.0	111.0	88.8	74.0
0.8	0.222	76.8	51.2	38.4	30.7	25.6	110.5	73.7	55.3	44.2	36.8
	0.200	82.0	54.7	41.0	32.8	27.3	118.1	78.7	59.0	47.2	39.4
	0.150	94.4	62.9	47.2	37.8	31.5	135.9	90.6	68.0	54.4	45.3
	0.100	107.6	71.7	53.8	43.0	35.9	155.0	103.3	77.5	62.0	51.7
	0.050	121.7	81.1	60.9	48.7	40.6	175.3	116.8	87.6	70.1	58.4
0.6	0.188	57.3	38.2	28.6	22.9	19.1	82.4	55.0	41.2	33.0	27.5
	0.150	65.3	43.6	32.7	26.1	21.8	94.1	62.7	47.0	37.6	31.4
	0.100	76.9	51.2	38.4	30.7	25.6	110.7	73.8	55.3	44.3	36.9
	0.050	89.3	59.5	44.6	35.7	29.8	128.5	85.7	64.3	51.4	42.8
0.4	0.143	37.7	25.1	18.8	15.1	12.6	54.2	36.1	27.1	21.7	18.1
	0.100	46.1	30.7	23.1	18.4	15.4	66.4	44.3	33.2	26.6	22.1
	0.050	56.8	37.9	28.4	22.7	18.9	81.8	54.5	40.9	32.7	27.3
0.2	0.083	18.3	12.2	9.1	7.3	6.1	26.3	17.5	13.2	10.5	8.8
	0.050	24.3	16.2	12.2	9.7	8.1	35.1	23.4	17.5	14.0	11.7

$$\frac{T_{Rd2}}{h^2} = 2\left(\frac{b}{h} - \frac{t}{h}\right)\left(1 - \frac{t}{h}\right)\frac{f_{ywk}}{\gamma_s}\frac{A_{sw}}{s}k_{\theta,2}$$ (with $f_{ywk} = 0.5$ kN/mm²; $\gamma_s = 1.15$; $\theta = 45°$ and so $k_{\theta,2} = 1.0$) in kN/m

- If $\gamma_s \neq 1.15$, multiply by $1.15/\gamma_s$
- If $f_{ywk} \neq 0.5$ kN/mm², multiply by $f_{ywk}/0.5$
- If $\theta \neq 45°$, multiply by $k_{\theta,2} = \cot\theta$

For t/h and θ, the same values as in Table 9.6 should be used.

9.2.3b $\frac{T_{Rd2}}{h^2}$ [4.43]

Table 9.8

$\frac{b}{h}$	$\frac{t}{h}$	$\frac{T_{Rd2}}{h^2}$ in kN/m per configuration of stirrups with one leg									
		Stirrups ϕ 8 - s					Stirrups ϕ 10 - s				
		100	150	200	250	300	100	150	200	250	300
		A_{sw}/s (mm²/m)					A_{sw}/s (mm²/m)				
		503	335	251	201	168	785	524	393	314	262
1.0	0.250	246.0	164.0	123.0	98.4	82.0	384.4	256.2	192.2	153.7	128.1
	0.200	279.9	186.6	139.9	112.0	93.3	437.3	291.5	218.7	174.9	145.8
	0.150	316.0	210.6	158.0	126.4	105.3	493.7	329.1	246.8	197.5	164.6
	0.100	354.2	236.1	177.1	141.7	118.1	553.5	369.0	276.7	221.4	184.5
	0.050	394.7	263.1	197.3	157.9	131.6	616.7	411.1	308.3	246.7	205.6
0.8	0.222	196.5	131.0	98.3	78.6	65.5	307.1	204.7	153.5	122.8	102.4
	0.200	209.9	139.9	105.0	84.0	70.0	328.0	218.7	164.0	131.2	109.3
	0.150	241.6	161.1	120.8	96.6	80.5	377.5	251.7	188.8	151.0	125.8
	0.100	275.5	183.7	137.8	110.2	91.8	430.5	287.0	215.2	172.2	143.5
	0.050	311.6	207.7	155.8	124.6	103.9	486.8	324.6	243.4	194.7	162.3
0.6	0.188	146.6	97.7	73.3	58.6	48.9	229.0	152.7	114.5	91.6	76.3
	0.150	167.3	111.5	83.6	66.9	55.8	261.4	174.2	130.7	104.5	87.1
	0.100	196.8	131.2	98.4	78.7	65.6	307.5	205.0	153.7	123.0	102.5
	0.050	228.5	152.3	114.2	91.4	76.2	357.0	238.0	178.5	142.8	119.0
0.4	0.143	96.4	64.3	48.2	38.6	32.1	150.6	100.4	75.3	60.2	50.2
	0.100	118.1	78.7	59.0	47.2	39.4	184.5	123.0	92.2	73.8	61.5
	0.050	145.4	96.9	72.7	58.2	48.5	227.2	151.5	113.6	90.9	75.7
0.2	0.083	46.8	31.2	23.4	18.7	15.6	73.1	48.7	36.5	29.2	24.4
	0.050	62.3	41.5	31.2	24.9	20.8	97.4	64.9	48.7	38.9	32.5

$$\frac{T_{Rd2}}{h^2} = 2\left(\frac{b}{h} - \frac{t}{h}\right)\left(1 - \frac{t}{h}\right)\frac{f_{ywk}}{\gamma_s}\frac{A_{sw}}{s}k_{\theta,2}$$ (with $f_{ywk} = 0.5$ kN/mm²; $\gamma_s = 1.15$; $\theta = 45°$ and so $k_{\theta,2} = 1.0$) in kN/m

- If $\gamma_s \neq 1.15$, multiply by $1.15/\gamma_s$
- If $f_{ywk} \neq 0.5$ kN/mm², multiply by $f_{ywk}/0.5$
- If $\theta \neq 45°$, multiply by $k_{\theta,2} = \cot\theta$

For t/h and θ, the same values as used in Table 9.6 should be used.

9.2.3c $\frac{T_{Rd2}}{h^2}$ **[4.43]**

Table 9.9

$\frac{b}{h}$	$\frac{t}{h}$	$\frac{T_{Rd2}}{h^2}$ in kN/m per configuration of stirrups with one leg									
		Stirrups ϕ 12 - s					Stirrups ϕ 16 - s				
		100	150	200	250	300	100	150	200	250	300
		A_{sw}/s in mm²/m					A_{sw}/s in mm²/m				
		1131	754	565	452	377	2011	1340	1005	804	670
1.0	0.250	553.5	369.0	276.7	221.4	184.5	983.9	656.0	492.0	393.6	328.0
	0.200	629.7	419.8	314.9	251.9	209.9	1119.5	746.3	559.8	447.8	373.2
	0.150	710.9	473.9	355.5	284.4	237.0	1263.8	842.6	631.9	505.5	421.3
	0.100	797.0	531.3	398.5	318.8	265.7	1416.9	944.6	708.4	566.8	472.3
	0.050	888.0	592.0	444.0	355.2	296.0	1578.7	1052.5	789.3	631.5	526.2
0.8	0.222	442.2	294.8	221.1	176.9	147.4	786.1	524.1	393.0	314.4	262.0
	0.200	472.3	314.9	236.1	188.9	157.4	839.6	559.8	419.8	335.9	279.9
	0.150	543.6	362.4	271.8	217.5	181.2	966.5	644.3	483.2	386.6	322.2
	0.100	619.9	413.3	309.9	248.0	206.6	1102.0	734.7	551.0	440.8	367.3
	0.050	701.1	467.4	350.5	280.4	233.7	1246.3	830.9	623.2	498.5	415.4
0.6	0.188	329.8	219.9	164.9	131.9	109.9	586.3	390.8	293.1	234.5	195.4
	0.150	376.4	250.9	188.2	150.5	125.5	669.1	446.1	334.5	267.6	223.0
	0.100	442.8	295.2	221.4	177.1	147.6	787.2	524.8	393.6	314.9	262.4
	0.050	514.1	342.7	257.1	205.6	171.4	914.0	609.3	457.0	365.6	304.7
0.4	0.143	216.9	144.6	108.4	86.7	72.3	385.5	257.0	192.8	154.2	128.5
	0.100	265.7	177.1	132.8	106.3	88.6	472.3	314.9	236.1	188.9	157.4
	0.050	327.2	218.1	163.6	130.9	109.1	581.6	387.7	290.8	232.6	193.9
0.2	0.083	105.2	70.2	52.6	42.1	35.1	187.1	124.7	93.5	74.8	62.4
	0.050	140.2	93.5	70.1	56.1	46.7	249.3	166.2	124.6	99.7	83.1

$$\frac{T_{Rd2}}{h^2} = 2\left(\frac{b}{h} - \frac{t}{h}\right)\left(1 - \frac{t}{h}\right)\frac{f_{ywk}}{\gamma_s}\frac{A_{sw}}{s}k_{\theta,2}$$ (with $f_{ywk} = 0.5$ kN/mm²; $\gamma_s = 1.15$; $\theta = 45°$ and so $k_{\theta,2} = 1.0$) in kN/m

- If $\gamma_s \neq 1.15$, multiply by $1.15/\gamma_s$
- If $f_{ywk} \neq 0.5$ kN/mm², multiply by $f_{ywk}/0.5$
- If $\theta \neq 45°$, multiply by $k_{\theta,2} = \cot\theta$

For t/h and θ, the same values as in Table 9.6 should be used.

9.2.4 $\dfrac{T_{Rd2}}{h^3}$ [4.43]

Table 9.10

$\frac{b}{h}$	$\frac{t}{h}$	$\frac{T_{Rd2}}{h^3}$ (N/mm²) — $\frac{100A_{sl}}{bh}$ (%) 0.1	0.2	0.3	0.4	0.5	0.6	0.8	1.0	1.2	max
1.0	0.250	0.163	0.326	0.489	0.653	0.816	0.979	1.305	1.631	1.958	1.641
	0.200	0.174	0.348	0.522	0.696	0.870	1.044	1.392	1.740		1.493
	0.150	0.185	0.370	0.555	0.740	0.924	1.109	1.479			1.264
	0.100	0.196	0.392	0.587	0.783	0.979					0.945
	0.050	0.207	0.413	0.620							0.526
0.8	0.222	0.115	0.231	0.346	0.461	0.577	0.692	0.923	1.154	1.384	1.165
	0.200	0.119	0.239	0.358	0.477	0.597	0.716	0.955	1.193		1.120
	0.150	0.128	0.256	0.385	0.513	0.641	0.769	1.025			0.967
	0.100	0.137	0.274	0.411	0.548	0.685	0.822				0.735
	0.050	0.146	0.292	0.438							0.416
0.6	0.188	0.071	0.143	0.214	0.286	0.357	0.428	0.571	0.714	0.857	0.733
	0.150	0.077	0.154	0.230	0.307	0.384	0.461	0.614	0.768		0.669
	0.100	0.084	0.168	0.252	0.336	0.419	0.503	0.671			0.525
	0.050	0.091	0.182	0.273	0.364						0.305
0.4	0.143	0.034	0.069	0.103	0.138	0.172	0.207	0.275	0.344	0.413	0.367
	0.100	0.039	0.078	0.117	0.157	0.196	0.235	0.313	0.392		0.315
	0.050	0.045	0.089	0.134	0.178	0.223					0.194
0.2	0.083	0.009	0.018	0.027	0.036	0.045	0.054	0.072	0.090	0.108	0.104
	0.050	0.011	0.023	0.034	0.045	0.056	0.068	0.090			0.083

$$\frac{T_{Rd2}}{h^3} = \frac{A_{sl}}{bh}\frac{f_{ylk}}{\gamma_s}\frac{\frac{b}{h}\left(\frac{b}{h}-\frac{t}{h}\right)\left(1-\frac{t}{h}\right)}{\frac{b}{h}-2\frac{t}{h}+1}k_{\theta,3}$$

(with f_{ylk} = 500 N/mm²; γ_s = 1.15; θ = 45° and so $k_{\theta,3}$ = 1.0) in N/mm²

- If $\gamma_s \neq 1.15$, multiply by $1.15/\gamma_s$
- If $f_{ylk} \neq 500$, N/mm² multiply by $f_{ylk}/500$
- If $\theta \neq 45°$, multiply by $k_{\theta,3} = 1/\cot\theta$

For t/h and θ, the same values as in Table 9.6 should be used.

9.3 Combination of torsion and shear

Torsion and shear 4.3.3.2.2(3)

Determine T_{Rd1} according to Table 9.6, V_{Rd2} according to Table 9.7, 9.8, 9.9 or 9.10 and check whether the following condition is satisfied:

$$\left(\frac{T_{Sd}}{T_{Rd1}}\right)^2 + \left(\frac{V_{Sd}}{V_{Rd2}}\right)^2 \leq 1 \qquad [4.47]$$

Torsion and shear for solid, approximately rectangular sections 4.3.3.2.2(5)

$$\frac{T_{Sd}}{V_{Sd} b_w} \leq \frac{1}{4.5} \quad \text{for} \quad \frac{V_{Rd1}}{V_{Sd}} > 2 \qquad [4.48]$$

$$\frac{T_{Sd}}{V_{Sd} b_w} = \frac{1}{4.5}\left(\frac{V_{Rd1}}{V_{Sd}} - 1\right) \quad \text{for} \quad 1 \leq \frac{V_{Rd1}}{V_{Sd}} \leq 2 \qquad [4.49]$$

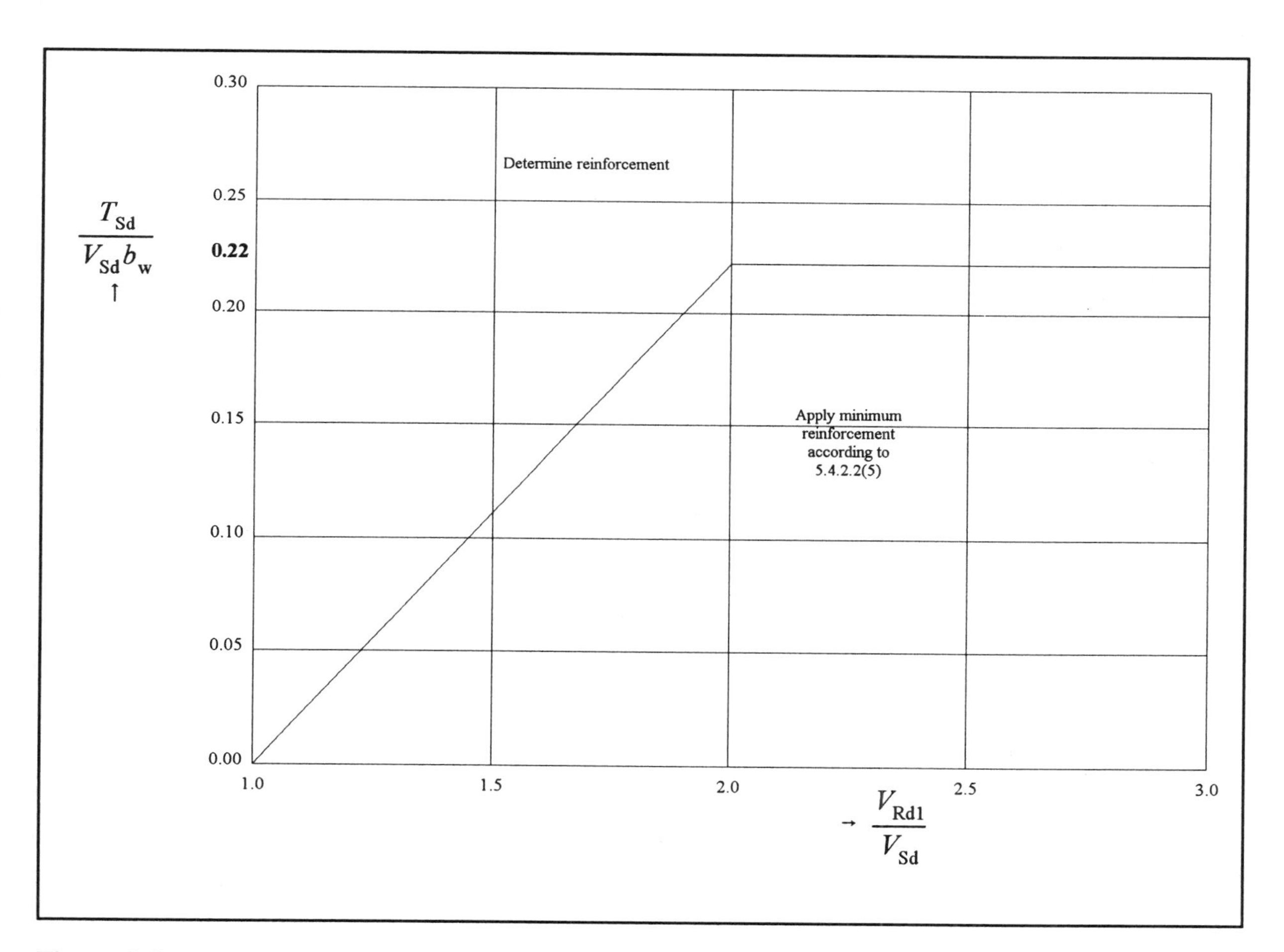

Figure 9.3

10 Punching

10.1 General

Punching 4.3.4

Slabs without punching shear reinforcement 4.3.4.5.1

Requirement:

$v_{Sd} \leq v_{Rd1} \quad \rightarrow \quad \frac{v_{Sd}}{d} \leq \frac{v_{Rd1}}{d}$ 4.3.4.3(2)

with:

$$\frac{v_{Sd}}{d} = \frac{V_{Sd}\,\beta}{d\,u} = \frac{\frac{V_{Sd}}{d^2}\,\beta}{\frac{u}{d}}$$ according to Table 10.2a or 10.2b

$$\frac{v_{Rd1}}{d} = \tau_{Rd}\,k\,(1.2 + 40\rho_1)$$ according to Table 10.3

Slabs with punching shear reinforcement 4.3.4.5.2

Requirements:

$v_{Sd} \leq v_{Rd3} \quad \rightarrow \quad \frac{v_{Sd}}{d} \leq \frac{v_{Rd3}}{d}$ 4.3.4.3(3)

and

$v_{Sd} \leq [1.6]\,v_{Rd1} \quad \rightarrow \quad \frac{v_{Sd}}{d} \leq [1.6]\,\frac{v_{Rd1}}{d}$ [4.57]

with:

$$\frac{v_{Sd}}{d} = \frac{V_{Sd}\,\beta}{d\,u} = \frac{\frac{V_{Sd}}{d^2}\,\beta}{\frac{u}{d}}$$ according to Table 10.2a or 10.2b

$$\frac{v_{Rd1}}{d} = \tau_{Rd}\,k\,(1.2 + 40\rho_1)$$ according to Table 8.3

$$\frac{v_{Rd3}}{d} = \frac{v_{Rd1}}{d} + \frac{\sum A_{sw} f_{ywk} \sin\alpha}{u\,d\,\gamma_s}$$ according to Table 10.4a or 10.4b

Apply minimum punching shear reinforcement by taking [60%] of the appropriate value of Table 5.5 (EC2). 4.3.4.5.2(4)

For the boxed values, apply the values given in the appropriate NAD.

Loaded area, critical perimeter and critical section 4.3.4.2.1-4

The critical perimeter is defined as a perimeter surrounding the perimeter of the loaded area at a defined distance of $1.5d$ (Figures 10.1 and 10.2).

For a circular loaded area with diameter a, the perimeter of the loaded area is πa.

The critical perimeter for a circular loaded area located away from unsupported edges is:

$u = \pi(a + 3d) \quad \rightarrow \quad u/d = \pi(a/d + 3)$

Limiting value: $a/d \not> 3.5$

For a rectangular loaded area with dimensions a and b the perimeter of the loaded area is $2(a + b)$.

The critical perimeter for a rectangular loaded area located away from unsupported edges is:

$u = 2(a + b) + 3\pi d \quad \rightarrow \quad u/d = 2(a/d + b/d) + 3\pi$

Limiting values: $a/d + b/d \not> 5.5$ and $a/b \not> 2$

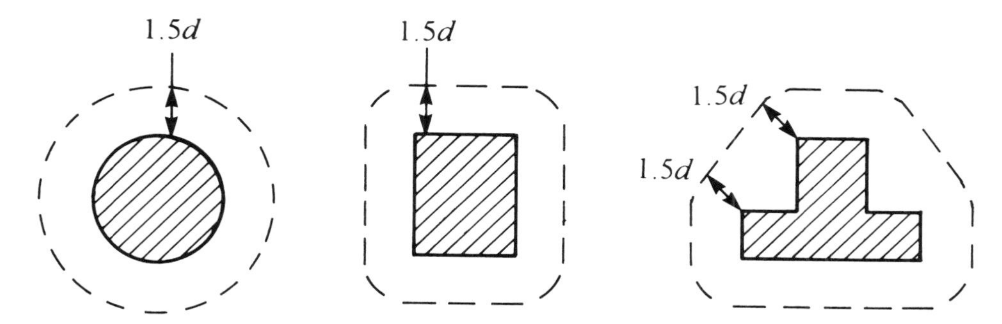

Figure 10.1 Critical perimeter round loaded areas located away from an unsupported edge.

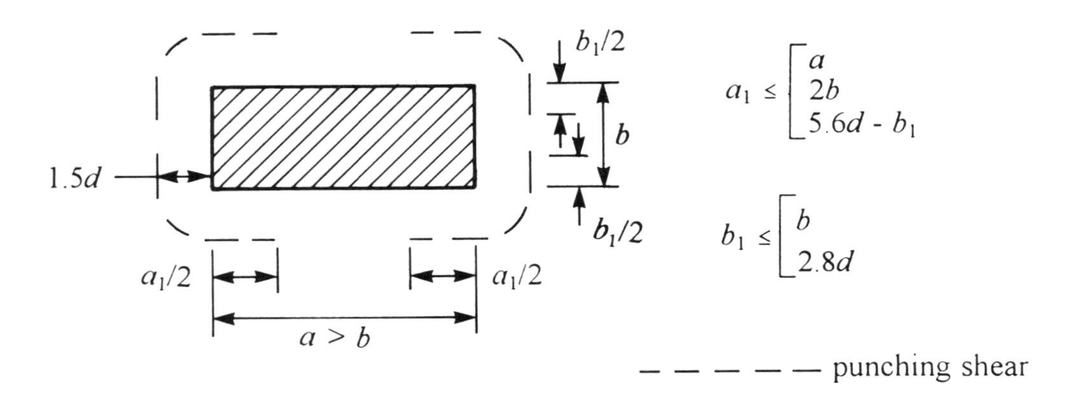

Figure 10.2 Application of punching provisions in non-standard cases.

For openings, determine the critical perimeter according to Figure 10.3.

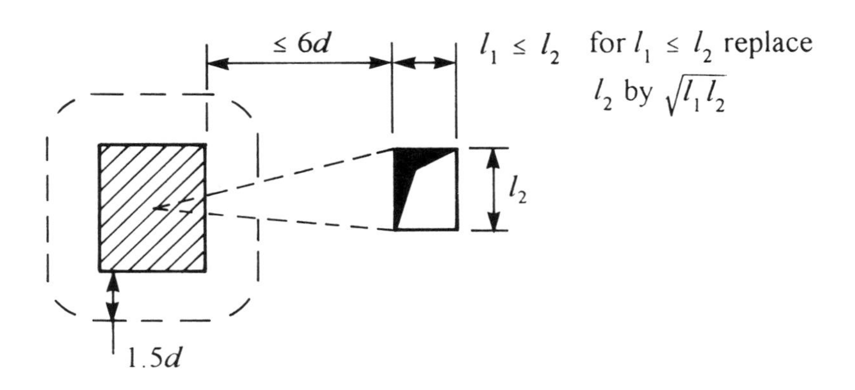

Figure 10.3 Critical perimeter near an opening.

For loaded areas near or on an unsupported edge or corner, determine the critical perimeter according to Figure 10.4.

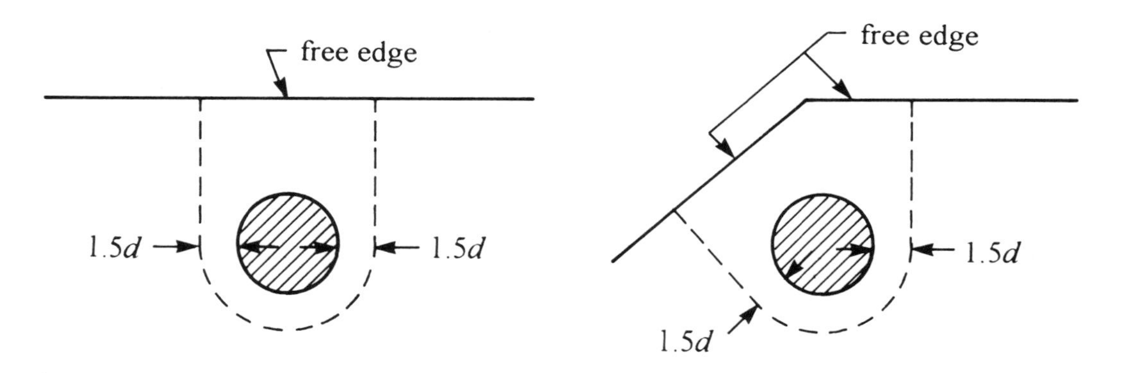

Figure 10.4 Critical sections near unsupported edges.

For slabs with column heads where $l_H < 1.5h_H$, determine critical sections according to Figure 10.5.

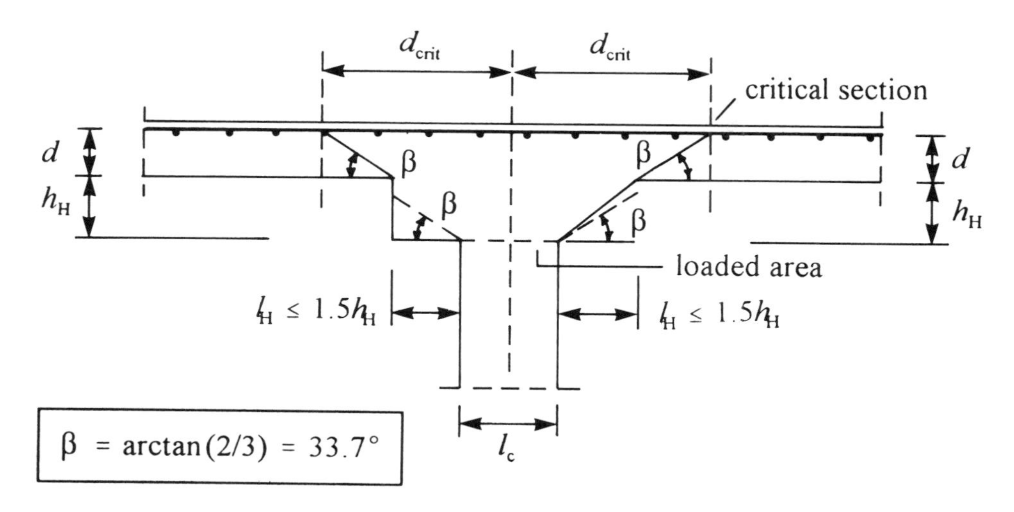

Figure 10.5 Slab with column heads where $l_H \leq 1.5h_H$.

For slabs with enlarged column head where $l_H > 1.5(d + h_H)$, determine critical sections according to Figure 10.6.

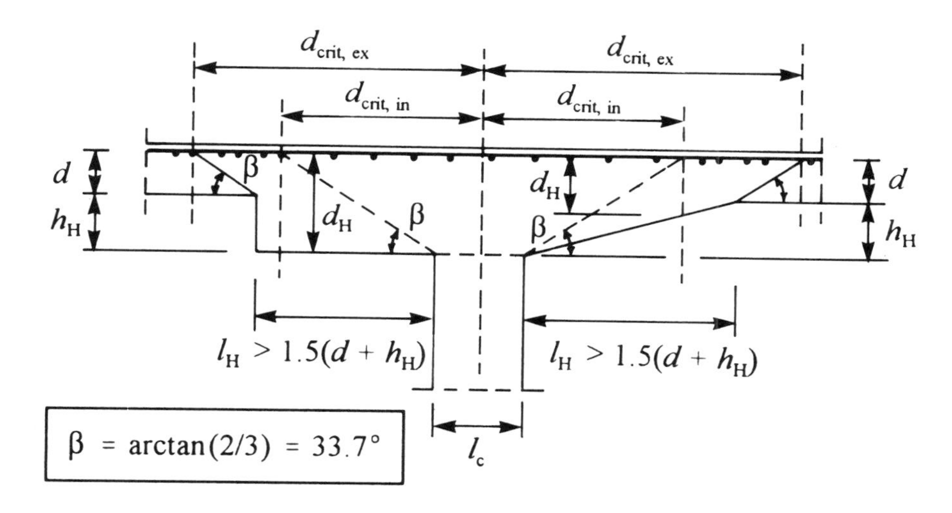

Figure 10.6 Slabs with enlarged column head where $l_H > 1.5(d + h_H)$.

For column heads where $1.5h_H < l_H < 1.5(d + h_H)$, the distance from the centroid of the column to the critical section may be taken as:

$$d_{crit} = 1.5l_H + 1.5l_c$$

Coefficient β 4.3.4.3(4)

β is a coefficient which takes account of the effects of eccentricity of loading. In cases where no eccentricity of loading is possible, β may be taken as 1.0. In other cases, the values given in Figure 10.7 may be adopted (4.3.4.3(4)).

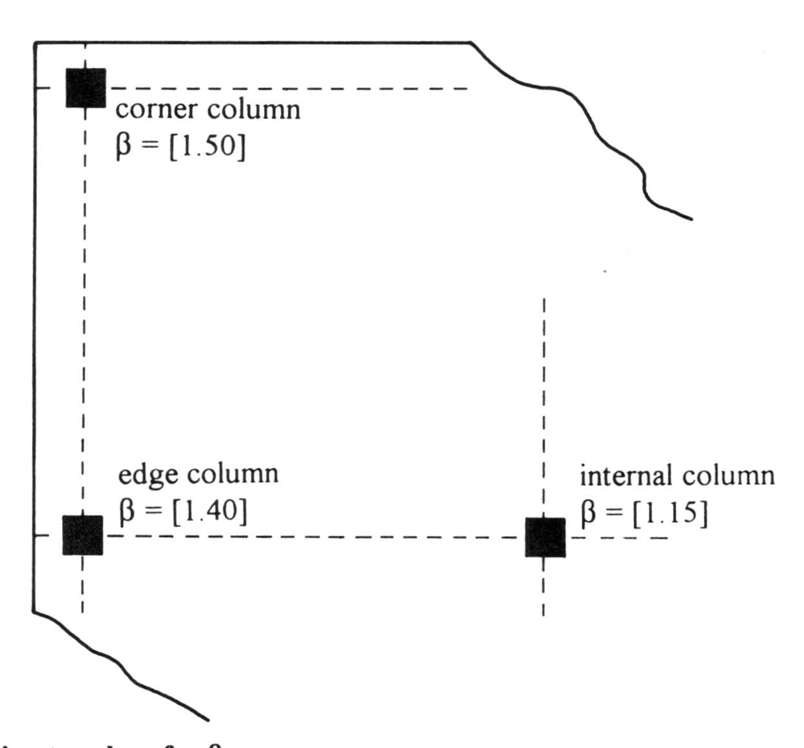

Figure 10.7 Approximate values for β.

For the boxed values, apply the values given in the appropriate NAD.

Minimum design moments

4.3.4.5.3

Design slab for minimum bending moments per unit width, m_{Sdx} and m_{Sdy} in the x- and y-direction, unless structural analysis leads to higher values according to:

$$m_{Sdx} \text{ (or } m_{Sdy}) \geq nV_{Sd} \qquad [4.25]$$

Take n into account according to Table 10.1 and Figure 10.8.

Position of column	n for m_{Sdx}			n for m_{Sdy}		
	Top	Bottom	Effective width	Top	Bottom	Effective width
Internal column	-0.125	0	$0.30\ l_y$	-0.125	0	$0.3\ l_x$
Edge columns, edge of slab parallel to x-axis	-0.250	0	$0.15\ l_y$	-0.125	+0.125	(per m)
Edge columns, edge of slab parallel to y-axis	-0.125	+0.125	(per m)	-0.250	0	$0.15\ l_x$
Corner column	-0.500	0	(per m)	+0.500	-0.500	(per m)

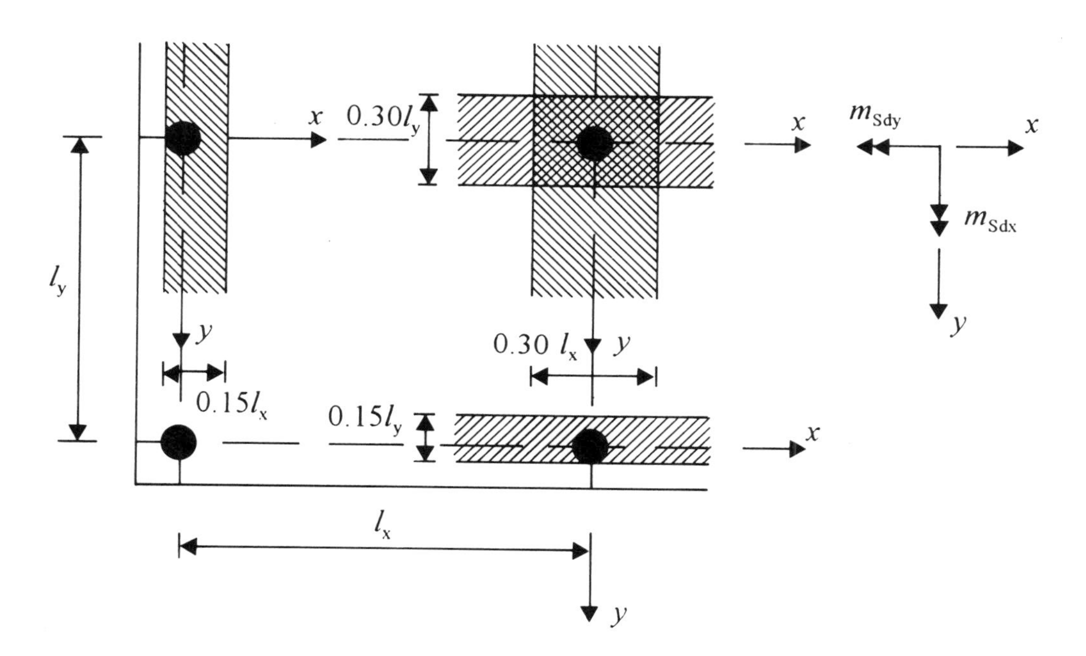

Figure 10.8 Bending moments m_{Sdx} and m_{Sdy} in slab-column joints subjected to eccentric loading, and effective width for resisting these moments.

10.2a $\frac{v_{Sd}}{d}$ for circular loaded areas **[4.50]**

Table 10.2

$\frac{v_{Sd}}{d}$ in N/mm^2 for circular loaded areas

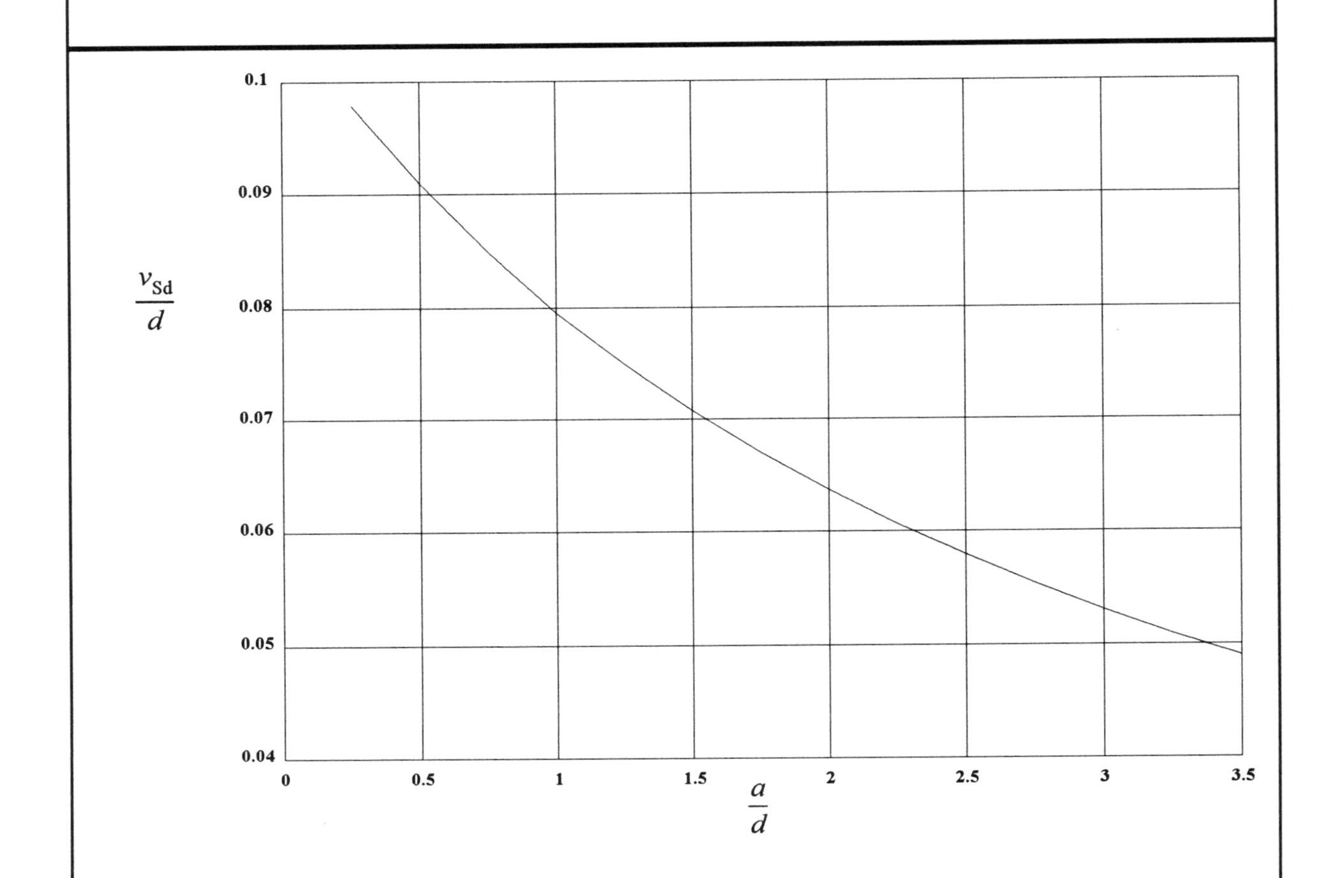

$$\frac{v_{Sd}}{d} = \frac{\frac{V_{Sd}}{d^2}\beta}{\pi\left(\frac{a}{d} + 3\right)} \quad \text{with } \frac{V_{Sd}}{d^2} = 1.0\ \text{N/mm}^2 \text{ and } \beta = 1.0$$

- If $\frac{V_{Sd}}{d^2} \neq 1.0$ N/mm^2 , multiply by $\frac{V_{Sd}}{d^2}$
- If $\beta \neq 1.0$, multiply by β according to Figure 10.7

10.2b $\frac{v_{Sd}}{d}$ for rectangular loaded areas [4.50]

Table 10.3

$\frac{v_{Sd}}{d}$ in N/mm² for rectangular loaded areas

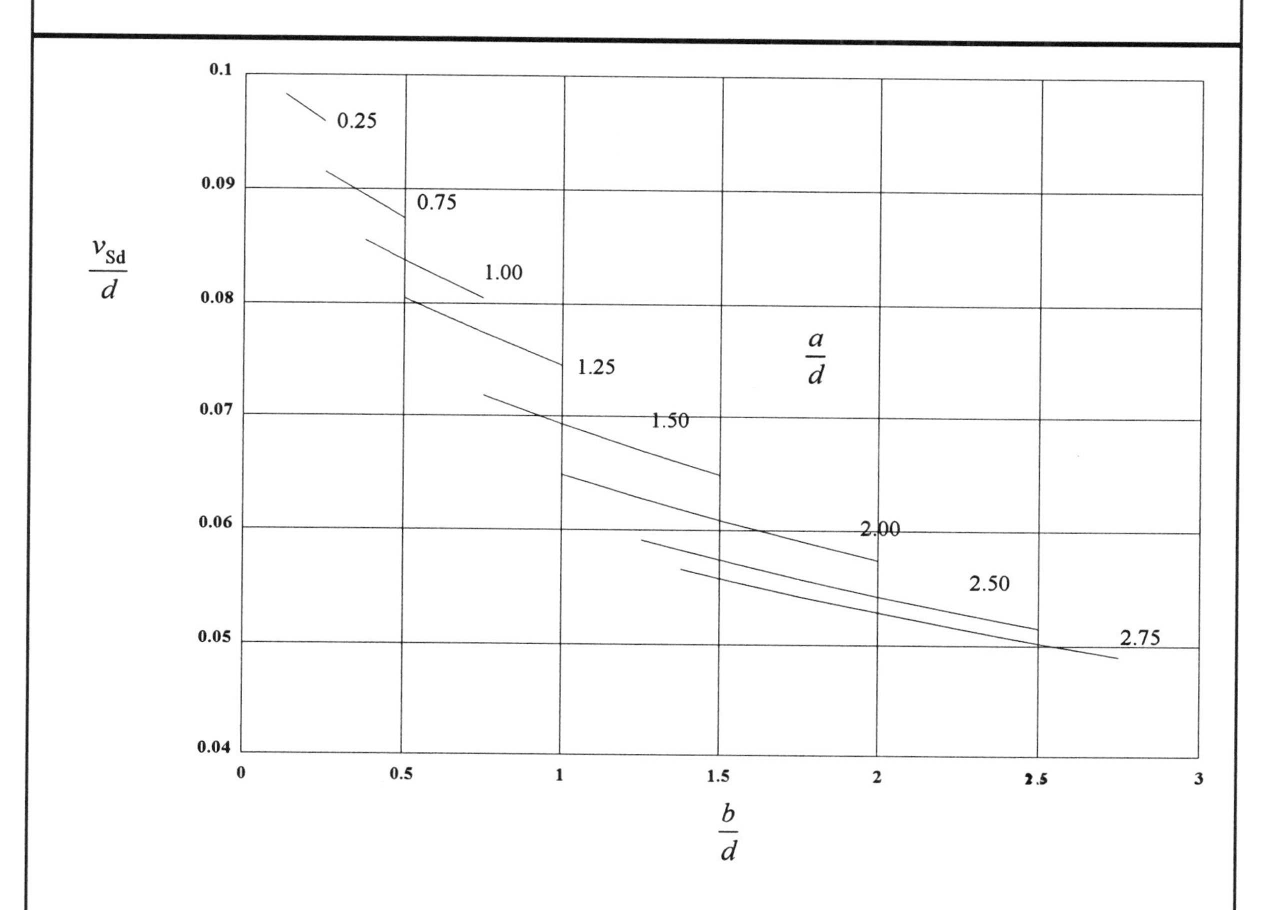

$$\frac{V_{Sd}}{d} = \frac{\frac{V_{Sd}}{d^2}\beta}{2\left(\frac{a}{d}+\frac{b}{d}\right)+3\pi} \quad \text{with} \quad \frac{V_{Sd}}{d^2} = 1.0 \text{ N/mm}^2 \text{ and } \beta = 1.0$$

- If $\frac{V_{Sd}}{d^2} \neq 1.0$ N/mm², multiply by $\frac{V_{Sd}}{d^2}$
- If $\beta \neq 1.0$, multiply by β according to Figure 10.7

10.3 $\frac{v_{Rd1}}{d}$ [4.56]

Table 10.4

$100\rho_1$ (%)	$\frac{v_{Rd1}}{d}$ (with $\gamma_c = 1.5$ and $k = 1.0$) in N/mm² per concrete class								
	C12/15	C16/20	C20/25	C25/30	C30/37	C35/45	C40/50	C45/55	C50/60
0.5	0.252	0.308	0.364	0.420	0.476	0.518	0.574	0.616	0.672
0.6	0.259	0.317	0.374	0.432	0.490	0.533	0.590	0.634	0.691
0.7	0.266	0.326	0.385	0.444	0.503	0.548	0.607	0.651	0.710
0.8	0.274	0.334	0.395	0.456	0.517	0.562	0.623	0.669	0.730
0.9	0.281	0.343	0.406	0.468	0.530	0.577	0.640	0.686	0.749
1.0	0.288	0.352	0.416	0.480	0.544	0.592	0.656	0.704	0.768
1.1	0.295	0.361	0.426	0.492	0.558	0.607	0.672	0.722	0.787
1.2	0.302	0.370	0.437	0.504	0.571	0.622	0.689	0.739	0.806
1.3	0.310	0.378	0.447	0.516	0.585	0.636	0.705	0.757	0.826
1.4	0.317	0.387	0.458	0.528	0.598	0.651	0.722	0.774	0.845
1.5	0.324	0.396	0.468	0.540	0.612	0.666	0.738	0.792	0.864
$\frac{v_{Rd1}}{d} = \tau_{Rd} k(1.2 + 40\rho_1)$ (with $\gamma_c = 1.5$ and $k = 1.0$) in N/mm²									
τ_{Rd} (N/mm²)	0.18	0.22	0.26	0.30	0.34	0.37	0.41	0.44	0.48

- If $\gamma_c \neq 1.5$, multiply by $1.5/\gamma_c$
- If $d < 0.6$ m, multiply by $k = 1.6 - d$ where d is in metres
- If $100\rho_1 < 0.5$ %, with $\rho_1 = \sqrt{\rho_{1x} + \rho_{1y}}$, apply $100\rho_1 = 0.5\%$
- If $100\rho_1 > 1.5$ %, with $\rho_1 = \sqrt{\rho_{1x} + \rho_{1y}}$, take $100\rho_1 = 1.5\%$ into account

10.4a $\frac{v_{Rd3}}{d} - \frac{v_{Rd1}}{d}$ circular loaded areas **[4.58]**

Table 10.5

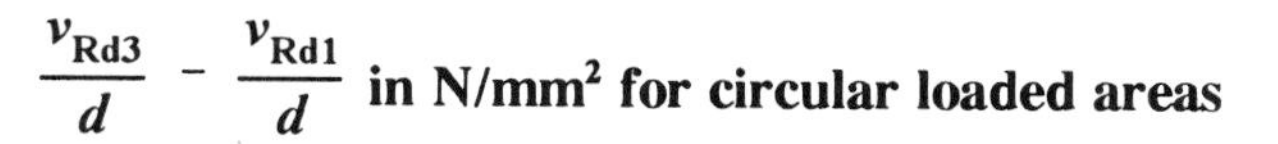

$\frac{v_{Rd3}}{d} - \frac{v_{Rd1}}{d}$ in N/mm² for circular loaded areas

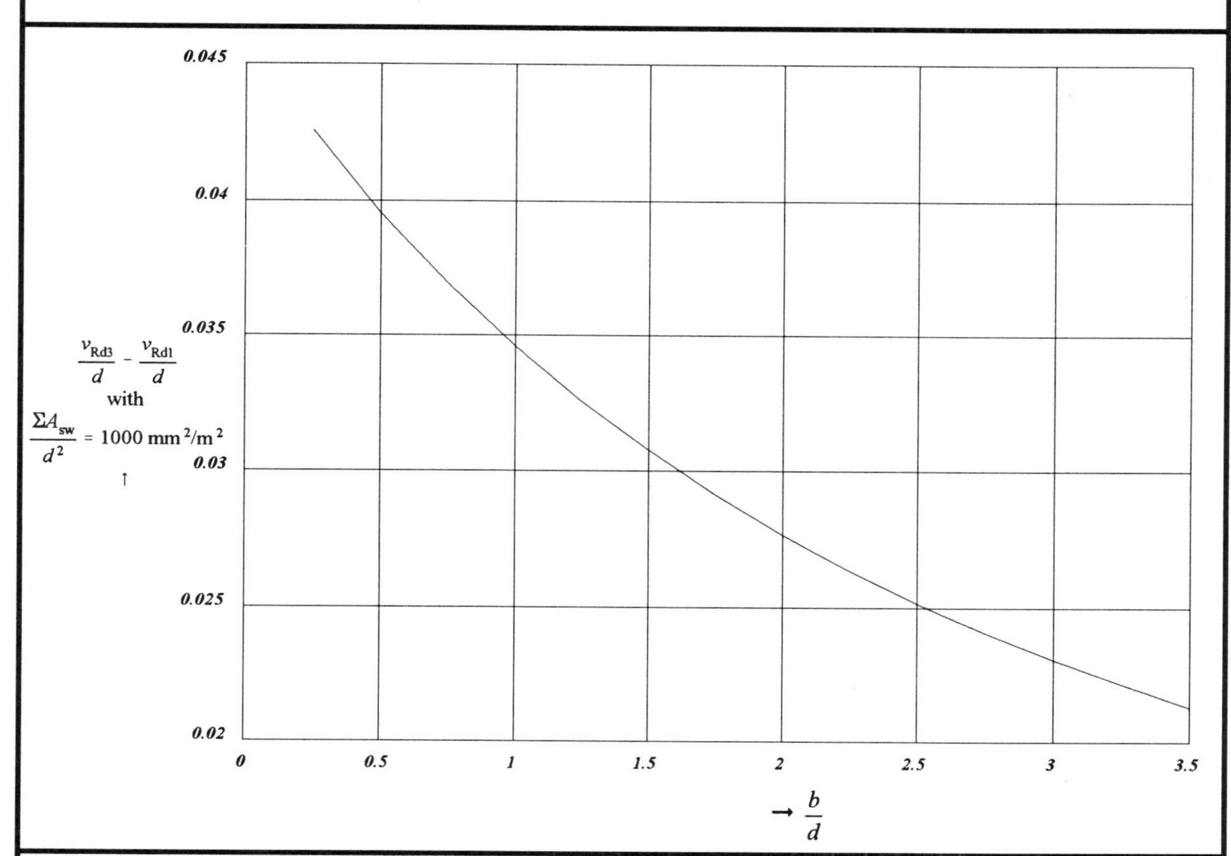

$$\frac{V_{Rd3}}{d} - \frac{V_{Rd1}}{d} = \frac{\Sigma A_{sw}}{d^2} \frac{1}{\pi\left(\frac{a}{d} + 3\right)} \frac{f_{ywk}}{\gamma_s} \sin\alpha$$

(with $\frac{\Sigma A_{sw}}{d^2} = 1000$ mm²/m²; $\gamma = 1.15$; $f_{ywk} = 500$ N/mm²; $\alpha = 90°$)

- If $\frac{\Sigma A_{sw}}{d^2} \neq 1000$ mm²/m², multiply by $\frac{\Sigma A_{sw}}{1000\, d^2}$
- If $\gamma_s \neq 1.15$, multiply by $1.15/\gamma_s$
- If $f_{ywk} \neq 500$, multiply by $f_{ywk}/500$
- If $\alpha \neq 90°$, multiply by $\sin\alpha$

10.4b $\frac{v_{Rd3}}{d} - \frac{v_{Rd1}}{d}$ rectangular loaded areas **[4.58]**

Table 10.6

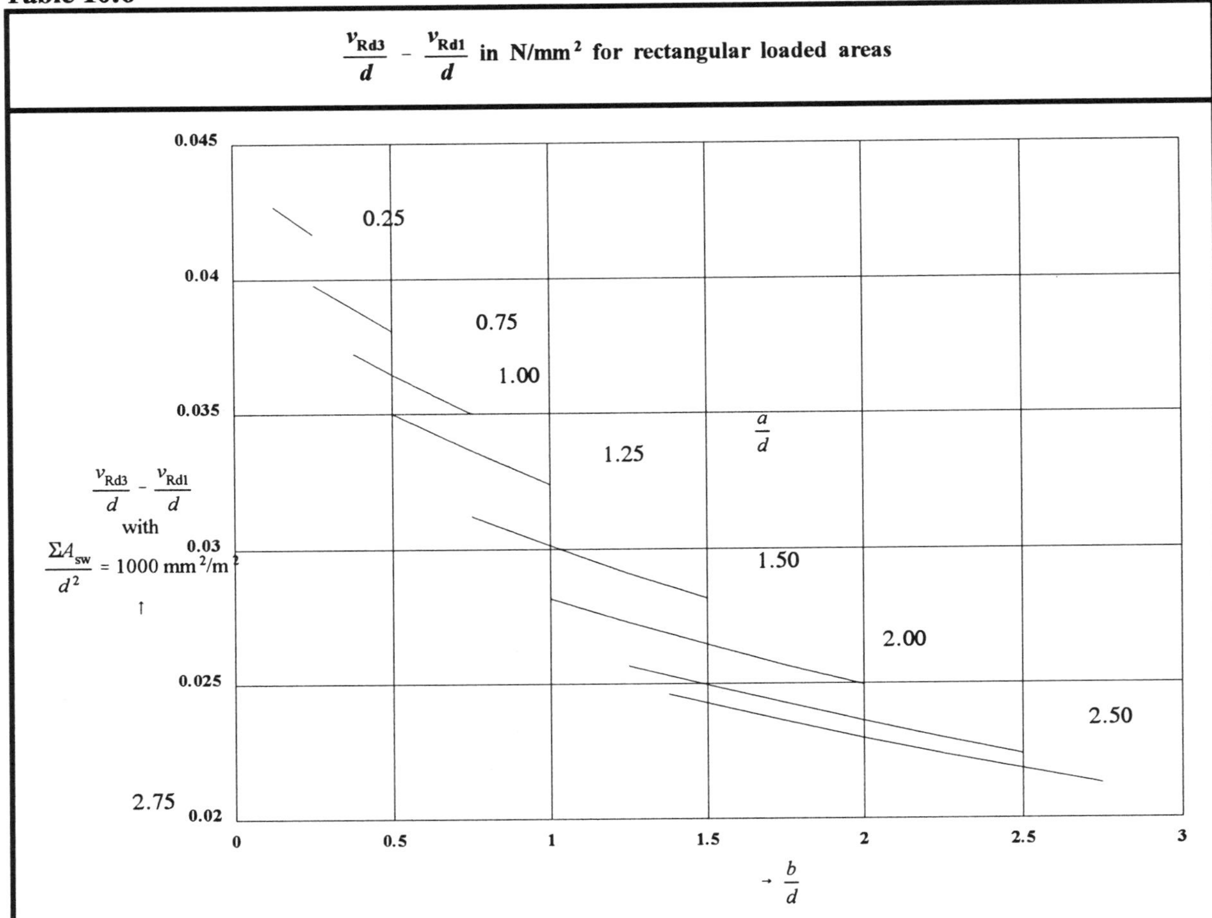

$$\frac{V_{Rd3}}{d} - \frac{V_{Rd1}}{d} = \frac{\Sigma A_{sw}}{d^2} \frac{1}{2\left(\frac{a}{d} + \frac{b}{d}\right) + 3\pi} \frac{f_{ywk}}{\gamma_s} \sin\alpha$$

(with $\frac{\Sigma A_{sw}}{d^2} \neq 1000$ mm²/m²; γ_s 1.15; f_{ywk} 500 N/mm² and α 90°)

- If $\frac{\Sigma A_{sw}}{d^2} \neq 1000$ mm²/m², multiply by $\frac{A_{sw}}{1000} d^2$
- If $\gamma_s \neq 1.15$, multiply by $1.15/\gamma_s$
- If $f_{ywk} \neq 500$, multiply by $f_{ywk}/500$
- If $\alpha \neq 90°$, multiply by $\sin\alpha$

11 Elements with second order effects

11.1 Determination of effective length of columns

The effective length of a column depends on the stiffness of the column relative to the stiffness of the structure connected to either end of the column. The effective length may be estimated from the relation:

$$l_{eff} = \beta \, l_{col}$$

where β may be obtained from Figure 11.1.

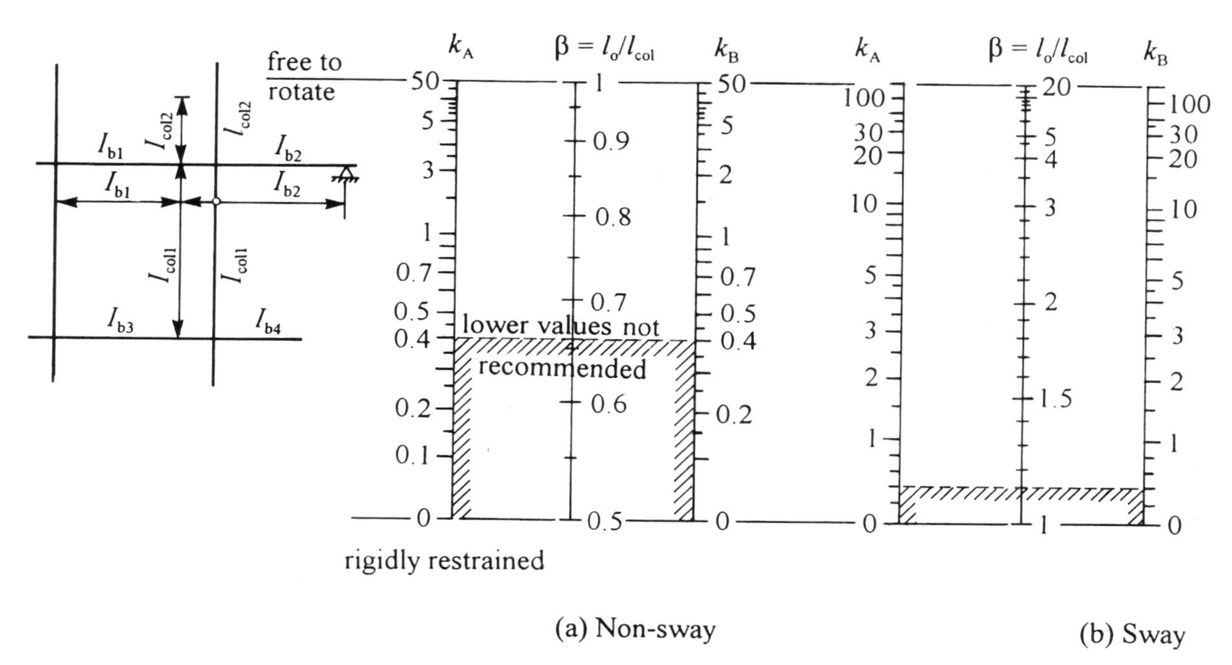

Figure 11.1: Nomogram for assessing effective lengths.

coefficients K_A and K_B denote the rigidity of restraint at the column ends:

$$\frac{\Sigma E_{cm} I_{col} / l_{col}}{\Sigma E_{cm} \alpha I_b / l_{eff}}$$

where

E_{cm}	=	modulus of elasticity of the concrete
I_{col}, I_b	=	moment of inertia (gross section) of the column or beam respectively
l_{col}	=	height of the column measured between centres of restraint
l_{eff}	=	effective span of the beam
α	=	factor taking into account the conditions of restraint of the beam at the opposite end:
	=	1.0 opposite end elastically or rigidly restrained
	=	0.5 opposite end free to rotate
	=	0 for a cantilever beam

Alternatively, for columns in braced frames, the effective height for framed structures may be taken as the lesser of:

$$l_e = l_o[0.7 + 0.05(\alpha_{c,1} + \alpha_{c,2})] < l_o$$

$$l_e = l_o[0.85 + 0.05\alpha_{c,min}] < l_o$$

The effective height for unbraced framed structures may be taken as the lesser of:

$l_e = l_o[1.0 + 0.15(\alpha_{c,1} + \alpha_{c,2})]$

$l_e = l_o[2.0 + 0.3\alpha_{c,min}]$

where

l_e = effective height of a column in the plane of bending considered

l_o = height between end restraints

$\alpha_{c,1}$ = ratio of the sum of the column stiffnesses to the sum of the beam stiffnesses at the lower end of a column

$\alpha_{c,2}$ = ratio of the sum of the column stiffnesses to the sum of the beam stiffnesses at the upper end of a column

$\alpha_{c,min}$ = lesser of $\alpha_{c,1}$ and $\alpha_{c,2}$

Where creep may significantly affect the performance of a member (e.g. where members are not well restrained at the ends by monolithic connections), this can be allowed for by increasing the effective length by a factor:

$$\left(1 + \frac{M_{qp}}{M_{sd}}\right)^{½}$$

where

M_{qp} is the moment under the quasi-permanent load
M_{sd} is the design first order moment.

Table 11.1 Simplified assessment of β for non-sway frames

(A) Assess K for each end of column using the following method:

(i) $K = 0.5$
(ii) If there is a column continuing beyond the joint, $K = K * 2$
(iii) If there is a beam on only one side of the joint, $K = K * 2$
(iv) If the span of the beam is more than twice the height of the columns, $K = K * 1.5$
(v) If the beams or slabs framing into the column are shallower than the column dimension, $K = K * 2$
(vi) If the joint nominally carries no moment (e.g. connection with a pad footing), $K = 10$

(B) Obtain β from the following:

***K* for lower joint**	***K* for upper joint**							
	0.5	0.75	1.0	1.5	2	3	10	PIN
0.5	0.69	0.70	0.74	0.75	0.77	0.8	0.81	0.84
0.75	0.70	0.74	0.75	0.77	0.80	0.81	0.84	0.85
1.0	0.74	0.75	0.77	0.80	0.81	0.84	0.85	0.86
1.5	0.75	0.77	0.80	0.81	0.84	0.85	0.86	0.90
2	0.77	0.80	0.81	0.84	0.85	0.86	0.90	0.92
3	0.80	0.81	0.84	0.85	0.86	0.90	0.92	0.95
10	0.81	0.84	0.85	0.86	0.90	0.92	0.95	0.98
PIN	0.84	0.85	0.86	0.90	0.92	0.95	0.98	1.00

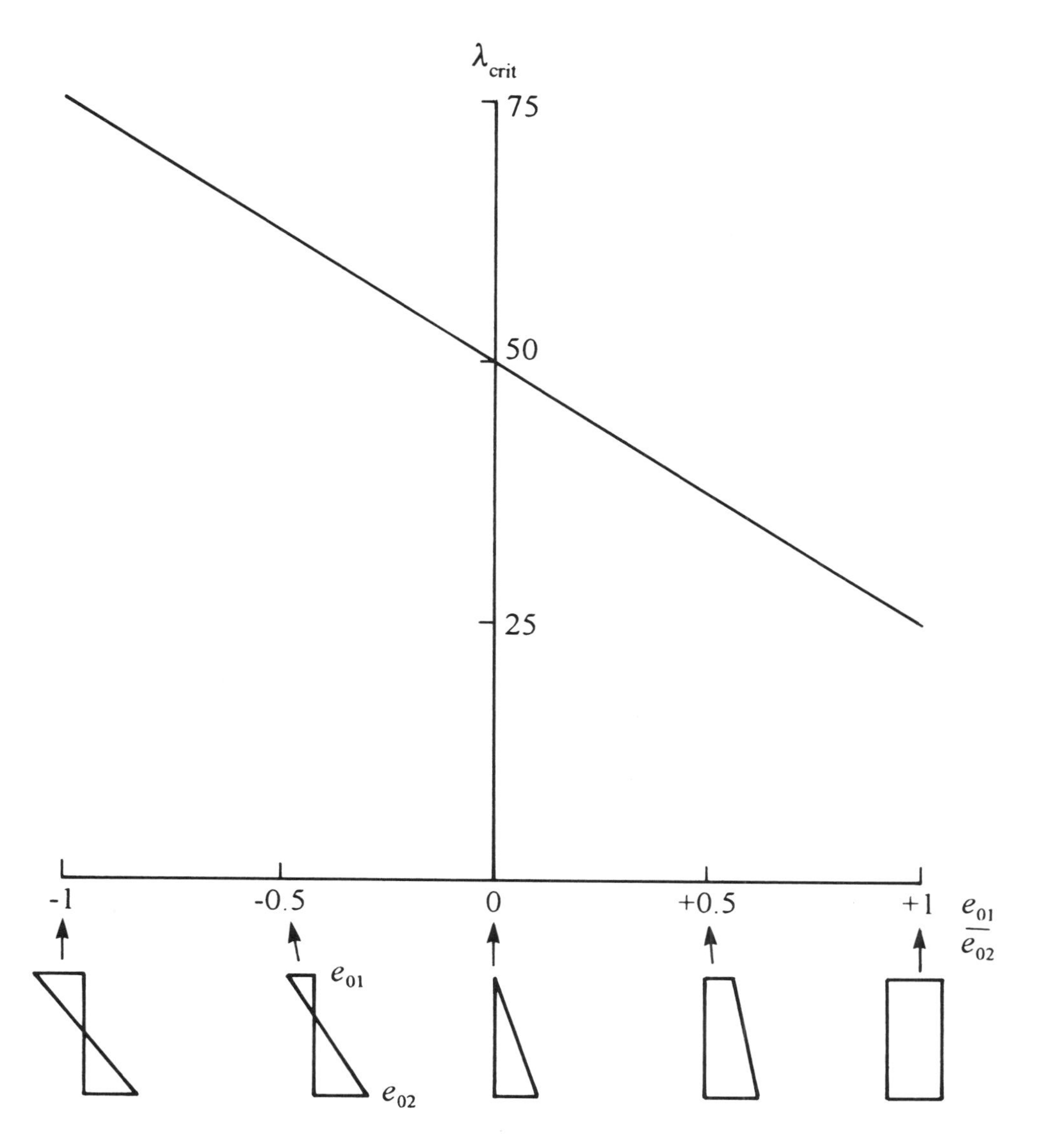

Figure 11.2 Critical slenderness ratio for isolated column.

Table 11.2 Model column method for isolated non-sway columns

If $\lambda = \lambda_{crit}$, $N_{Rd} = N_{sd}$

$M_{Rd} = N_{sd}h/20$

If $\lambda > \lambda_{crit}$, $N_{Rd} = N_{sd}$

$M_{Rd} = N_{sd}e_{tot}$ $< N_{sd}h/20$

$< N_{sd}e_{0.2}$

e_{tot} = $e_o + e_a + e_2$

e_o = first order eccentricity

= $0.4e_{0.1} + 0.6e_{0.2} < 0.4e_{0.2}$

e_a = accidental eccentricity

= $vl_o/2$

where

$$v = \frac{\alpha_n}{\sqrt{100l}} \geq \frac{l}{200}$$

α_n = $\sqrt{(1 + l/n)/2}$

l = total height of structure in metres

n = number of vertical elements acting together

e_2 = second order eccentricity

$$= \frac{k_1 k_2 l_o^2 f_{yd}}{90{,}000}$$

$$k_1 = \frac{\lambda}{20} - 0.75 \text{ for } 15 < \lambda \leq 35$$

$$= 1 \text{ for } \lambda > 35$$

$$k_2 = (N_{ud} - N_{sd})/(N_{ud} - N_{bal}) \leq 1$$

Table 11.3 Detailing requirements for columns (EC2 Clause 5.4.1)

Minimum dimensions:

200 mm	vertical columns, cast in-situ
140 mm	precast columns cast horizontally

Minimum area of longitudinal reinforcement

$$A_{s.min} = \frac{0.15N_{sd}}{f_{yd}} < 0.003A_c$$

Maximum area of longitudinal reinforcement

$$A_{s.max} = 0.08A_c$$

Transverse reinforcement:

Minimum diameter of links: $\dfrac{\text{maximum longitudinal bar size}}{4} < 6\text{ mm}$

Maximum spacing: the smallest of:

- 12 times minimum diameter of longitudinal bars
- the least dimension of the column
- 300 mm

The resulting maximum spacing should be multiplied by 0.6

- in sections immediately above or below a beam or slab over a height equal to the larger dimension of the column
- near lapped joints where the size of the longitudinal bars exceed 14 mm

12 Control of cracking

It should be clearly understood that there are many causes of cracking and that only certain of these lead to cracks that will be controlled by the provisions of chapter 4.4.2 of EC2. Chapter 4.4.2 is concerned with cracks that form in hardened concrete either from restrained imposed deformations, such as shrinkage or early thermal movements, or from the effects of loads.

The fundamental principle behind the provisions of the code is as follows. Crack control is only possible where spread cracking can occur (i.e. the tensile strain is accommodated in multiple cracks, or a crack accommodates only tensile strains that arise near the crack). For this to occur, there must be sufficient reinforcement in the section to ensure that the reinforcement does not yield on first cracking. The rules for minimum reinforcement areas in 4.4.2.2 are aimed at ensuring that this requirement is met. Provided this minimum is present, crack widths can normally be controlled by simple detailing rules.

Table 12.1 Minimum areas of reinforcement

$A_s \geq K_c K f_{ct.eff} A_c / \sigma_s$

where:

A_c	=	the area of concrete in tension immediately before the formation of the first crack
$f_{ct.eff}$	=	the tensile strength of the concrete effective at the time when the cracks first form. Except where the cracks can be guaranteed to form at an early age, it is suggested that the value chosen should not be less than 3 N/mm^2
σ_s	=	the stress in the reinforcement, which may be taken as the yield strength of the reinforcement
K	=	a coefficient that takes account of the effects of non-linear stress distribution. See Table 12.2 for values for K
K_c	=	a coefficient taking account of the form of loading causing the cracks. See Table 12.2 for values of K_c

Table 12.2 Values of K and K_c

(1) Values of K
(a) Extrinsic, or external deformations imposed on a member: $K = 1.0$
(b) Internal deformations (e.g. restrained shrinkage or temperature change):
for members with least dimension ≤ 300 $K = 0.8$
for members with least dimension ≥ 800 $K = 0.5$
Interpolation may be used between these values
(2) Values of K_c
(a) Pure tension $K_c = 1.0$
(b) Pure flexure: $K_c = 0.4$
(c) Section in compression with zero stress at least compressed fibre (under rare load combination) $K_c = 0$
(d) Sections where the neutral axis depth calculated on the basis of a cracked section under the cracking load is less than the lesser of $h/2$ or 500 mm: $K_c = 0$
(e) Box sections
Webs: $K_c = 0.4$
Tension chords: $K_c = 0.8$
(f) Parts of sections in tension distant from main reinforcement
$0.5 < K_c < 1.0$

To help interpolation between (a), (b), (c), Fig. 12.1 may be used. Checking the crack width requires (a) crack width criteria and (b) an estimate of the stress in the reinforcement under the quasi-permanent load. The criteria are given in Table 12.3.

Table 12.3: Crack width criteria

(1) Reinforced concrete: 0.3 m. If acceptable, a greater value may be used in exposure Class 1.
(2) Prestressed members.

Exposure class	Design crack width, w_k, under the frequent load combination (mm)	
	Post-tensioned	Pre-tensioned
1	0.2	0.2
2	0.2	Decompression
3 4	Decompression or coating of the tendons and $w_k = 0.2$	Decompression

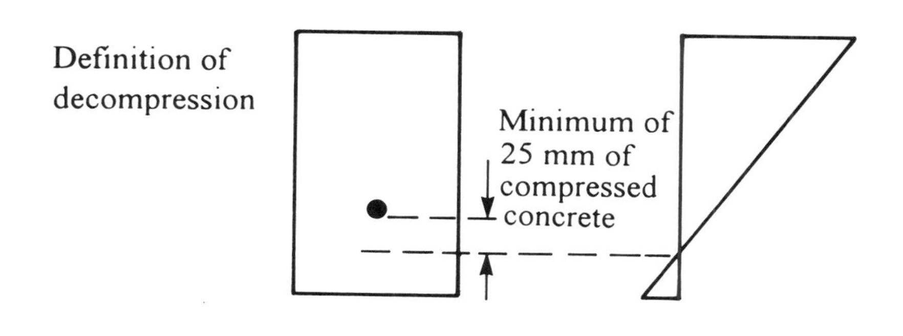

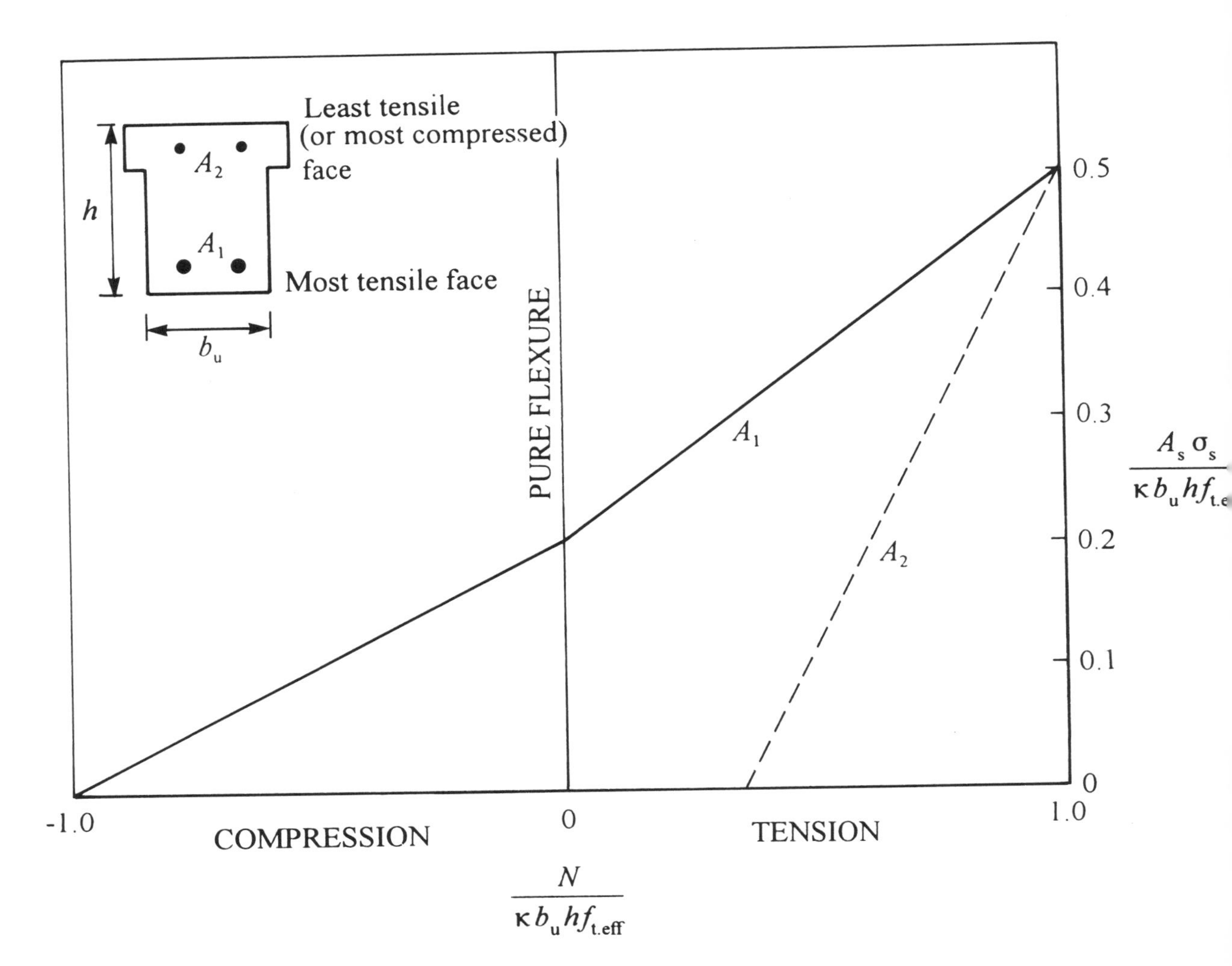

Figure 12.1 Calculation of minimum reinforcement areas.

The steel stress may be calculated on the basis of a cracked section under the quasi-permanent load. Creep may be allowed for by taking the modular ratio as 15. Table 12.4 and Figures 12.2 and 12.3 may be used to estimate the properties of a cracked section.

Alternatively, an approximate estimate of the stress may be obtained for reinforced concrete using the formula:

$$\sigma_s \frac{f_{yk}\delta}{\gamma_s}\frac{M_{qp}}{M_{sd}}\frac{A_{s.req}}{A_{s.prov}}$$

where

δ = the ratio of the design ultimate moment after redistribution to the elastically calculated value under the ultimate loads

γ_s = the partial safety factor on the reinforcement (i.e. 1.15)

M_{qp} = the moment under the quasi-permanent load

M_{sd} = the design ultimate load

$A_{s.req}$ = the reinforcement area required for the ultimate limit state

$A_{s.prov}$ = the area of tension reinforcement provided

M_{qp}/M_{sd} may be taken approximately as N_{qp}/N_{sd} where N_{qp} and N_{ad} are, respectively, the quasi-permanent and design ultimate loads on the member.

Where the stress in the reinforcement is dominantly due to imposed deformations, the value of σ_s used in Table 12.1 should be adopted.

Crack control may be achieved either by satisfying the provisions of either Table 12.5 or Table 12.6 or by direct calculation of crack widths. This is covered in Table 12.7.

Table 12.4 Neutral axis depths and moments of inertia for flanged beams (a) with *hf/d* = 0.2

αρ	*br/b* = 1		*br/b*= 0.5		*br/b* = 0.4		*br/b* = 0.3		*br/b* = 0.2	
	x/d	I/bd^3	*x/d*	I/bd^3	*x/d*	I/bd^3	*x/d*	I/bd^3	*x/d*	I/bd^3
0.02	0.181	0.015								
0.03	0.217	0.022	0.217	0.022	0.217	0.022	0.217	0.022	0.217	0.022
0.04	0.246	0.028	0.248	0.028	0.248	0.028	0.249	0.028	0.249	0.028
0.05	0.270	0.033	0.274	0.033	0.275	0.033	0.276	0.033	0.278	0.033
0.06	0.292	0.038	0.298	0.038	0.300	0.038	0.302	0.038	0.304	0.038
0.07	0.311	0.043	0.320	0.043	0.322	0.043	0.325	0.043	0.327	0.043
0.08	0.328	0.048	0.340	0.047	0.343	0.047	0.346	0.047	0.349	0.047
0.09	0.344	0.052	0.358	0.052	0.361	0.052	0.365	0.051	0.369	0.051
0.10	0.358	0.057	0.375	0.056	0.379	0.056	0.383	0.055	0.388	0.055
0.11	0.372	0.061	0.390	0.060	0.395	0.059	0.400	0.059	0.406	0.059
0.12	0.384	0.064	0.405	0.063	0.410	0.063	0.416	0.063	0.422	0.062
0.13	0.396	0.068	0.418	0.067	0.424	0.066	0.430	0.066	0.437	0.065
0.14	0.407	0.072	0.431	0.070	0.437	0.070	0.444	0.069	0.452	0.069
0.15	0.418	0.075	0.443	0.073	0.450	0.073	0.457	0.072	0.466	0.071
0.16	0.428	0.078	0.455	0.076	0.462	0.076	0.470	0.075	0.478	0.074
0.17	0.437	0.082	0.466	0.079	0.473	0.078	0.481	0.078	0.491	0.077
0.18	0.446	0.085	0.476	0.082	0.484	0.081	0.493	0.080	0.502	0.079
0.19	0.455	0.088	0.486	0.085	0.494	0.084	0.503	0.083	0.513	0.082
0.20	0.463	0.091	0.495	0.087	0.504	0.086	0.513	0.085	0.524	0.084
0.21	0.471	0.094	0.504	0.090	0.513	0.089	0.523	0.088	0.534	0.086
0.22	0.479	0.096	0.513	0.092	0.522	0.091	0.532	0.090	0.543	0.089
0.23	0.486	0.099	0.521	0.094	0.531	0.093	0.541	0.092	0.552	0.091
0.24	0.493	0.102	0.529	0.097	0.539	0.095	0.549	0.094	0.561	0.093

Table 12.4 Neutral axis depths and moments of inertia for flanged beams (b) with $hf/d = 0.3$

$\alpha\rho$	$br/b = 1$		$br/b = 0.5$		$br/b = 0.4$		$br/b = 0.3$		$br/b = 0.2$	
	x/d	I/bd^3	x/d	I/bd^3	x/d	I/bd^3	x/d	I/bd^3	x/d	I/bd^3
0.02	0.181	0.015								
0.03	0.217	0.022								
0.04	0.246	0.028								
0.05	0.270	0.033								
0.06	0.292	0.038								
0.07	0.311	0.043	0.311	0.043	0.311	0.043	0.311	0.043	0.311	0.043
0.08	0.328	0.048	0.328	0.048	0.329	0.048	0.329	0.048	0.329	0.048
0.09	0.344	0.052	0.345	0.052	0.345	0.052	0.345	0.052	0.346	0.052
0.10	0.358	0.057	0.360	0.056	0.361	0.056	0.361	0.056	0.362	0.056
0.11	0.372	0.061	0.375	0.060	0.375	0.060	0.376	0.060	0.377	0.060
0.12	0.384	0.064	0.388	0.064	0.389	0.064	0.390	0.064	0.391	0.064
0.13	0.396	0.068	0.401	0.068	0.402	0.068	0.403	0.068	0.404	0.068
0.14	0.407	0.072	0.413	0.071	0.414	0.071	0.416	0.071	0.417	0.071
0.15	0.418	0.075	0.425	0.075	0.426	0.075	0.428	0.075	0.430	0.075
0.16	0.428	0.078	0.436	0.078	0.437	0.078	0.439	0.078	0.441	0.078
0.17	0.437	0.082	0.446	0.081	0.448	0.081	0.450	0.081	0.452	0.081
0.18	0.446	0.085	0.456	0.084	0.458	0.084	0.461	0.084	0.463	0.084
0.19	0.455	0.088	0.466	0.087	0.468	0.087	0.471	0.087	0.473	0.087
0.20	0.463	0.091	0.475	0.090	0.477	0.090	0.480	0.090	0.483	0.089
0.21	0.471	0.094	0.483	0.093	0.486	0.092	0.489	0.092	0.493	0.092
0.22	0.479	0.096	0.492	0.095	0.495	0.095	0.498	0.095	0.502	0.095
0.23	0.486	0.099	0.500	0.098	0.503	0.098	0.507	0.097	0.511	0.097
0.24	0.493	0.102	0.508	0.100	0.511	0.100	0.515	0.100	0.519	0.099

Table 12.4 Neutral axis depths and moments of inertia for flanged beams (c) with hf/d = 0.4

αp	br/b = 1		br/b = 0.5		br/b = 0.4		br/b = 0.3		br/b = 0.2	
	x/d	I/bd^3	x/d	I/bd^3	x/d	I/bd^3	x/d	I/bd^3	x/d	I/bd^3
0.02	0.181	0.015								
0.03	0.217	0.022								
0.04	0.246	0.028								
0.05	0.270	0.033								
0.06	0.292	0.038								
0.07	0.311	0.043								
0.08	0.328	0.048								
0.09	0.344	0.052								
0.10	0.358	0.057								
0.11	0.372	0.061								
0.12	0.384	0.064								
0.13	0.396	0.068								
0.14	0.407	0.072	0.407	0.072	0.407	0.072	0.407	0.072	0.407	0.072
0.15	0.418	0.075	0.418	0.075	0.418	0.075	0.418	0.075	0.418	0.075
0.16	0.428	0.078	0.428	0.078	0.428	0.078	0.428	0.078	0.428	0.078
0.17	0.437	0.082	0.438	0.082	0.438	0.082	0.438	0.082	0.438	0.082
0.18	0.446	0.085	0.447	0.085	0.447	0.085	0.448	0.085	0.448	0.085
0.19	0.455	0.088	0.456	0.088	0.457	0.088	0.457	0.088	0.457	0.088
0.20	0.463	0.091	0.465	0.091	0.465	0.091	0.466	0.091	0.466	0.091
0.21	0.471	0.094	0.473	0.094	0.474	0.094	0.474	0.094	0.474	0.093
0.22	0.479	0.096	0.481	0.096	0.482	0.096	0.482	0.096	0.483	0.096
0.23	0.486	0.099	0.489	0.099	0.490	0.099	0.490	0.099	0.491	0.099
0.24	0.493	0.102	0.496	0.101	0.497	0.101	0.498	0.101	0.498	0.101

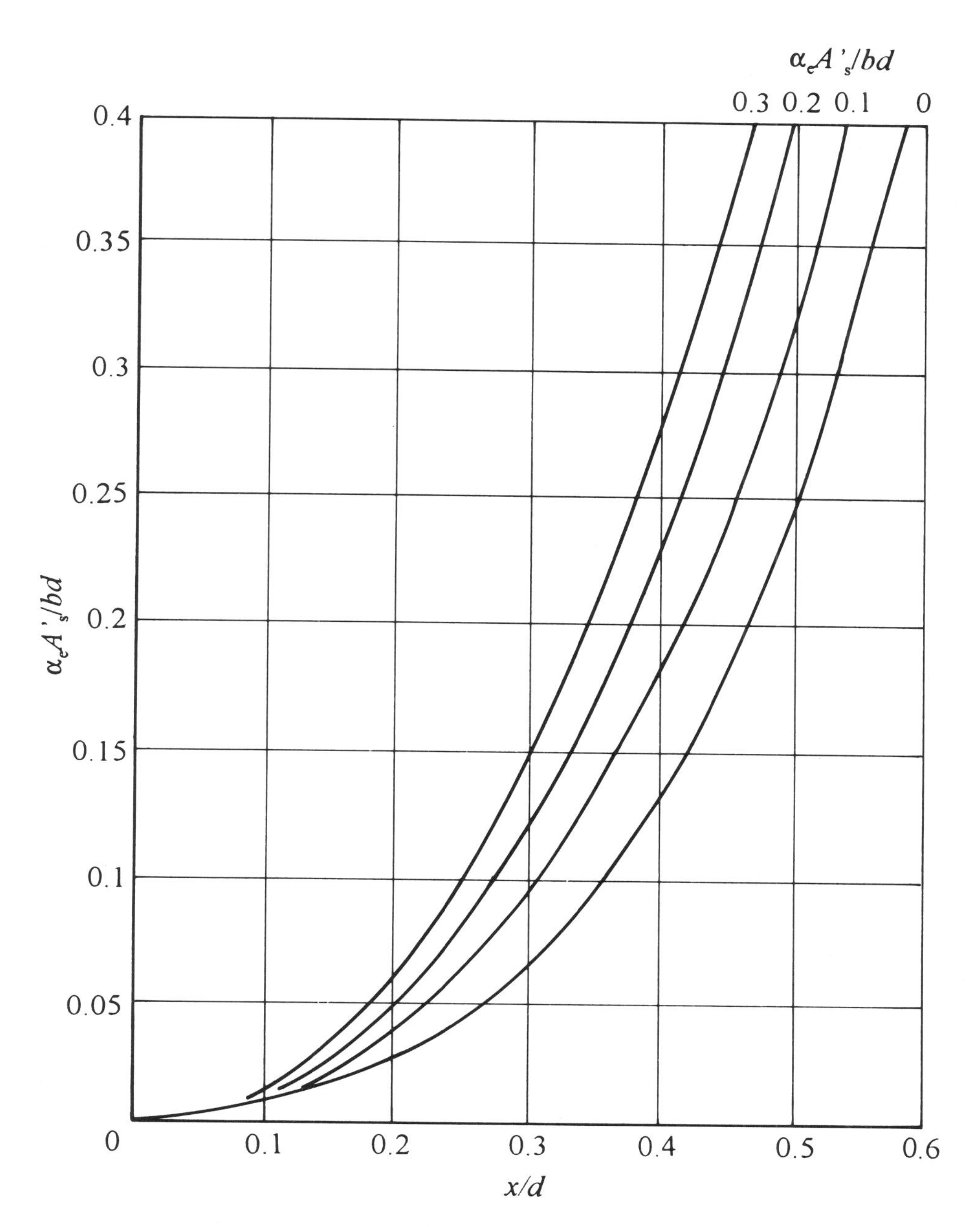

Figure 12.2 Neutral axis depths for rectangular sections.

Table 12.5 Maximum bar diameters

Steel stress (MPa)	Maximum bar size (mm)	
	Reinforced sections	Prestressed sections
160	32	25
200	25	16
240	20	12
280	16	8
320	12	6
360	10	5
400	8	4
450	6	

For reinforced concrete, the maximum bar diameter may be modified as follows:

$ø_s = ø_s^* f_{ctm}/(2.5)h/[10(h-d)] \geq ø_s^* (f_{ctm}/2.5)$ for restraint cracking

$ø_s = ø_s^* \dfrac{h}{10(h-d)} \geq ø_s^*$ for load-induced cracking

where:

$ø_s$ = the adjusted maximum bar diameter

$ø_s^*$ = the maximum bar size in Table 12.5

h = the overall depth of the section

Table 12.6 Maximum bar spacings for high bond bars

Steel stress (MPa)	Maximum bar spacing (mm)		
	Pure flexure	Pure tension	Prestressed sections (bending)
160	300	200	200
200	250	150	150
240	200	125	100
280	150	75	50
320	100	-	-
360	50	-	-

Table 12.7 Crack width by direct calculation

Design crack width, $w = \beta\left(50 + 0.25 k_1 k_2 \left(\frac{ø}{\rho_r}\right)\right) \varepsilon_{sm}$

β	=	coefficient relating the maximum crack spacing to the average value.
	=	1.7 for load-induced cracking and for restraint cracking in members with a minimum dimension greater than 800 mm
	=	1.3 for sections with a minimum dimension less than 300 mm. Intermediate values may be interpolated
ø	=	bar size in mm. For a mixture of bar sizes in a section, take the average
k_1	=	a coefficient that takes account of the bond properties of the bars; $k_1 = 0.8$ for high bond bars and 1.6 for plain bars. In the case of imposed deformations, k_1 should be replaced by k_1, k, with k being in accordance with Table 12.2
k_2	=	a coefficient that takes account of the form of the strain distribution
	=	0.5 for bending and 1.0 for pure tension
		For cases of eccentric tension or for local areas, intermediate values of k_2 should be used which can be calculated from the relation: $k_2 = \frac{\varepsilon_1 + \varepsilon_2}{2\varepsilon_1}$ where ε_1 is the greater and ε_2 the lesser tensile strain at the boundaries of the section considered, assessed on the basis of a cracked section
ρ_r	=	the effective reinforcement ratio, $A_s/A_{c.eff}$, where A_s is the area of reinforcement contained within the effective tension area $A_{c.eff}$

The effective tension area is generally the area of concrete surrounding the tension reinforcement of depth equal to 2.5 times the distance from the tension face of the section to the centroid of the reinforcement (see Figure 12.4). For slabs,

ε_{sm} is the mean strain allowing for the effects of tension stiffening, shrinkage, etc. under the relevant load combinations, and may be calculated from the relation:

$$\varepsilon_{sm} = \frac{\sigma_s}{E_s}\left[1 - \beta_1\beta_2\left(\frac{\sigma_{sr}}{\sigma_s}\right)^2\right]$$

where

σ_s	=	the stress in the tension reinforcement calculated on the basis of a cracked section
σ_{sr}	=	the stress in the tension reinforcement calculated on the basis of a cracked section under the loading conditions causing first cracking
β_1	=	a coefficient that takes account of the bond properties of the bars
	=	1.0 for high bond bars
	=	0.5 for plain bars
β_2	=	a coefficient that takes account of the duration of the loading or of repeated loading
	=	1.0 for a single, short-term load
	=	0.5 for a sustained load or for many cycles of repeated loading

For members subjected only to intrinsic imposed deformations, σ_s may be taken as equal to σ_{sr}.

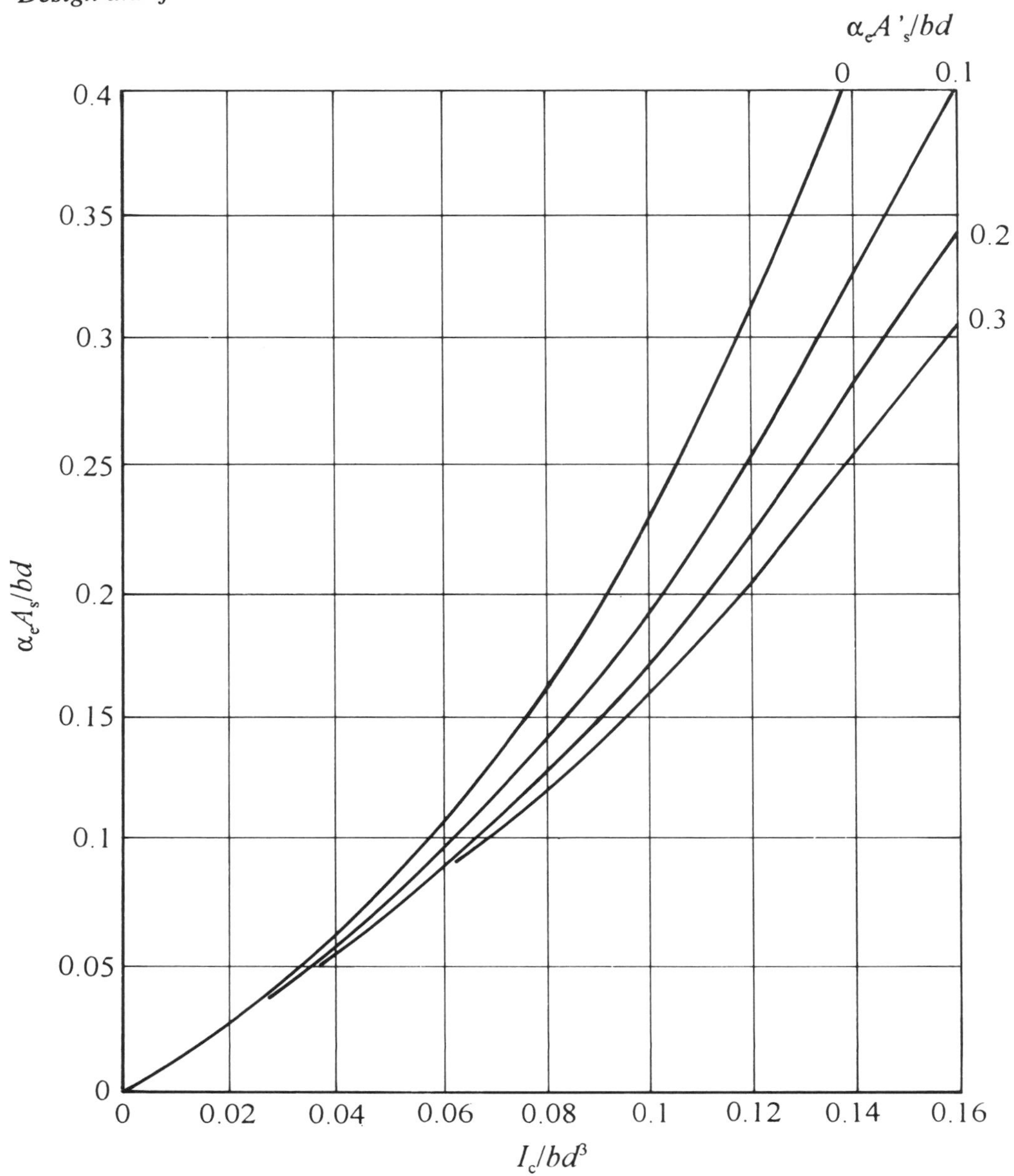

Figure 12.3 Second moments of area of rectangular sections based on a cracked transformed section.

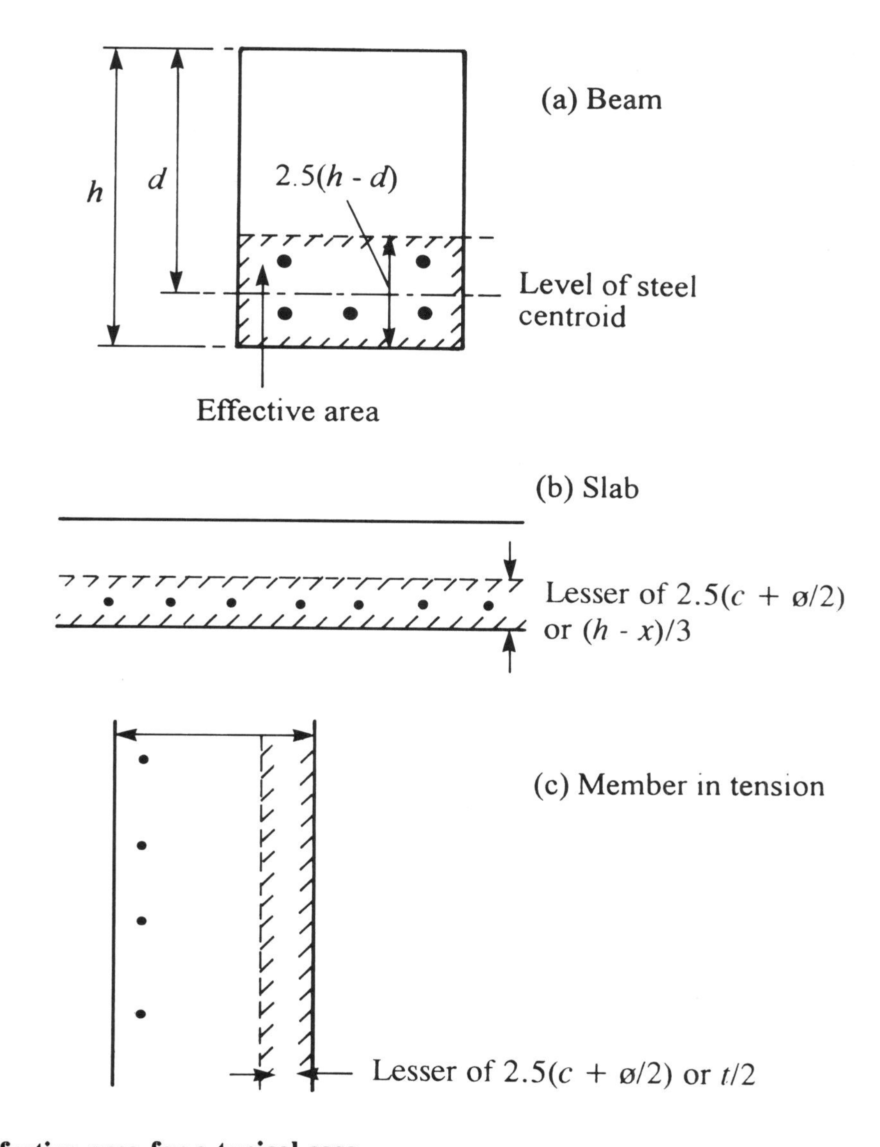

Figure 12.4 Effective area for a typical case.

13 Deflections

13.1 General

Eurocode 2 assumes that deflections will generally be checked using span/effective depth ratios, though a calculation method is also given (in Appendix 4). Owing particularly to uncertainties about the likely tensile strength of the concrete, calculation of deflection in the design stage for reinforced concrete members is likely to be very approximate. Hence direct calculation, rather than use of simple checks, is generally inappropriate.

Limits to deflection should be considered in the light of the intended function of the structure and the nature of finishes and partitions. The limits given in the code are intended only as guidance. They are: (1) Limit to overall total deflection: span/250; (2) Limit to deflection after construction of partitions and finishes where these are susceptible to damage: span/500

13.2 Ratios of span to effective depth

The span/effective depth ratios should generally ensure that these limits are met. The ratios depend upon: the nature of the structural system; the stress in the tension reinforcement; the reinforcement ratio; the geometry of the section (whether flanged or rectangular).

Figure 13.1 gives permissible ratios on the assumption that f_{yk} is 500 N/mm^2 and hence that the service stress at the critical section is approximately 250 N/mm^2. The values in Figure 13.1 should be adjusted according to those in Table 13.1. The critical section for assessing the reinforcement ratio and the steel stress is at mid-span for all members but cantilevers where the support section is used. For two-way spanning slabs supported on beams on all sides, the span/effective depth ratios should be based on the shorter span. For flat slabs, the longer span should be used.

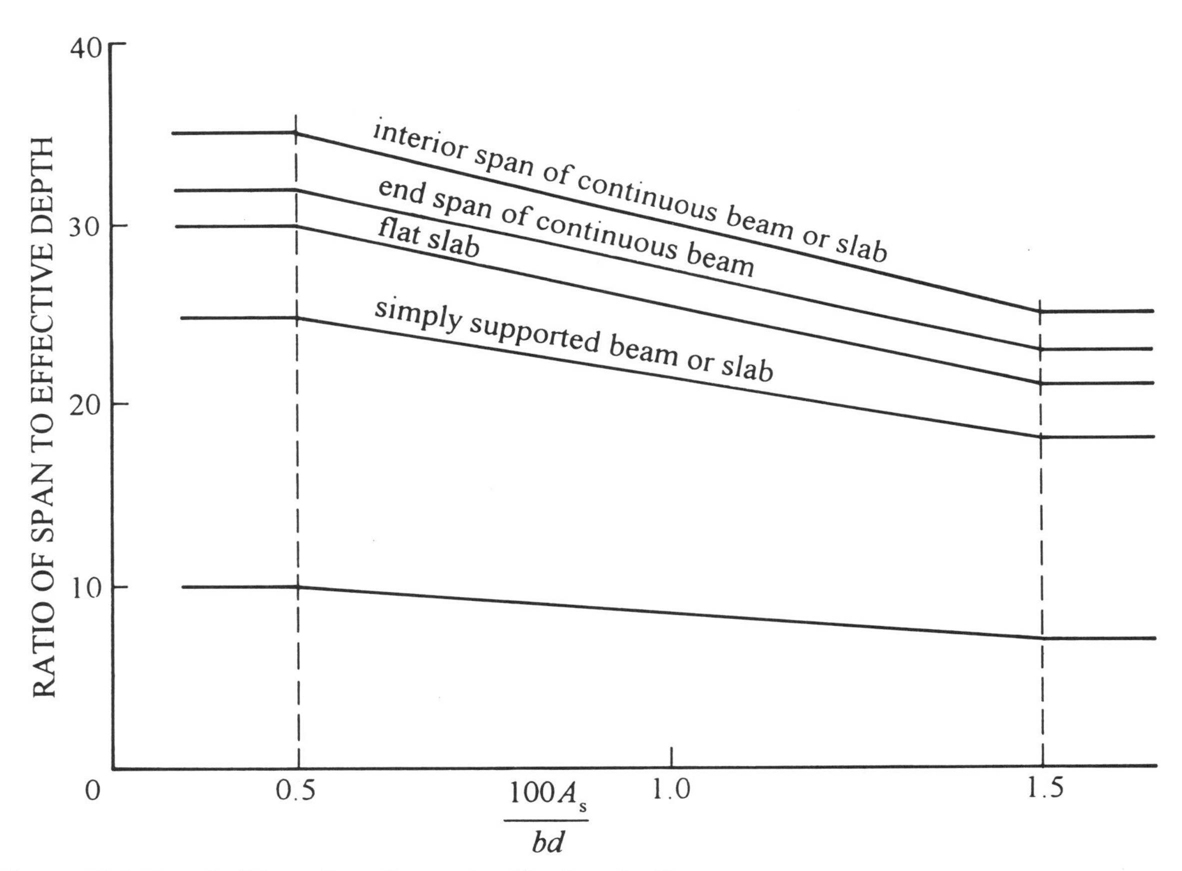

Figure 13.1 Permissible ratios of span to effective depth.

Table 13.1 Adjustment factors for span/effective depth ratios

(1) Different levels of stress in tension reinforcement

Multiply ratios by $250/f_s$

f_s = stress under quasi-permanent load. This may be estimated approximately from:

$$f_s = \frac{f_{yk}}{\gamma_s}\frac{M_{qp}}{M_{sd}}\frac{A_{s.req}}{A_{s.prov}}\frac{1}{\delta}$$

where

γ_s = partial safety factor for reinforcement
M_{qp} = moment at critical section under the quasi-permanent load
M_{sd} = design ultimate load
$A_{s.req}$ = area of tension reinforcement required at critical section
$A_{s.prov}$ = area of tension reinforcement provided
δ = ratio of design moment after redistribution to the elastically calculated moment

(2) Flanged beams where $b_t/b < 0.3$

Multiply ratios by 0.8

(3) Long spans

(a) Members other than flat slabs with spans > 7 m

Multiply by 7/span

(b) Flat slabs with spans > 8.5 m

Multiply by 8.5/span

(c) Other deflection limits: for total deflections other than span/250

Multiply by $250/k$ where new deflection limit is span/k

13.3 Calculation of deflection

There are two ways of approaching the calculation of deflections: one rigorous, the other more approximate. In the more rigorous approach, the curvature is calculated at a reasonable number of sections along the beam and then the deflection is calculated by numerical double integration.

The curvature may be calculated from:

$$\frac{1}{r} = \xi(1/r)_{II} + (1 - \xi)(1/r)_{I}$$

where

$(1/r)_I$ = the curvature calculated assuming the section is uncracked
$(1/r)_{II}$ = the curvature calculated assuming the section to be fully cracked

ξ = a distribution factor = $-\beta_1\beta_2\left(\frac{\tau_{sr}}{\tau_s}\right)^2$ where

β_1 = a coefficient that takes account of the bond properties of the bars
= 1 for high bond bars
= 0.5 for plain bars
β_2 = a coefficient that takes account of the duration of the loading or of repeated loading
= 1 for a single short-term loading
= 0.5 for sustained loads or many cycles of repeated loading
σ_s = the stress in the tension steel calculated on the basis of a cracked section
σ_{sr} = the stress in the tension steel calculated on the basis of a cracked section under the loading which will just cause cracking at the section being considered
(Note: σ_s/σ_{sr} can be replaced by M/M_{cr} for flexure or N/N_{cr} for pure tension.)
ξ = zero for uncracked sections

The curvature may be calculated from the relation:

$$\frac{1}{r} = \frac{M}{EI_I} \text{ or } \frac{M}{EI_{II}}$$ as appropriate or, for cracked sections, from the relation:

$$\frac{1}{r} = \frac{\varepsilon_{sII}}{(d - x)}$$

where ε_{sII} is the strain in the reinforcement calculated on the basis of a cracked section. Values of I_{II} can be obtained from Table 12.4 or Figure 12.3.

The second method is to use the relation:

$$a = \xi a_I + (1 - \xi)a_{II}$$

where
a = the deflection
a_I = the deflection calculated on the basis of an uncracked section
a_{II} = the deflection calculated on the basis of a cracked section

Standard elastic formulae may be used for obtaining a_I and a_{II}, using the appropriate values of II and III.

The calculation of a_I and a_{II} may be obtained from the relation:

$$a = \frac{kL^2M}{EI}$$

where k is a coefficient that depends on the shape of the bending moment diagram. Values for k are given in Table 13.2, taken from the UK code, BS8110, Part 2.

Table 13.2 Values of κ for various bending moment diagrams

Loading	Bending moment diagram	κ
M M	M	0.125
al W	$M = Wa(1 - a)l$	$\frac{3 - 4a^2}{48(1 - a)}$ if $a = \frac{1}{2}$, $\kappa = \frac{1}{12}$
	M	0.0625
al $W/2$ $W/2$ al	$M = Wal/2$	$0.125 - \frac{a^2}{6}$
q	$ql^2/8$	0.104
q	$ql^2/15.6$	0.102
q	M_A M_C M_B	$\kappa = 0.104(1 - \frac{\beta}{10})$ $\beta = \frac{M_A + M_B}{M_C}$
al W	Wal	end deflection = $\frac{a(3 - a)}{6}$ load at end $\kappa = 0.333$
al q	$qa^2l^2/2$	$\frac{a(4 - a)}{12}$ if $a = l$, $\kappa = 0.25$
	M_A M_B M_C	$\kappa = 0.083(1 - \frac{\beta}{4})$ $\beta = \frac{M_A + M_B}{M_C}$
al al	$(Wl^2/24)(3 - 4a^2)$	$\frac{1}{80}\frac{(5 - 4a^2)^2}{3 - 4a^2}$

14 Detailing

14.1 Bond conditions

The bond conditions affect the anchorage and lap lengths. Good and poor bond conditions are illustrated in Figure 14.1.

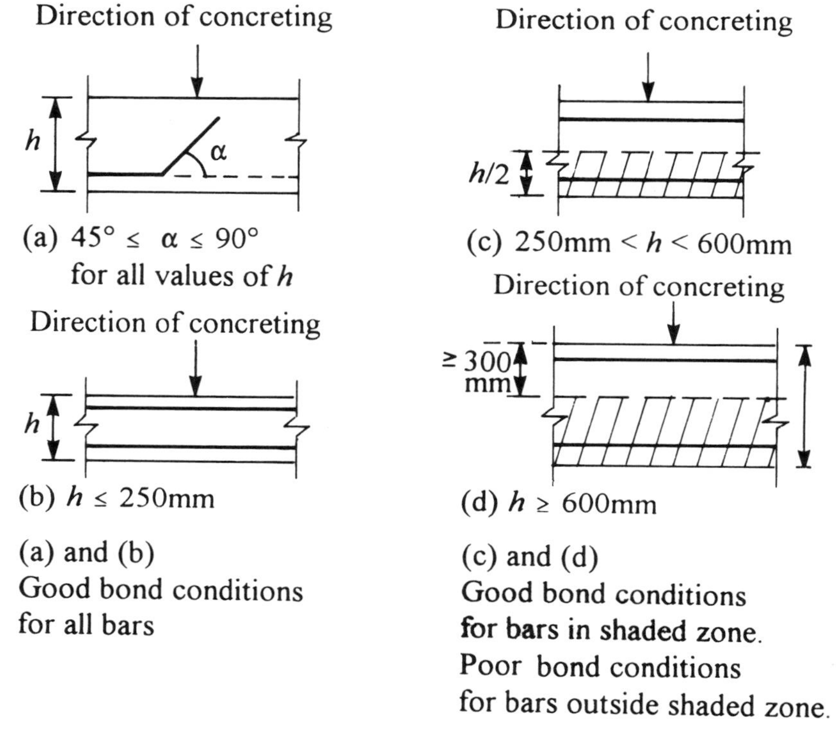

Figure 14.1 Good bond conditions.

14.2 Anchorage and lap lengths

Anchorage and lap lengths should be obtained from Table 14.1 for high-bond bars and Table 14.2 for weld mesh fabric made with high-bond bars.

14.3 Transverse reinforcement

(a) Anchorage zones

Transverse reinforcement should be provided for all anchorages in compression. In the absence of transverse compression caused by support reactions, transverse reinforcement should also be provided for anchorage in tension.

The minimum total area of transverse reinforcement required within the anchorage zone is 25% of the area of the anchored bar.

The transverse reinforcement should be evenly distributed in tension anchorages and concentrated at the ends of compression anchorages.

(b) Laps

No special transverse reinforcement is required if the size of bars lapped is less than 16 mm or fewer than 20% of the bars in the section are lapped. When required, the transverse reinforcement should be placed as shown in Figure 14.2.

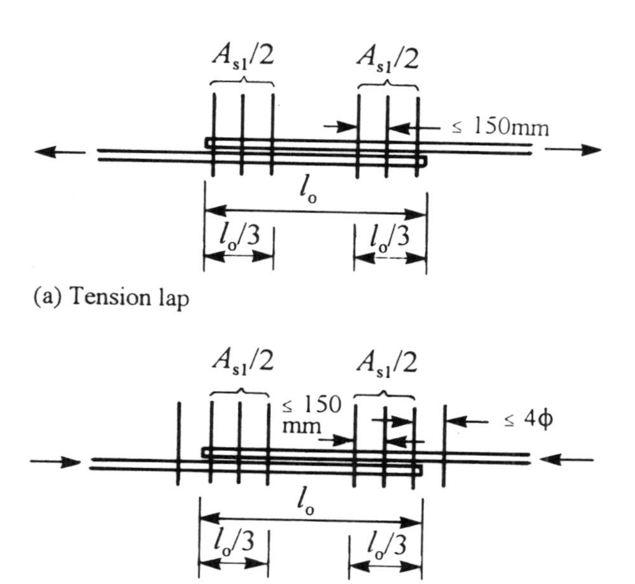

Figure 14.2 Transverse reinforcement at laps.

14.4 Curtailment of bars in flexural members

When a bar is curtailed in a flexural member, it should be anchored beyond the point when it is no longer required, for a length of $l_{b,\,net}$ or d, whichever is the greater.

In determining the location when a bar is no longer required, force in bars should be calculated taking into account (a) the bending moment and (b) the effect of truss modal for resisting shear.

A practical method for curtailment is as follows:

(a) Determine where the bar can be curtailed based on bending moment alone; and
(b) Anchor this bar beyond this location for a distance $l_{b,net} + a_l$, where $a_l = 0.45d$ for beams and $1.0d$ for slabs.

This procedure is diagrammatically illustrated in Figure 14.3.

At simply supported ends, the bars should be anchored beyond the line of contact between the member and its support by

$0.67\ l_{b,net}$ at a direct support and
$1.00\ l_{b,net}$ at an indirect support.

This requirement is illustrated in Figure 14.4.

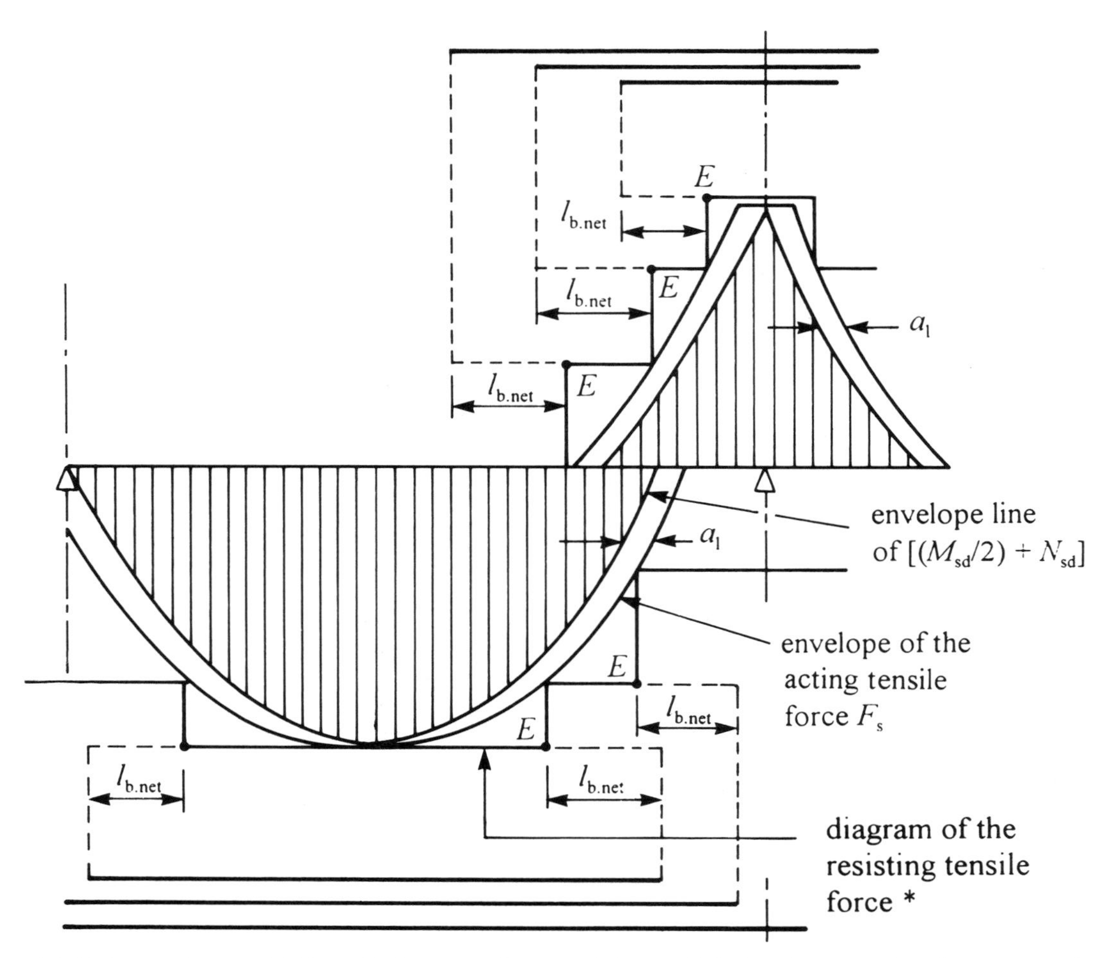

Figure 14.3 Illustration of 'shift-rule' for curtailment of bars. (* It is also permitted to use a diagram in which the resisting tensile force progressively decreases along the length $l_{b,\,net}$.)

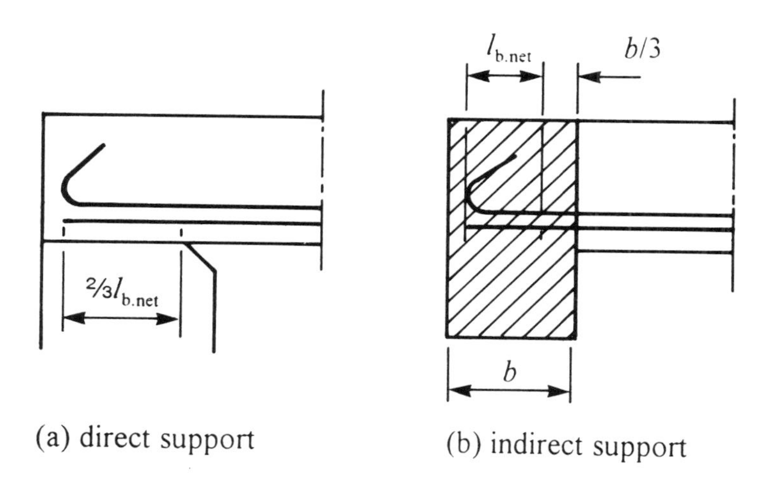

(a) direct support (b) indirect support

Figure 14.4 Anchorage of bottom reinforcement on end supports.

Table 14.1 Anchorage and lap lengths as multiples of bar size: high bond bars f_{yk} = 500 N/mm²

Concrete strength (N/mm²)	f_{ck}	20	25	30	35	40
	f_{cu}	25	30	37	45	50
Anchorage - straight bars compression and tension		48	40	37	33	29
Anchorage - curved bars[4] tension		34	28	26	23	21
Laps - compression - tension [5]		48	40	37	33	29
Laps - tension [6]		67	57	52	46	41
Laps - tension [7]		96	80	74	65	59

NOTES:
General
1. For bars with f_{yk} other than 500 N/mm², the values should be multiplied by (f_{yk}/500).
2. The values in the Table apply to (a) good bond conditions (see Fig. 14.1) and (b) bar size ≤ 32.
3. For poor bond conditions (see Figure 14.1), the Table values should be divided by 0.7.
4. For bar size > 32, the values should be divided by [(132 - ø)/100], where ø is the bar diameter in mm.

Specific conditions
5. In the anchorage region, cover perpendicular to the plane of curvature should be at least 3ø.
6. Proportion of bars lapped at the section < 30% and clear spacing between bars ≥ 10ø and side cover to the outer bar ≥ 5ø.
7. Proportion of bars lapped at the section > 30% or clear spacing between bars < 10ø or side cover to the outer bar < 5ø.
8. Proportion of bars lapped at the section > 30% and clear spacing between bars < 10ø or side cover to the outer bar < 5ø.

Table 14.2 Anchorage and lap lengths as multiples of bar size. Welded mesh fabric made with high-bond bars with f_{yk} = 500 N/mm²

Concrete strength f_{ck} (N/mm²)	20	25	30	35	40
Basic anchorage and lap lengths (mm)	48	40	37	33	29

Notes:

1. For bars with f_{yk} other than 500 N/mm², the values should be multiplied by (f_{yk}/500).
2. Where welded transverse bars are present in the anchorage zone, the Table values for anchorage may be multiplied by 0.7.
3. The values given in the Table apply to good bond conditions and to bar sizes ≤ 32mm.
4. For poor bond conditions, the values should be divided by 0.7.
5. For bar sizes > 32mm, the values should be divided by [(132 - ø)/100], where ø is the diameter of the bar in mm.
6. The Table values should be multiplied by the following factors corresponding to the different (A_s/S) values. A_s is the area of the main reinforcement (mm²) bar and S is the spacing of the bars forming the main reinforcement (m).

A_s/S	≤ 480	680	880	1080	1280
Multiplier	1.00	1.25	1.50	1.75	2.00

15 Numerical examples designed to ENV 1992-1-1

15.1 Introduction

Three types of building have been designed to Eurocode 2 (ENV 1992-1-1). Criteria for the choice of the buildings were:

- the type of structural members
- magnitude of vertical (imposed) loads
- character of the imposed loads (i.e. static or dynamic)
- the ultimate limit states to be considered (e.g. punching, fatigue).

The objectives of these calculations were to demonstrate the applicability of Eurocode 2 in practice. The main conclusion of these calculations therefore is that no basic difficulties have been observed when applying the new European Prestandard in a practical design process.

15.2 References

Reference	Title		Abbreviation
ENV 1991-1:	Eurocode 1: Basis of design and actions on structures. Part 1: Basis of design. Edition 1994.		EC1-1
ENV 1991-2-1:	Eurocode 1: Basis of design and actions on structures. Part 2.1: Densities, self-weight and imposed loads. Final draft April 1993.		EC1-2.1
ENV 1991-2-3:	Eurocode 1: Basis of design and actions on structures. Part 2.3: Snow loads. Final draft April 1993.		EC1-2.3
ENV 1991-2-4:	Eurocode 1: Basis of design and actions on structures. Part 2.4: Wind loads. Final draft April 1993.		EC1-2.4
ENV 1992-1:	Eurocode 2: Design of concrete structures. Part 1: General rules and rules for buildings. Edition 1991.	EC2	
ENV 1992-1-2:	Eurocode 2: Design of concrete structures. Part 1-2: Structural fire design. Draft August 1994.	EC2-1.2	
pr ENV 1992-2:	Eurocode 2: Design of concrete structures. Part 2: Concrete bridges. Draft June 1995.	EC2-2	
ENV 10 080:	Steels for the reinforcement of concrete; Weldable ribbed reinforcing steel grade B500; Technical delivery conditions for bars, coils and welded fabrics. Final draft April 1994.		ENV 10 080
ENV 206:	Concrete production, placing and compliance criteria. Edition 1990.		ENV 206
DIN 15 018:	Cranes; Principles for steel structures, stress analysis. Part 1. Edition November 1984.		

15.2 References

1. Litzner, H.-U.: *Design of Concrete Structures to ENV 1992-Eurocode 2. Concrete Structures-Euro-Design Handbook.* 1st volume 1994/1996. Berlin: Ernst & Sohn 1994.

2. Deutscher Ausschuß für Stahlbeton (DAfStb): *Bemessungshilfsmittel zu Eurocode 2 Teil 1 (DIN V ENV 1992 Teil 1-1, Ausgabe 06.92). 2. ergänzte Auflage. Heft 425 der DAfStb-Schriftenreihe.* Berlin, Köln: Beuth Verlag GmbH 1992.

3. British Cement Association: *Worked examples for the design of concrete buildings.* Crowthorne: British Cement Association 1994.

4. Deutscher Beton-Verein E.V.: *Beispiele zur Bemessung von Betontragwerken nach EC2.* Wiesbaden, Berlin: Bauverlag GmbH 1994.

5. Dieterle, H.: *Zur Bemessung quadratischer Stützenfundamente aus Stahlbeton unter zentrischer Belastung mit Hilfe von Bemessungsdiagrammen.* Heft 387 der DAfStb-Schriftenreihe 1987.

6. British Cement Association: *Concise Eurocode for the design of concrete buildings.* Crowthorne, 1993.

7. Betonvereniging: *GTB Deel 2. Grafieken en Tabellen voor Beton.* Gouda 1992.

15.3 Calculation for an office building

15.3.1 Floor plan, structural details and basic data

15.3.1.1 Floor plan of an office building

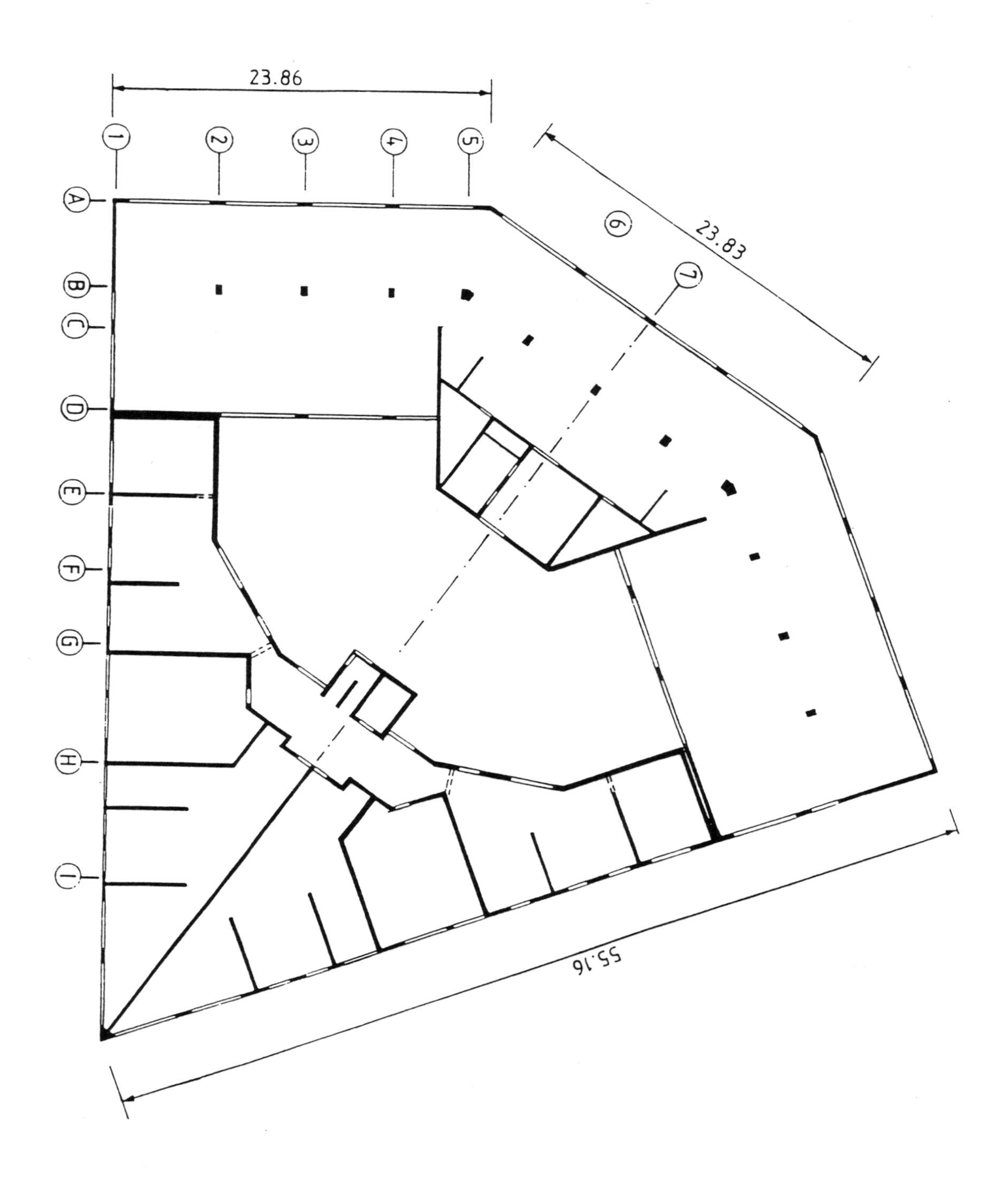

15.3.1.2 Structural details of an office building

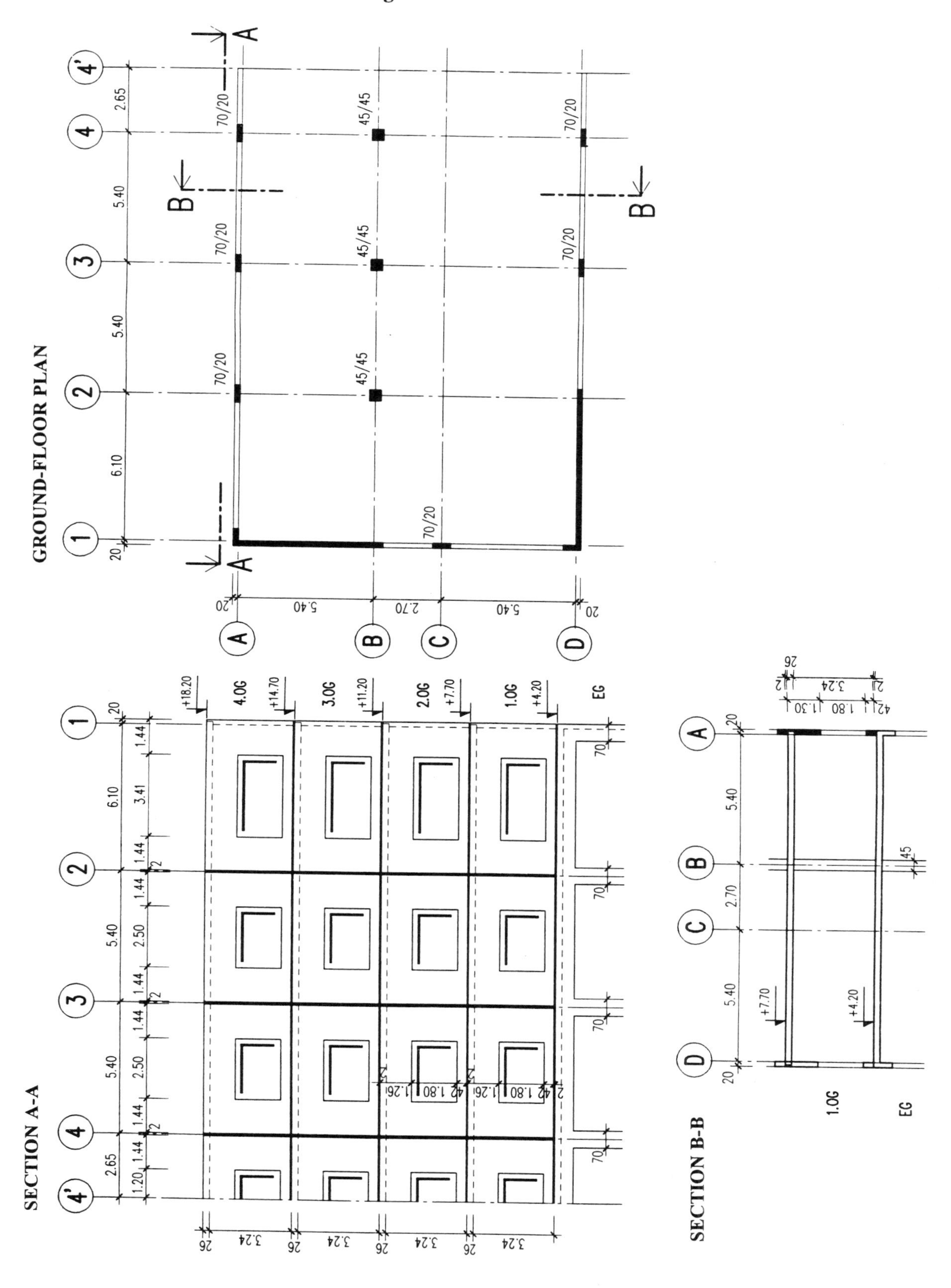

15.3.1.3 Basic data of structure, materials and loading

		Reference
Intended use:	Office block	see floor plan
Fire resistance:	1 hour for all elements	EC2-1.2, 1.3

Loading (excluding self-weight of structure):

					Reference
Flat slab:	- imposed:	Q_k	=	3 kN/m^2	EC1-2.1
	- finished:	$G_{k,2}$	=	1.25 kN/m^2	
Category B					
	- partitions:	$G_{k,3}$	=	1.25 kN/m^2	EC1-2.1

Combination factors:

				Reference
Frequent actions:	ψ_1	=	0.5	EC1-1, Table 9.3, Category B
Quasi-permanent actions:	ψ_2	=	0.3	

Exposure classes: (Reference: EC2, Table 4.1)

Flat slab:

Internal columns: Class 1 (indoors)

Façade elements: Class 2b (humid environment with frost)

Block foundation: Class 5a (slightly aggressive chemical environment)

Subsoil conditions:

Sand, gravel — Allowable pressure 300 kN/m^2 (Reference: from soil investigation)

Materials:

		Reference
Concrete grade	C 30/37	EC2, Table 3.1; ENV 206, Table 3 and Table 20;
Steel grade	B 500	ENV 10 080;
Self-weight of concrete	25 kN/m^3	EC1-2.1

15.3.2 Calculation of a flat slab
15.3.2.1 Actions

				Reference
Self-weight of slab: 0.26 * 25		=	6.50 kN/m²	h = 0.26 m
Finishes			1.25 kN/m²	
Partitions			1.25 kN/m²	
Permanent actions:	G_k	=	9.00 kN/m²	
Imposed load:	Q_k	=	3.00 kN/m²	
Design values of actions at the ultimate limit states: $\gamma_G G_k + \gamma_Q Q_k = 1.35 * 9.0 + 1.5 * 3.0$		=	16.65 kN/m²	EC2, Equation (2.7a), fundamental combination
Design values of actions at the serviceability limit states: Rare combination of actions:		=	12.00 kN/m²	EC2, 2.3.4
Frequent combination: $G_k + \psi_1 Q_k = 9.00 + 0.5 * 3.00$		=	10.50 kN/m²	
Quasi-permanent combination: $G_k + \psi_2 Q_k = 9.00 + 0.3 * 3.00$		=	9.90 kN/m²	

15.3.2.2 Structural model at the ultimate limit states (finite element grid)

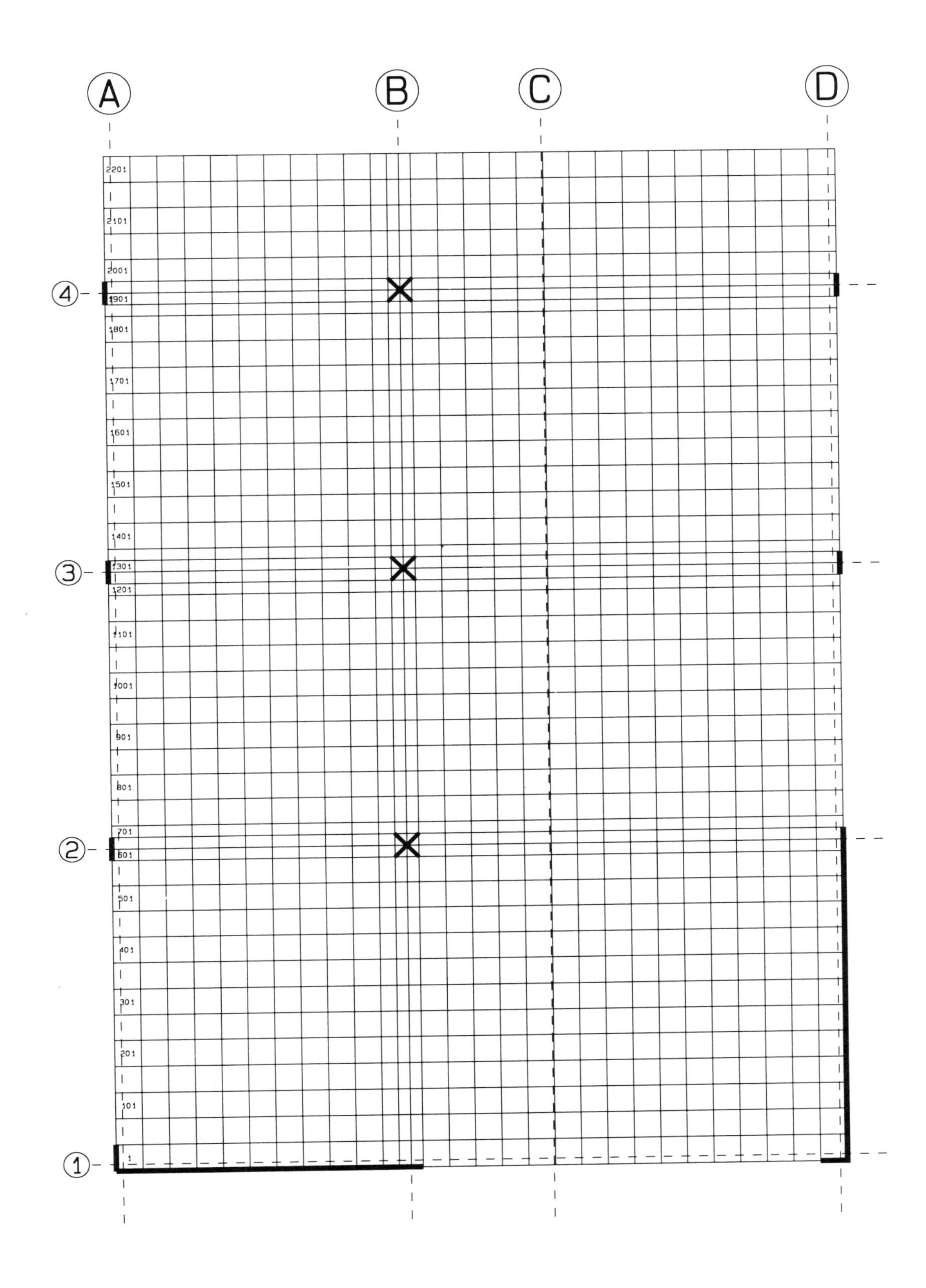

15.3.2.3 Design values of bending moments (example)

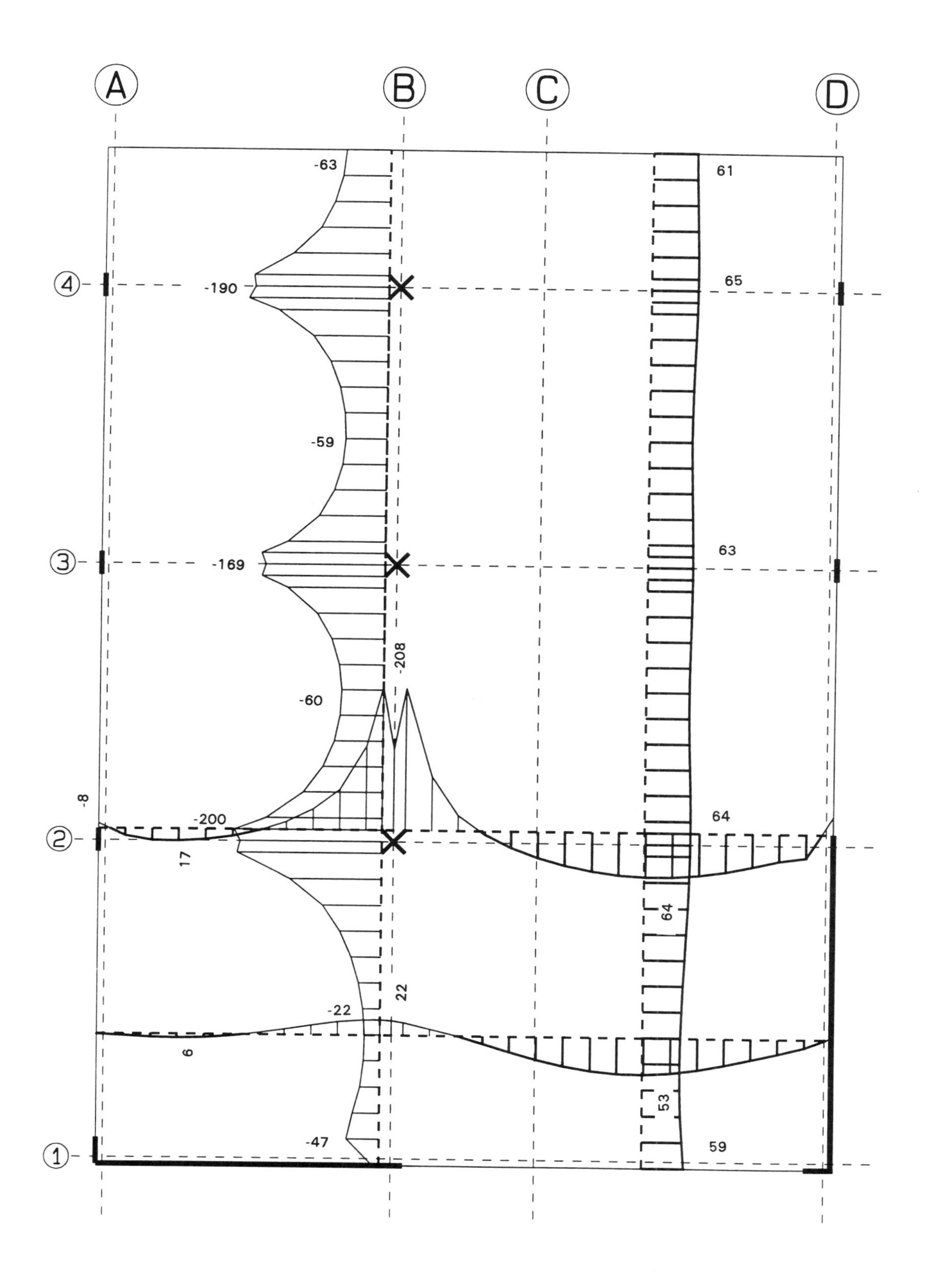

Table 15.1: Design bending moments at the ultimate limit states

Direction	Location	Section	Kind of moment	Moment (kNm/m)	Mean value of moment (kNm/m)
x	B/2	left of axis B	support: min m_{Sd}	- 184.59	
		centre of axis B		- 175.07	- 180.72
		right of axis B		- 182.50	
x	B - D/2	left of axis B	span: max m_{Sd}	63.26	
		centre of axis B		64.02	63.85
		right of axis B		64.26	
y	B/2	top of axis 2	support: m_{Sd}	- 208.26	
		centre of axis 2		- 200.02	- 204.16
		bottom of axis 2		- 204.21	
y	B/1-2	top of axis 2	span: m_{Sd}	93.14	
		centre of axis 2		93.00	92.95
		bottom of axis 2		92.71	

15.3.2.4 Design of bending at the ultimate limit states

Table 15.2: Design for bending

Direction	x		y	
Axis	B/1-3		2/A-D	
Location	Support	Span	Support	Span
m_{Sd} (kN/m)	180.72	63.85	204.16	92.95
d (m)	0.219	0.224	0.233	0.235
μ_{Sds}	0.188	0.063	0.188	0.084
ω	0.2163	0.0657	0.2163	0.0888
ξ	0.3142	0.1069	0.3142	0.1332
$A_{s,\,req}$ (mm²/m)	$21.78*10^2$	$6.80*10^2$	$23.17*10^2$	$9.59*10^2$
Selected B 500B(S)	2*ø 14-14.0		2*ø 14-14.0	
Selected B 500A(M)		2*ø 7.0-100		2*ø 8.0-100
$A_{s,prov}$ (mm²/m)	$22.00*10^2$	$7.70*10^2$	$23.68*10^2$	$10.05*10^2$

f_{cd} = 30/1.5 = 20 N/mm²
f_{yd} = 500/1.15 = 435 N/mm²

<u>Calculation for supports</u>

d_y = h - (min c + Δh) - ø/2
= 0.260 - (0.015 + 0.005) - 0.014/2 = 0.233 m
d_x = d_y - ø = 0.233 - 0.014 = 0.219 m

(S): Reinforcing bars
(M): Welded mesh fabric

15.3.2.5 Ultimate limit state for punching shear

Shear forces due to permanent actions

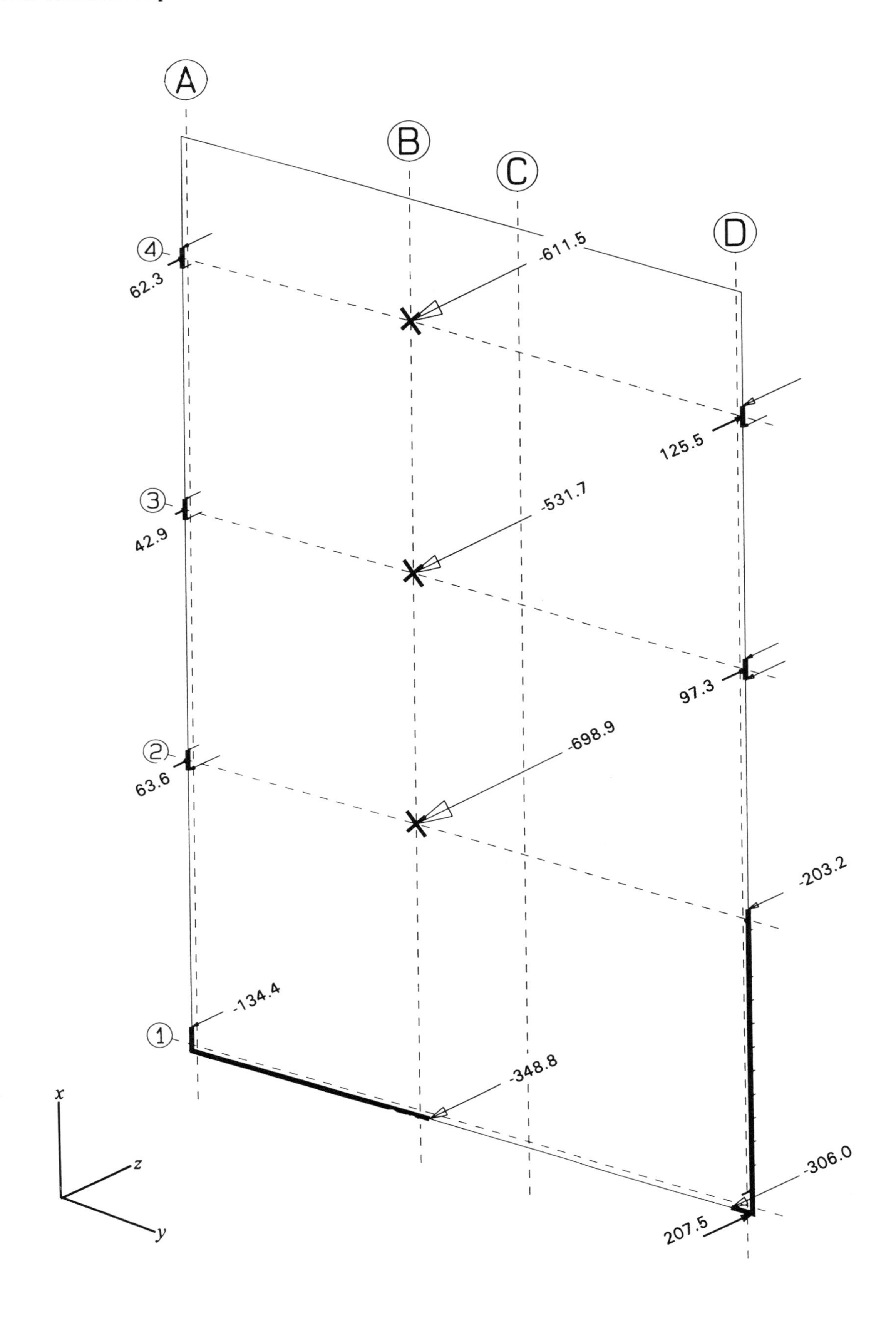

Shear forces due to variable actions

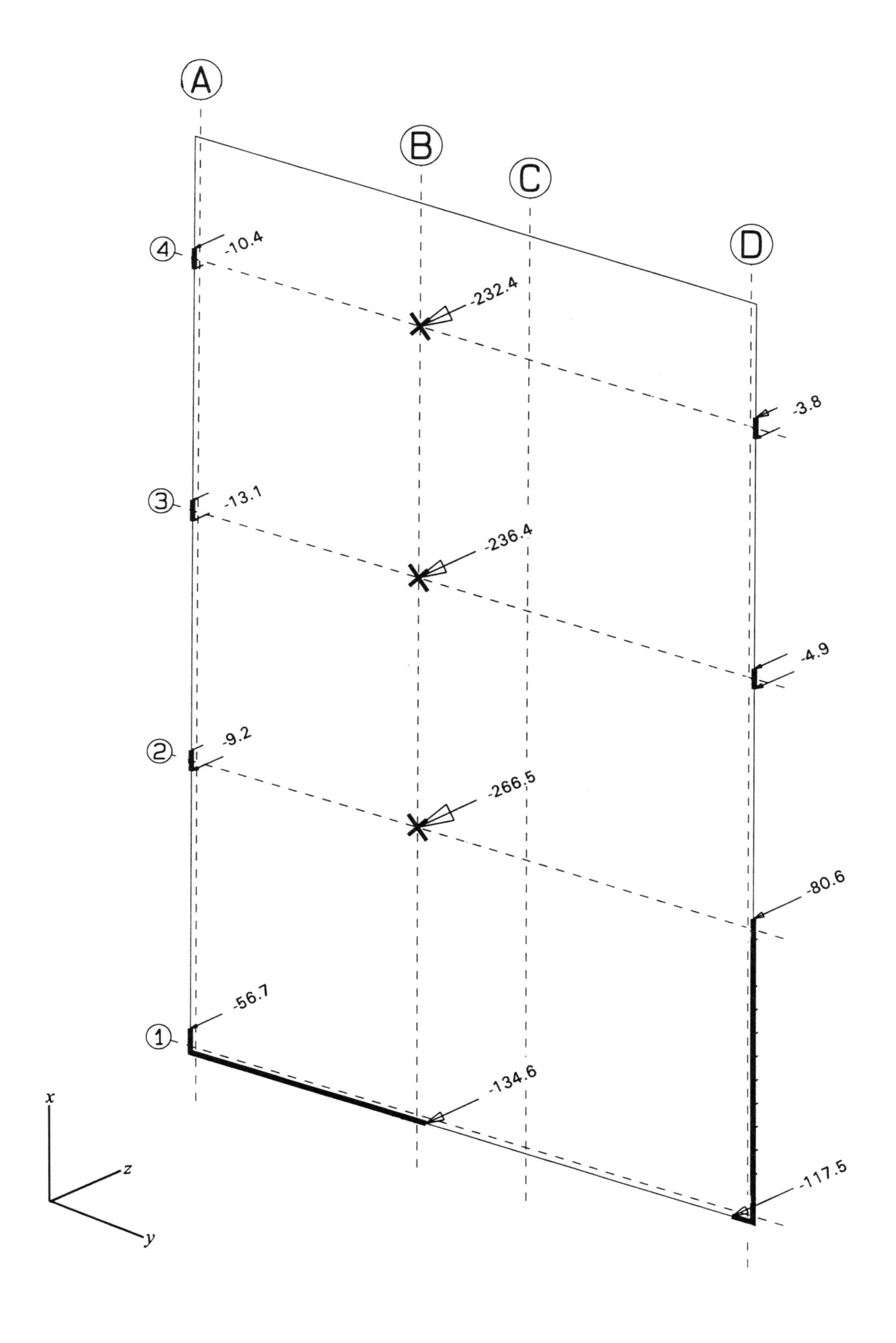

			Reference
Design for punching shear in axis B/2			EC2, 4.3.4
V_{sd} = 698.9 + 266.5	=	966 kN	see distribution of shear forces
d_m = (0.233 + 0.219)/2	=	0.226 m see Table 2	
Critical perimeter u = 4 * 0.45 + 2π * 1.5 * 0.226	=	3.93 m	EC2, 4.3.4.2.2
Acting shear force v_{Sd} = 966 * 1.15/3.93	=	283 kN/m	EC2, Eq.(4.50) for internal columns
Shear resistance of slabs without shear reinforcement ρ_l ≈ 0.01			EC2, 4.3.4.5.1 see Table 2; > 0.5%
v_{Rd1} = 0.34 * (1.6 - 0.226) * (1.2 + 40 * 0.01) * 0.226 * 10^3	= <	169 kN/m v_{Sd}	EC2, Eq.(4.56)
1.6 * v_{Rd1} = 1.6 * 169	= <	270 kN/m v_{Sd}	
ρ_l must be increased			
Required ρ_l = [283/(1.6*0.34*1.374*0.226*10^3) - 1.2]/40	=	1.19%	< 1.5% Table 2: ρ_{lx} = 1.19*21.9*10^2 = 26.0*10^2 mm²/m ρ_{ly} = 1.19*23.3*10^2 = 27.7*10^2 mm²/m
Calculation of shear reinforcement			EC2, Eq.(4.58)
v_{Sd} - v_{Rd1} = 283 - 283/1.6	=	107 kN/m	
$\sum A_{sw}$ = 107 * 10^3 * 3.93/(435 * sin 60°)	=	11.2 *10^2 mm²	
Selected four bent-up bars ø 14 mm provided 4.2.1.5.3 = 12.2*10^2 mm²			α = 60°
Minimum shear reinforcement			EC2, 4.3.4.5.2(4)
0.6 * min ρ_w = 0.6 * 0.11	=	0.066%	EC2, Table 5.5
The critical area minus the loaded area = = 4 * 0.45 * 1.5 * 0.226 + π*(1.5 * 0.226)²	=	0.97 m²	
$\sum A_{sw,min}$ = 0.066 * 10^{-2} * 0.97 * 10^6/sin 60°	=	7.9 * 10^2 mm² < 12.2 * 10^2 mm²	

Reference

Minimum design moment

EC2, 4.3.4.5.3

$m_{Sd,min}$ = -966 * 0.125 = -121 kNm/m < m_{Sd}

for internal columns see Table 15.1 above

15.3.2.6 Limitation of deflections

EC2, 4.4.3

It is assumed that, with regard to deflections under quasi-permanent actions, a limiting value of 25 mm was agreed with the client. The deflection diagram for cracked cross-section shows that this requirement is met between axes 1 and 4. The deflection of the cantilever slab accounting for creep deformations is about 34 mm.

EC2, 4.4.3.1P(2)

see following deflection Figures

EC2, Eq.(A.4.3)

Therefore, in order to ensure proper functioning and appearance of the structure, precamber of the cantilever slab of 10 mm is suggested.

Deformations of flat slab due to quasi-permanent actions, uncracked cross-sections assumed

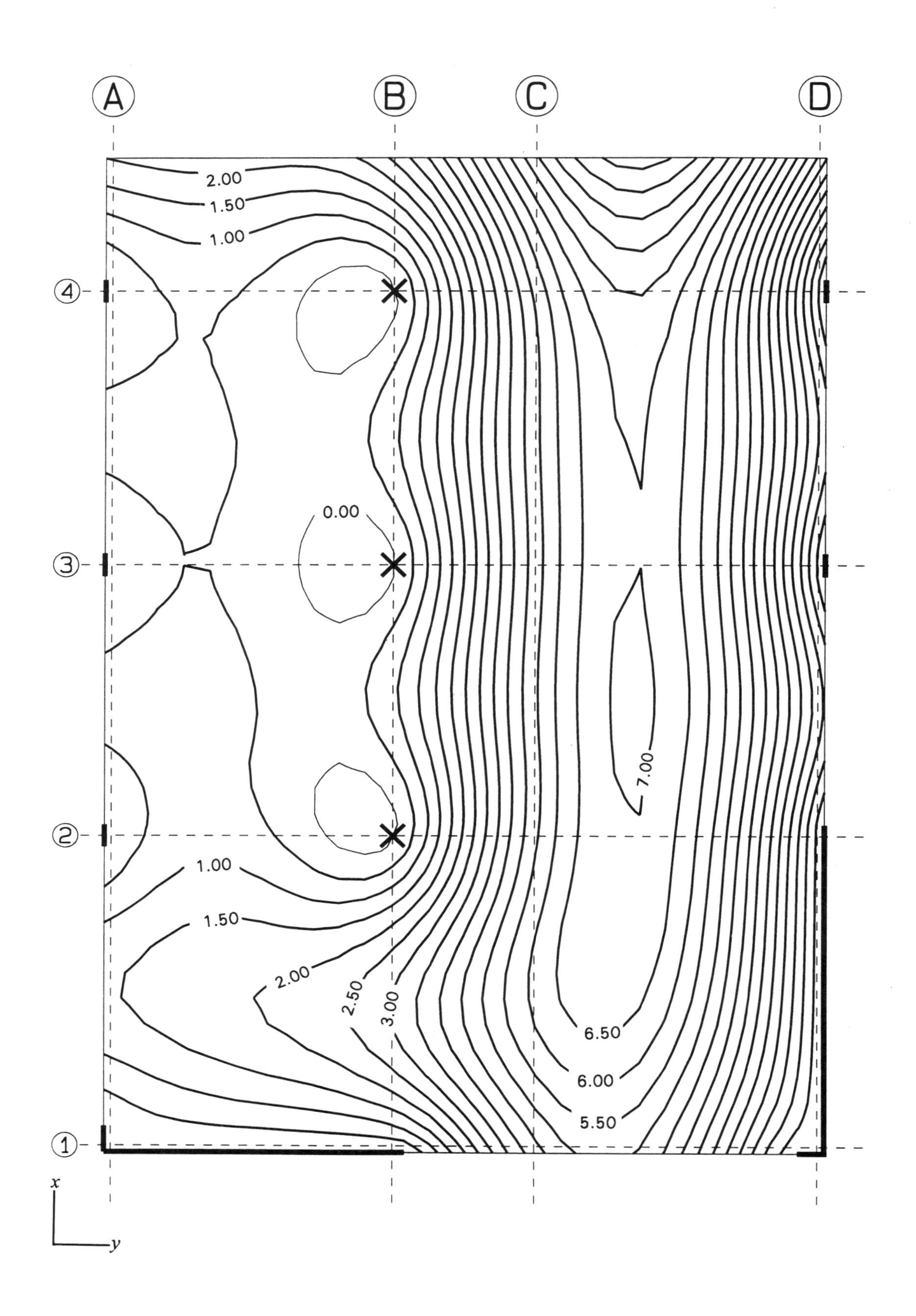

Deformations of flat slab due to quasi-permanent actions, cracked cross-sections assumed

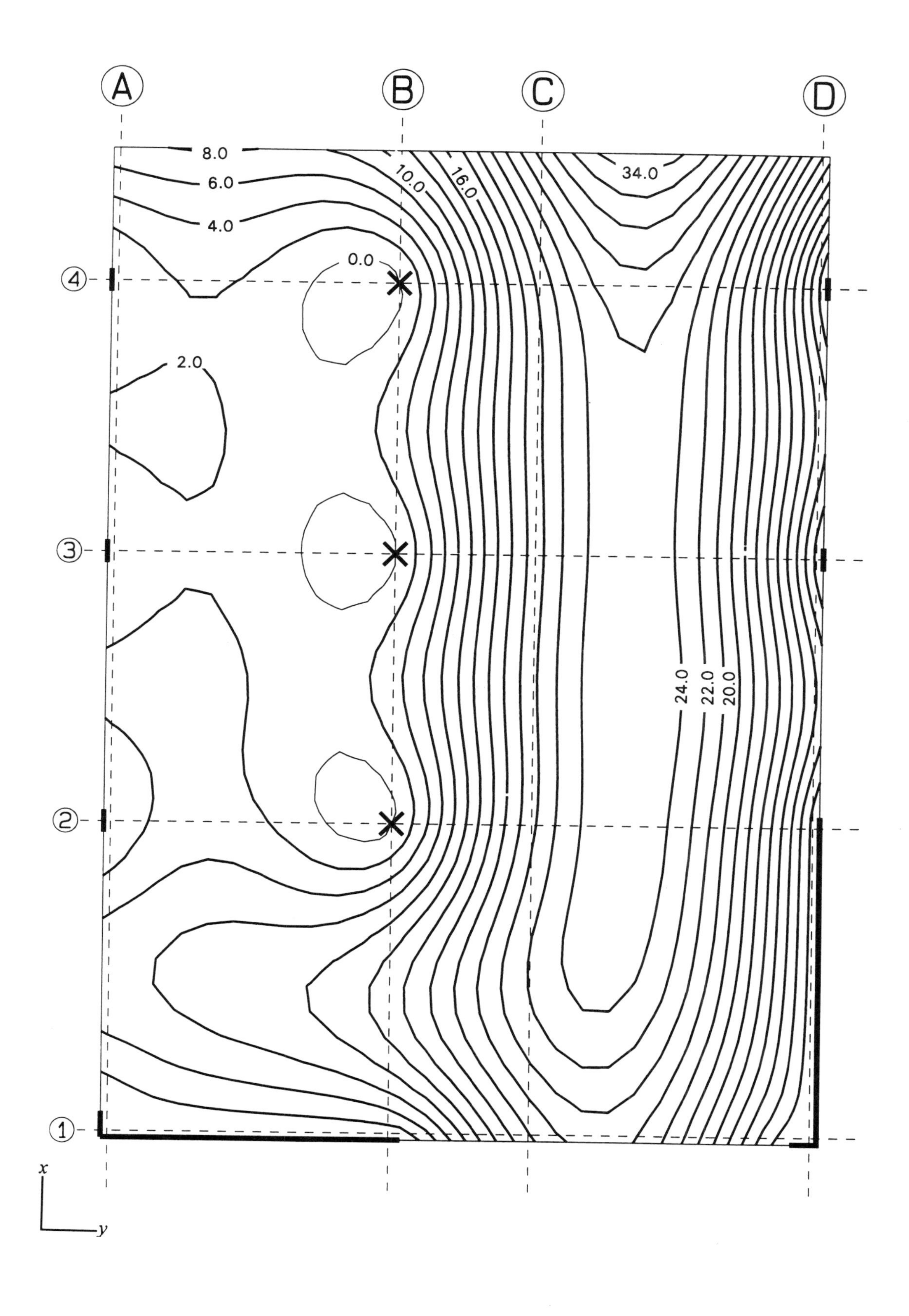

15.3.3 Internal column

Reference

Design model of the column

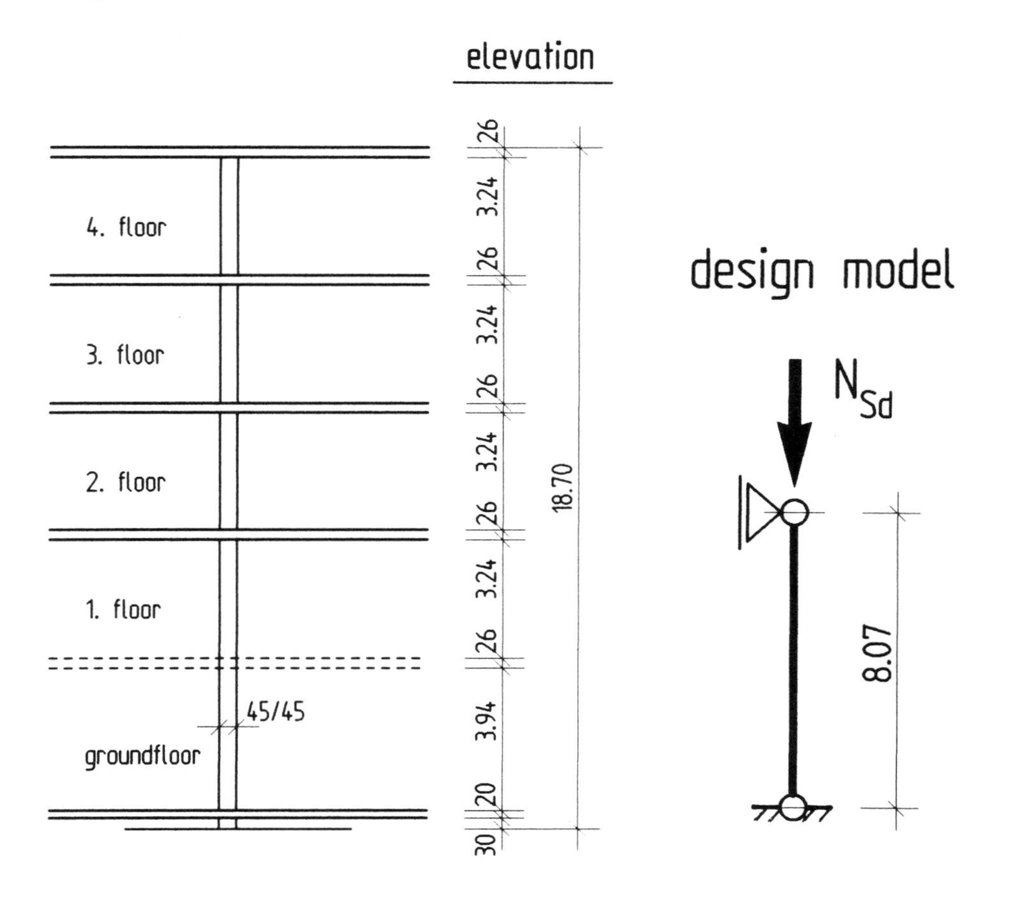

The column in the ground floor/first floor in axis B/2 will be designed to EC2. The structural model is shown in the Figure above. The column is analysed on the assumption that the adjacent slab and block foundation provide no rotational restraint.

see 15.3.2.5 above
EC2, 2.5.3.3(3)

Design value of the axial force N_{sd}:

see 15.3.2.5 above

On the roof, a uniformly distributed snow load is assumed:

$s \quad = \quad 0.9 \text{ kN/m}^2$

EC1-2.3

The combination factor for this load is taken as:

$\psi_0 \quad = \quad 1.0$

conservative assumption

$N_{sd} = -[4 * 698.9 + 3 * 266.5 + 266.5 * 0.9/3.0 + 0.45^2 * 18.7 * 25]$

$\approx \quad -3800 \text{ kN}$

see 15.3.2.5, Figures of shear forces

Reference

Design of the column

using the Figures in [2]

f_{cd} = 20 N/mm^2

C 30/37

f_{yd} = 435 N/mm^2

Creep deformations are neglected.

EC2, A.3.4(9)

Additional eccentricity e_a:

EC2, 4.3.5.4(3)

$e_a = l_0/400 = 8.07/400 = 0.02$ m

$e_a/h = 0.02/0.45 = 0.05$

$l_0/h = 8.07/0.45 = 18$

$\nu_u = -3.8/(0.45^2 * 20) \approx -1.0$

From the design diagram, ω is taken as:

$\omega = 0.40$

$A_{s,\,tot} = 0.40 * 450^2 * 20/435 = 37.3 * 10^2$ mm^2

Selected eight bars ø 25 mm
$A_{s,prov} = 39.3 * 10^2$ mm^2
Links ø 8 mm - 300 mm

EC2, 5.4.1.2.1(2):

$A_{s,min}$ not relevant here

Design diagram for the column

Reference
[2], diagram
R2-05

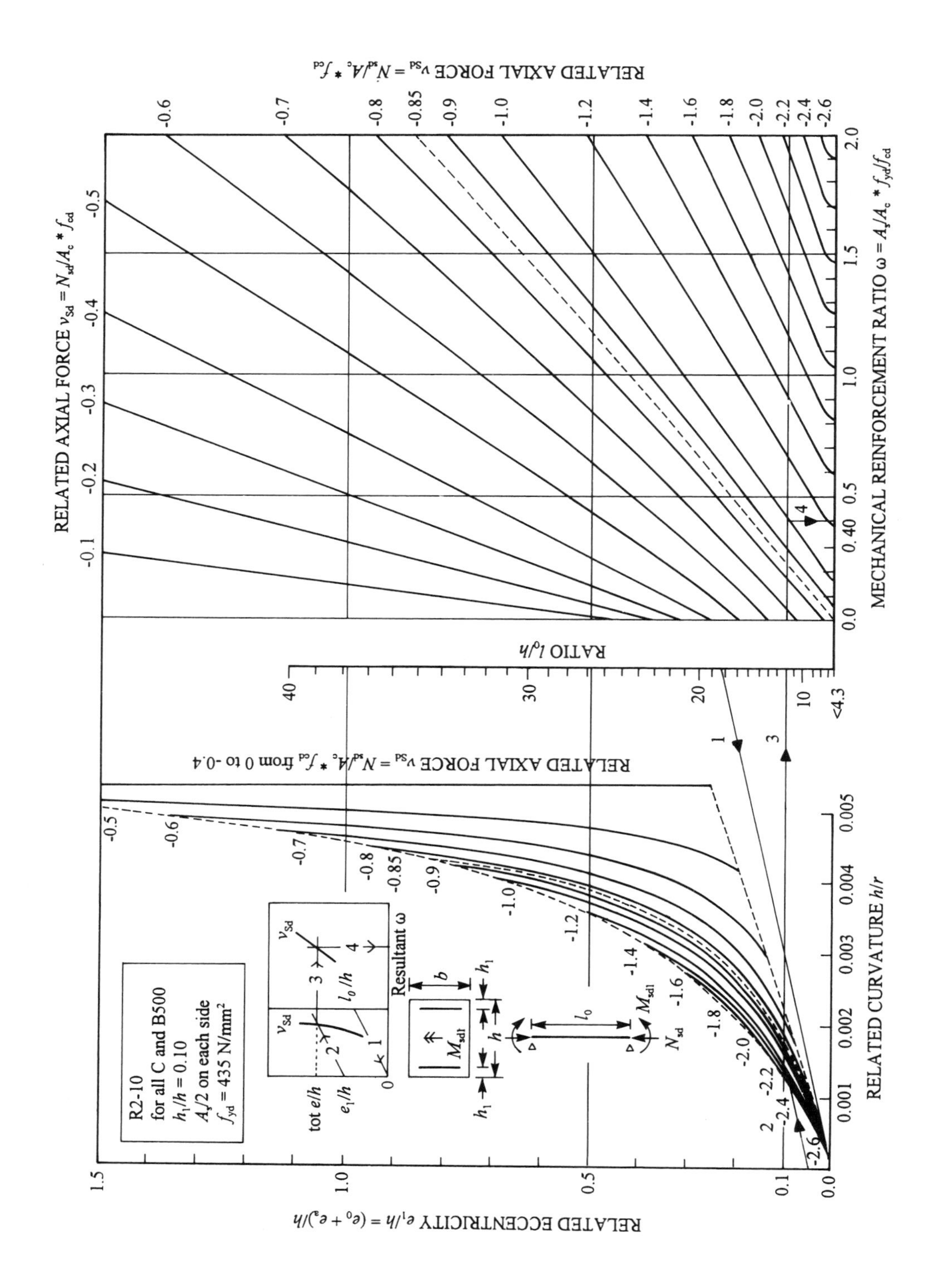

Detailing of reinforcement

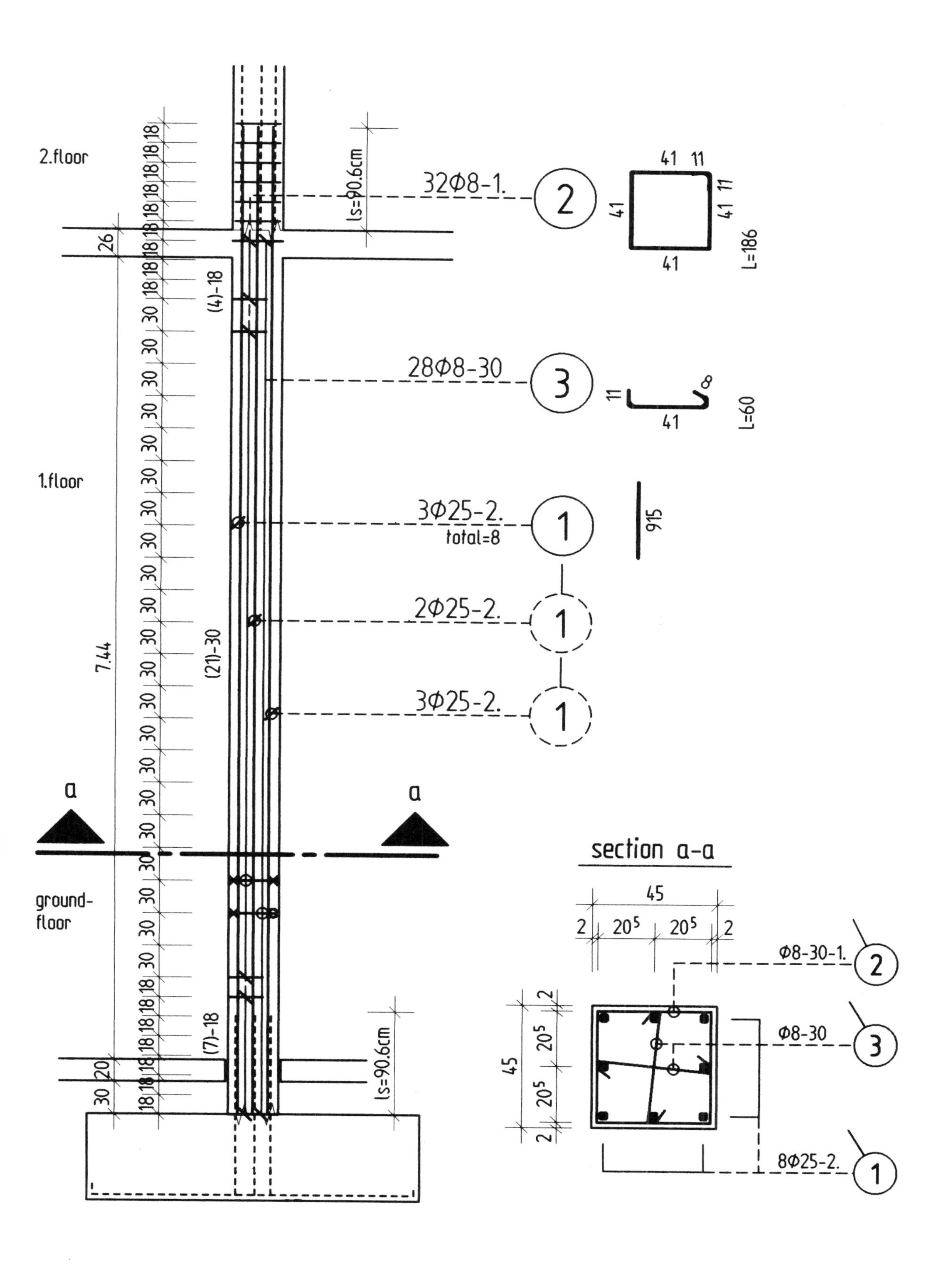

Reference

15.3.4 Facade element

The facade of the building consists of precast elements (see Figure below). As an example, the element between axis 2 and 3 will be designed to EC2. As model, strut and ties are used.

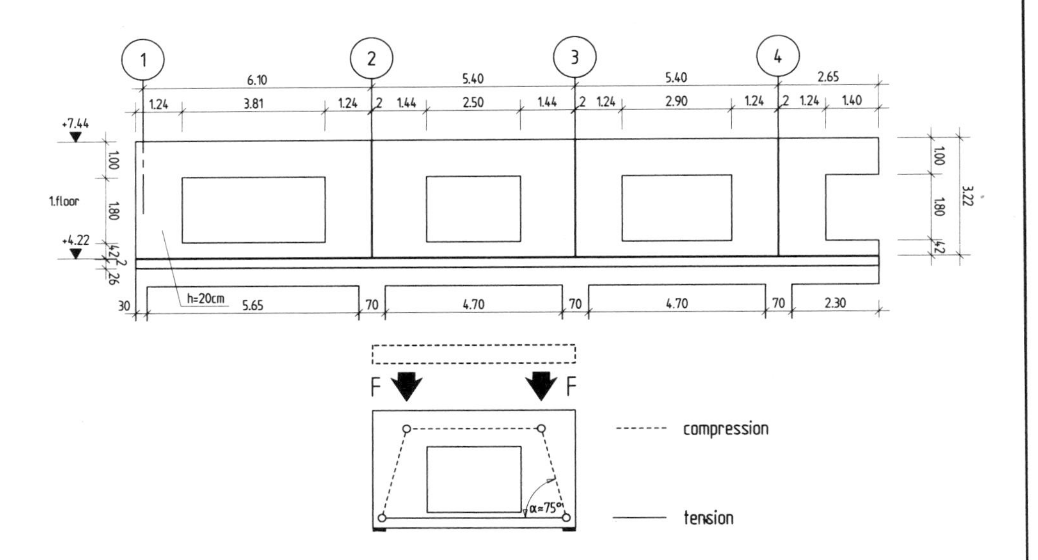

For simplification, the maximum shear forces, between axis 1 and 2 are considered

see 15.3.2.5, Figures for shear forces are considered

max F	=	135 + 70.6	=	206 kN

Maximum tie force:

max T	=	max F cos α		
	=	206 cos 75°	=	79 kN

$A_{s,req}$	=	$79 * 10^3/435$	=	$1.8 * 10^2$ mm^2
$A_{s,prov}$	=	2 ø 12 + 2 ø 10	=	$3.8 * 10^2$ mm^2

see details of reinforcement

Reinforcement details

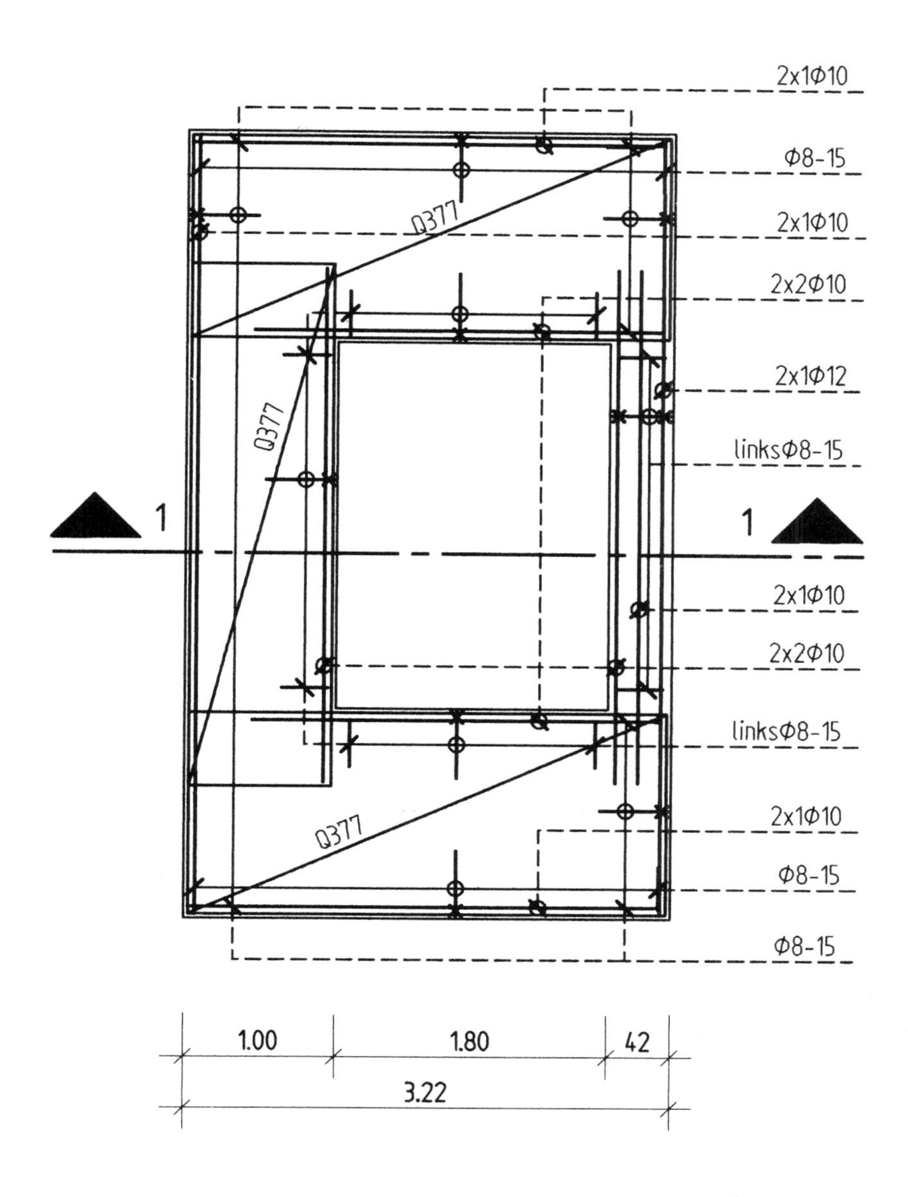

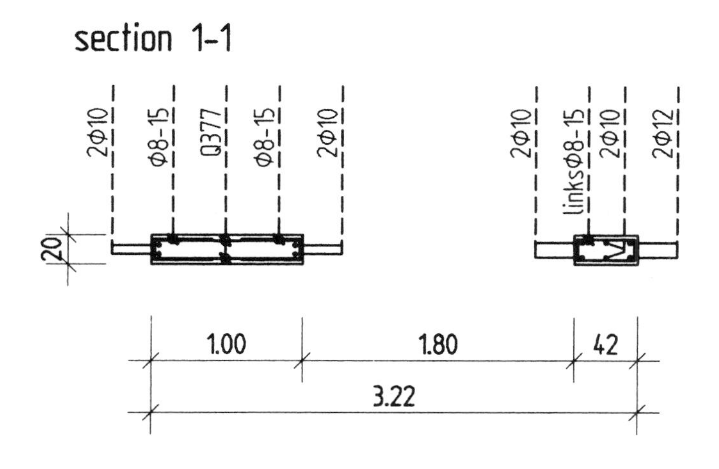

15.3.5 Block foundation

Reference

section I-I

80

3.00

plan

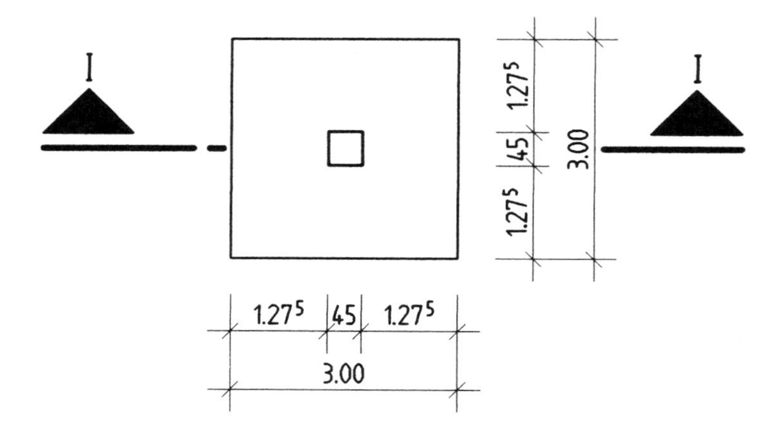

It is assumed that the foundation is subjected to an axial force N_{Sd} only which acts in the centre of gravity of the foundation slab. The axial force N_{Sd} results from the internal column in axis B/2 and is given by

see 15.3.3 above

N_{Sd} = = -3800 kN

Design value of bending moment:

M_{Sd} = $N_{Sd}\ a\ (1-h_{col}\ /a)^2/8$

page 97 in [5]

= $3800 * 3.0 * (1-0.45/3.0)^2/8$ = 1030 kNm

					Reference
Design for bending					EC2, 4.3.1
Effective depth:					
d_x	=	h_f - (min c + Δh + $ø_x/2$)			EC2, 4.1.3.3(9):
	=	0.800 - (0.040 + 0.010 + 0.012/2)	=	0.744 m	min c = 40 mm
					assumption:
d_y	=	d_x - ø = 0.744 - 0.012	=	0.732 m	ø = 12 mm
μ_{Sds}	=	$1.030/(3.0 * 0.732^2 * 20)$	=	0.032	[1], Table 7.1b
ω	=		=	0.033	
$A_{s,req}$	=	0.033 * 3000 * 732 * 20/435	=	$34 * 10^2$ mm²	

Provided in both directions:

$A_{s,\,prov} = 36\ ø\ 12 = 40.7 * 10^2\ mm^2$

spacing see reinforcement details below

Design for punching shear					EC2, 4.3.4
d_m	=	(0.744 + 0.732)/2	=	0.738 m	
Distance of the critical perimeter from the face of the column					EC2, 4.3.4.2.2
s	=	1.5 * d_m = 1.5 * 0.738	=	1.10 m	
Length of critical perimeter					
u	=	4 * 0.45 + 2π * 1.10	=	8.71 m	
Mean value of ground pressure due to N_{Sd}					
σ_g	=	N_{sd}/a^2 = 3800/9.00	=	422 kN/m²	
Area within critical perimeter					
	=	$0.45^2 + 4 * 0.45 * 1.10 + \pi * 1.10^2$	=	6.0 m²	
Critical force to be resisted					EC2, 4.3.4.1(5)
V_{Sd}	=	3800 - 6.0 * 422	=	1268 kN	
v_{Sd}	=	1268/6.0	=	212 kN/m	
Design shear resistance of slabs without punching shear reinforcement:					EC2, Eq.(4.56)
ρ_l	=	40.7/(300 * 73.8)	=	0.18%	
k	=	1.6 - 0.738	=	0.862	
			<	1.0	
τ_{Rd}	=		=	0.34 N/mm²	for C 30/37
v_{Rd1}	=	$1.0 * 0.34 * (1.2 + 40 * 0.0018) * 0.738 * 10^3$			
			=	319 kN/m	
			>	v_{Sd}	

Reinforcement details

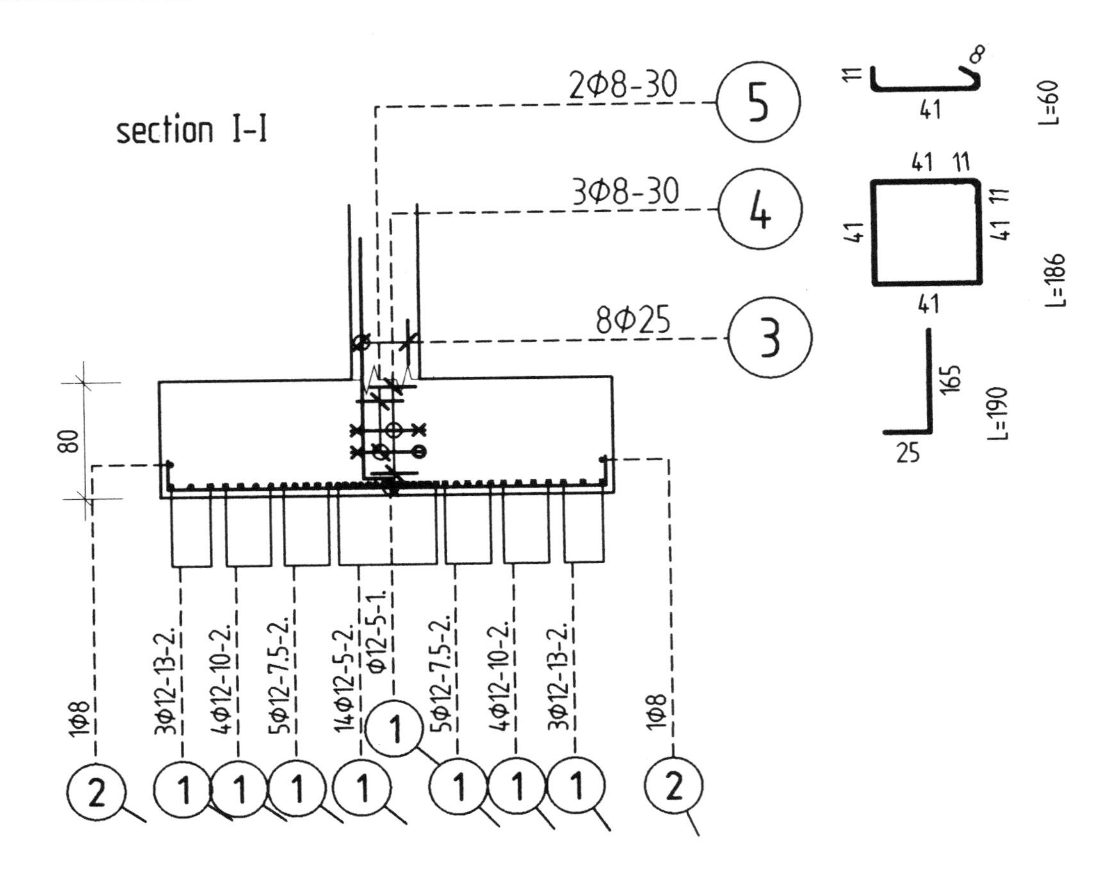

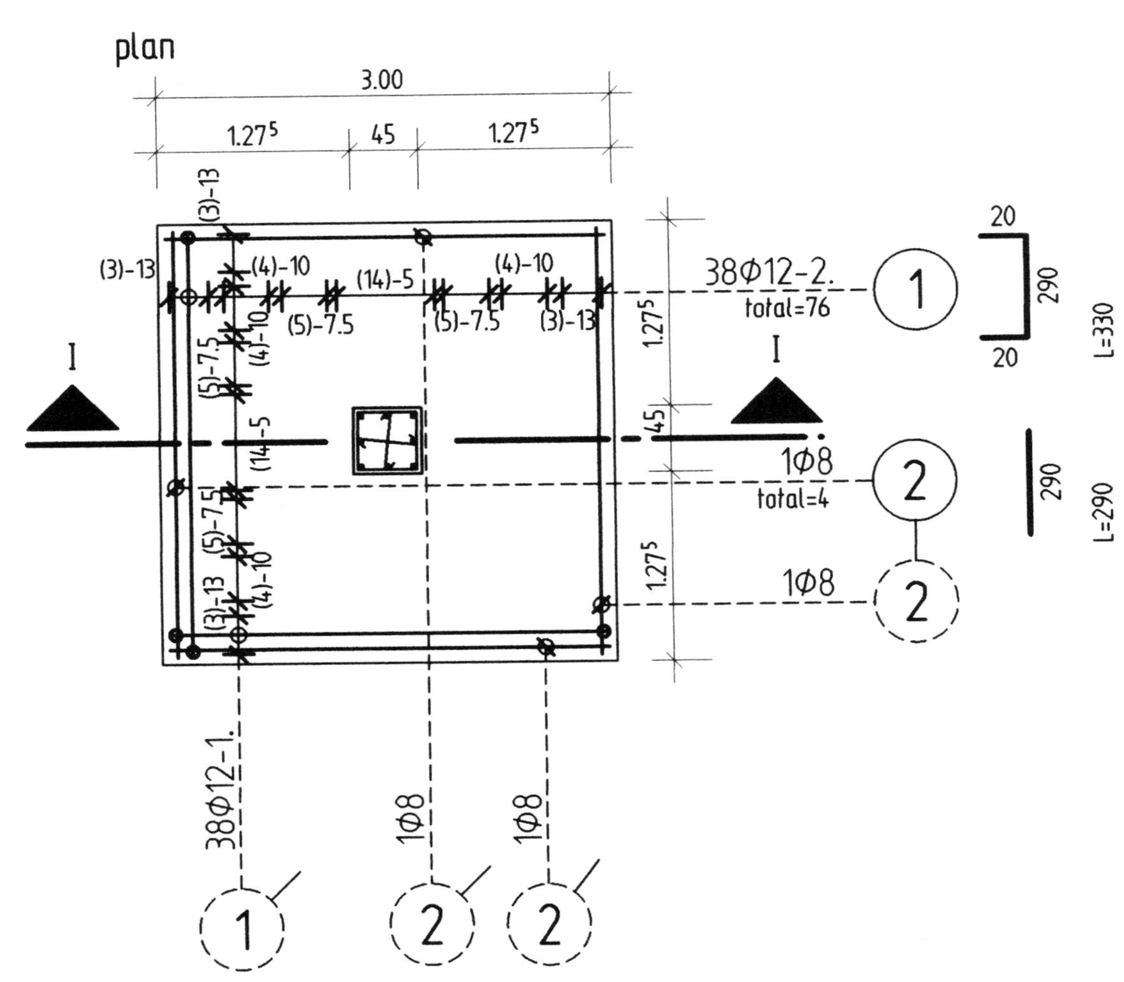

15.4 Calculation for a residential building

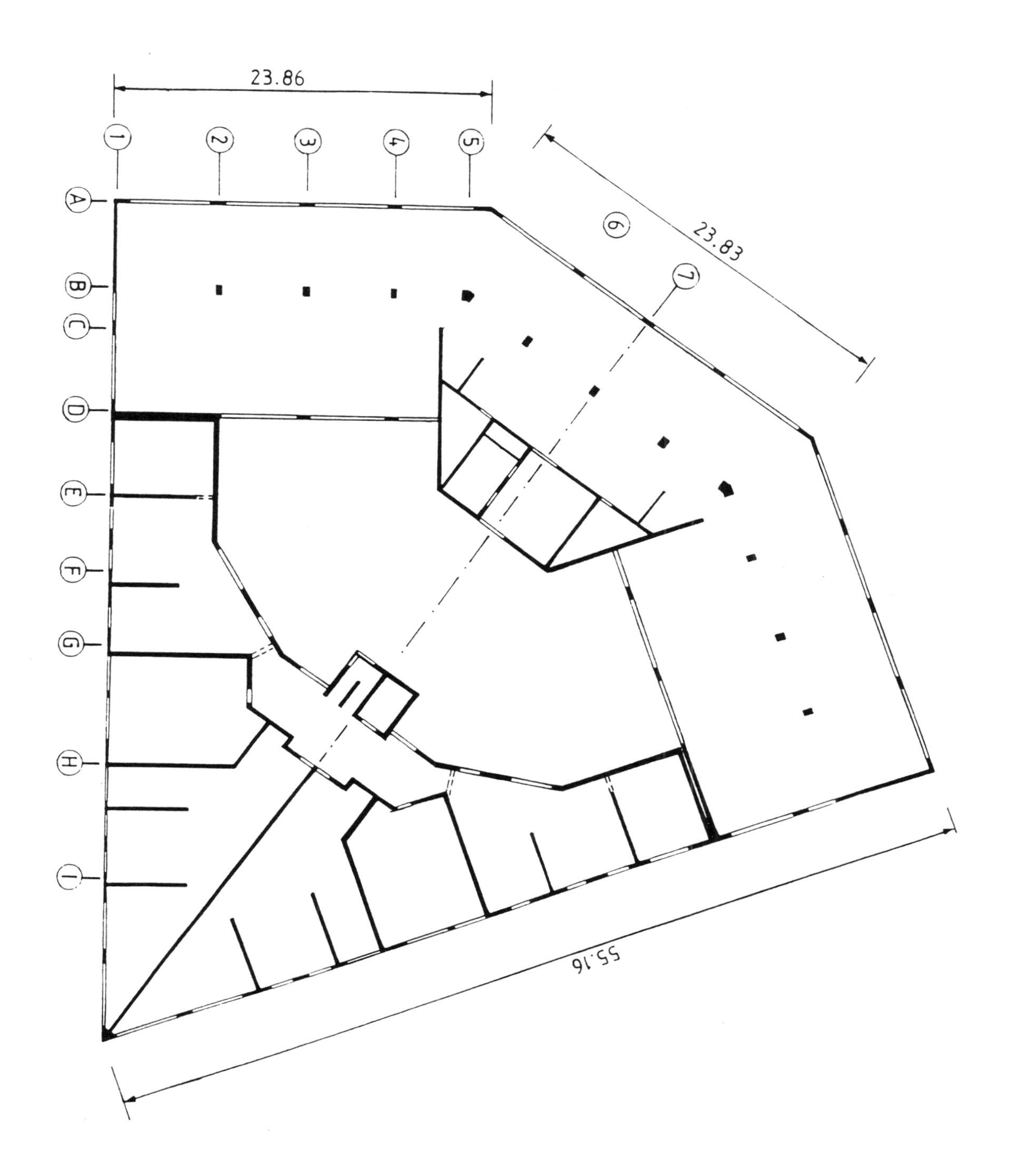

15.4.1.2 Basic data of structure, materials and loading		**Reference**
Intended use:	Residential building	see floor plan
Fire resistance:	1 hour for all elements	EC2-1.2, 1.3
Loading (excluding self-weight of structure):		
Continuous slab:	- imposed: 2.0 kN/m^2	EC1-2.1 for Category A
	- finishes: 1.5 kN/m^2	EC1-2.1
Combination factors:		
Serviceability limit states are not considered		
Exposure classes:		
Class 1 (indoor) for all members		EC2, Table 4.1
Materials:		
Concrete grade	C 30/37	EC2, Table 3.1
Steel grade	B 500	ENV 10 080
Self-weight of concrete	25 kN/m^3	EC1-2.1

15.4.2 Continuous slab (end span)

15.4.2.1 Floor span and idealization of the structure

Floor plan of the continuous slab

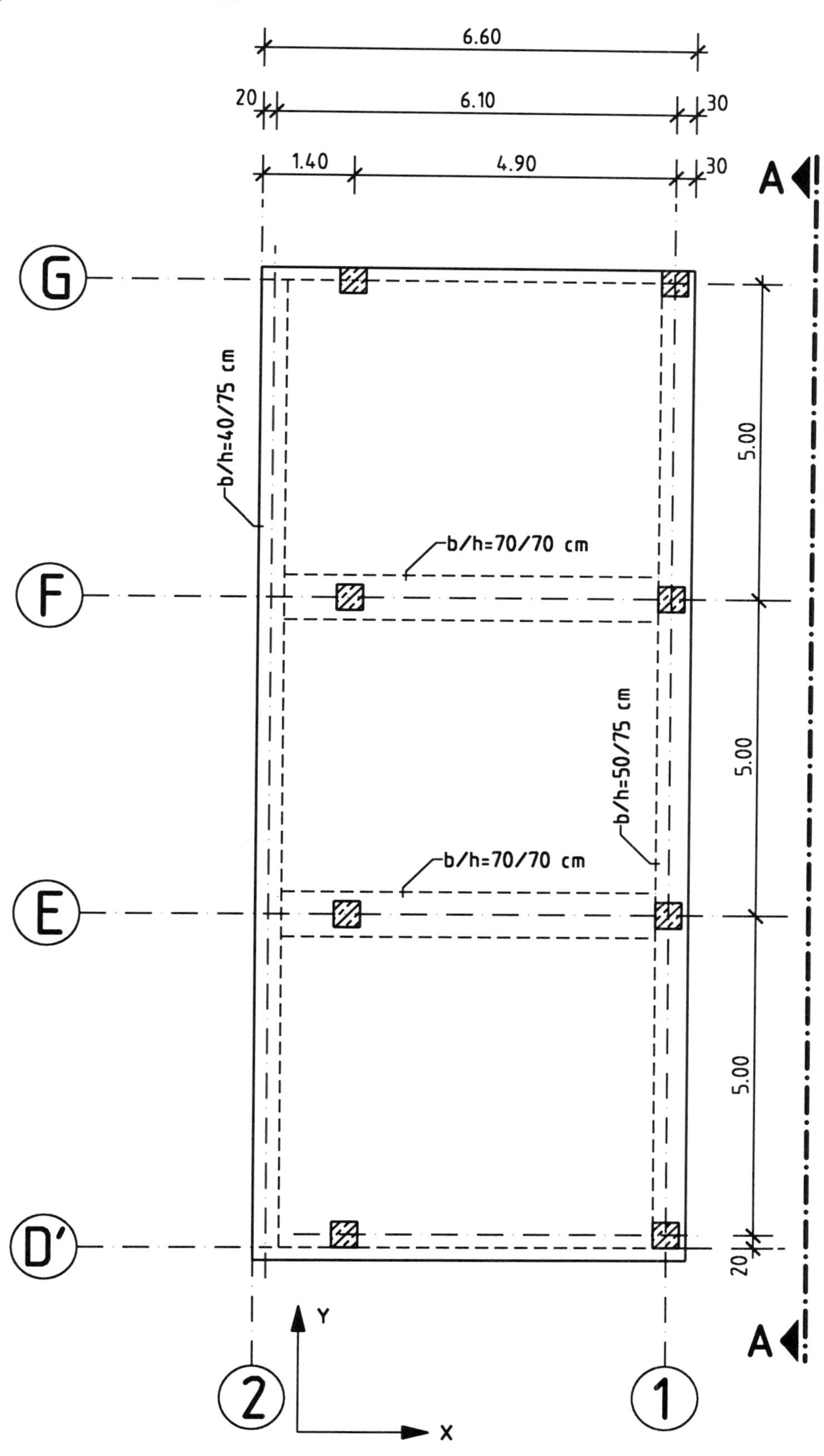

Reference

Idealization of structure:

EC2, 2.5.2.1

The end span of a two-way continuous slab is designed to Eurocode 2.
The effective spans are given as:

EC2, Eq.(2.15)

l_x = 6.10 + 0.2/3 + 0.3/3 = 6.27 m
l_y = = 5.00 m

a_i in axis 1 and 2

15.4.2.2 Limitation of deflections

EC2, 4.4.3

Assumptions:

- admissible deflection is given by $l_y/250$
- $\sigma_s \leq 250$ N/mm^2 in service conditions
- $\rho_l \leq 0.5\%$, i.e. concrete is considered as lightly stressed.

EC2, 4.4.3.1 and 4.4.3.2

From Table 4.14 in Eurocode 2 with ø assumed as 10 mm:

l_y/d = = 32
d_{req} = 5.00/32 = 0.156 m
h_{req} = d_{req} + nom c + ø/2 = 0.156 + 0.025 + 0.005 = 0.186 m

ø assumed as 10 mm

Selected: h = 0.25 m

15.4.2.3 Actions

Self-weight of slab: 0.25 * 25		=	6.25 kN/m^2
Finishes		=	1.50 kN/m^2
Permanent load	G_k	=	7.75 kN/m^2
Imposed load	Q_k	=	2.0 kN/m^2

15.4.2.4 Structural analysis

EC2, 2.5

In the present example, only the ultimate limit states are considered. The slab is analysed using the simplified yield-line method in [6], A3.2. This method is based on the following assumptions.

EC2, 2.5.3.2.2

- high ductility reinforcement is used

EC2, 3.2.4.2

- at the ultimate limit state for bending, the ratio $x/d \leq 0.25$

[6], A3.2(1)

- the spans in any one direction are approximately the same

l_x = 5.0 m for all span conditions

- the loadings on the adjacent panels are approximately the same.

Loading on the panels:

$\gamma_G\, G_k$ = 1.35 * 7.75 = 10.50 kN/m^2
$\gamma_Q\, Q_k$ = 1.5 * 2.00 = 3.00 kN/m^2

13.50 kN/m^2

					Reference
Design moment over the continuous edge (l_y/l_x = 6.27/5.0 = 1.25):					
m_0	=	13.50 * 5.0^2	=	337.5 kN	[6], A3.2(1), and Table A2
$m_{Sd,x}$	=	-337.5 * 0.0735	=	-24.81 kNm/m	three edges discontinuous, one edge continuous
Design span moment in x-direction:					
$m_{Sd,x}$	=	337.5 * 0.055	=	+18.56 kNm/m	[6], Eq.(A5)
Span moment in y-direction:					
$m_{Sd,y}$	=	337.5 * 0.044	=	14.85 kNm/m	three edges discontinuous, one edge continuous
Maximum shear force:					[6], A3.2.(2)
V_{sd}	=	0.52 *13.5 * 5.0	=	35.1 kN/m	

15.4.2.5 Design at ultimate limit states for bending and axial force

EC2, 4.3.1

f_{cd}	=	30/1.5	=	20 N/mm^2	
f_{yd}	=	500/1.15	=	435 N/mm^2	
d_x	=	0.25 - 0.03	=	0.22 m	see 15.4.2.2 above
Design of the continuous edge:					
μ_{Sds}	=	24.81 * 10^{-3} /(1.0 * 0.22^2 * 20)	=	0.026	
$A_{s,req}$	=	0.027 * 10^3 * 220 * 20/435	=	2.73 * 10^2 mm^2/m	
x/d	=	0.067	<	0.25	Table 7.1 (b) in [1]

Selected: Welded mesh with twin bars ø 6.0 mm — see 15.4.2.7 and 15.4.2.8.1 below

Steel B 500 B - R 377
2 * ø 6.0 - 150
$A_{s,prov}$ = 3.77 * 10^2 mm^2/m

Design for the span moments:
x-direction:

μ_{Sds}	=	18.56 * 10^{-3}/(1.0 * 0.22^2 * 20)	=	0.019	
$A_{s,req}$	=	0.020 * 10^3 * 220 * 20/435	=	2.02 * 10^2 mm^2/m	Table 7.1(b) in [1]

y-direction:

μ_{Sds}	=	14.85 * 10^{-3}/(1.0 * 0.21^2 * 20)	=	0.017
$A_{s,req}$	=	0.018 * 10^3 * 210 * 20/435	=	1.74 * 10^2 mm^2/m

Selected in x-direction:
Welded mesh with twin bars ø 5.5 mm — see 15.4.2.7 and 15.4.2.8.1 below

Steel B 500 B - R 317
2 * ø 5.5 - 150
$A_{s,prov}$ = 3.17 * 10^2 mm$^{2/}$m — in x-direction
$A_{s,prov}$ = 0.64 * 10^2 mm$^{2/}$m — in y-direction

					Reference
Additional span reinforcement in y-direction:					
Selected: Welded mesh fabric with bars ø 7.0					

Steel B 500 B - R 257
ø 7.0 - 150
$A_{s,prov}$ = 2.57 * 10² mm²/m — in x-direction
$A_{s,prov}$ = 0.64 * 10² mm²/m — in y-direction

Total reinforcement in y-direction:

$A_{s,prov}$	=	$0.64 * 10^2 + 2.57 * 10^2$	=	$3.21 * 10^2$ mm²/m	

15.4.2.6 Design for shear — EC2, 4.3.2

					Reference
τ_{Rd}	=		=	0.34 N/mm²	EC2, Table 4.8 for C30/37
ρ_l	=	3.77/(100 * 22)	=	0.17%	
k	=	1.6 - 0.22	=	1.38	
V_{Rd1}	=	0.34*1.38*(1.2+40*0.0017) * 0.22 * 10³			EC2, Eq.(4.18)
			=	130 kN/m	
			>	V_{sd}	

Design shear resistance of compression struts:

V_{Rd2}	=	0.5 * 0.575 * 20 * 0.9 * 0.22* 10³	=	1138 kN/m	EC2, Eq.(4.19)
			>	V_{sd}	

15.4.2.7 Minimum reinforcement for crack control — EC2, 4.4.2.2

A_s	=	$k_c\, k\, f_{ct,eff}\, A_{ct}/\sigma_s$			EC2, Eq.(4.78)
where					
k_c	=	0.4 for bending			
k	=	0.8 for h ≤ 300 mm			
$f_{ct,eff}$	=	3.0 N/mm²			
A_{ct}	=	0.25/2*1.0	=	0.125 m²	
σ_s	=	400 N/mm²			from EC2, Table 4.11, column 2, for ø ≤ 8 mm
A_s	=	0.4 * 0.8 * 3 * 0.125 * 10⁶/400	=	3.0 * 10² mm²/m	
$A_{s,prov}$	≥	3.21 * 10² mm²/m	>	3.0 * 10² mm²/m	

15.4.2.8 Detailing of reinforcement — EC2, 5

15.4.2.8.1 Minimum reinforcement areas for the avoidance of brittle failure — EC2, 5.4.2.1.1

$A_{s,min}$	=	0.0015 * 220 * 10³	=	3.3 * 10² mm²/m	EC2, Eq.(5.14)
$A_{s,prov}$	≥	3.17 * 10² + 0.64 *10²	=	3.81 * 10² mm²/m	in x-direction

15.4.2.8.2 Basic anchorage length — EC2, 5.2.2.2

l_b	=	$0.25\, ø\, f_{yd}/f_{bd}$ or			EC2, Eq.(5.3)
l_b	=	$0.25\, ø_n\, f_{yd}/f_{bd}$			for twin bars
f_{bd}	=		=	3.0 N/mm²	EC2, Table 5.3, for C30/37

Reference

Location		ø (mm)	$ø_n$ (mm)	l_b (mm)
Support		6.0	8.5	310
Span	x	5.5	7.8	290
	y	4.5	-	170

for mesh fabric R 317

15.4.2.8.3 Anchorage at the discontinuous edges

EC2, 5.4.3.2.1(5)

$F_s = V_{Sd}\, a_l/d + N_{sd}$ — EC2, Eq.(5.15)

$a_l = d$ — EC2, 5.4.3.2.1(1)

$F_s = 35.1 * 1.0 + 0 = 35.1$ kN/m

$A_{s,req} = 35.1 * 10^3/435 = 0.81 * 10^2$ mm²/m

Required anchorage length:

$l_{b,net} = 0.7 * 290 * 0.81/3.17 = 52$ mm — EC2, 5.2.3.4.1; EC2, Eq.(5.4), and 5.2.3.4.2(2)

Minimum values:

$l_{b,min} = 0.3 * 290 = 87$ mm

$= 10\,ø = 10 * 5.5 = 55$ mm

$= 100$ mm — The largest value of $l_{b,min}$ should be used

Anchorage length:

$l_{b,anch} = 2/3 * 100 = 70$ mm — EC2, 5.4.2.1.4(3)

15.4.2.8.4 Anchorage at the continuous edges

EC2, 5.4.3.2.1(5)

$l_{b,anch} = 10\,ø = 10 * 7.8 \approx 80$ mm

15.4.2.8.5 Lap lengths of mesh fabrics in y-direction

Lapping of mesh fabrics R 317 with bar diameter ø 4.5

$l_s = \alpha_2\, l_b\, A_{s,req}/A_{s,prov}$ — EC2, Eq.(5.9)

$\alpha_2 = 0.4 + 64/800 = 0.48 < 1.00$

$l_s = 1.0 * 170 * 1.0 = 170$ mm

$l_{s,min} = 0.3 * 1.0 * 170 = 51$ mm

$= s_t = 150$ mm

$= 200$ mm — The largest value of $l_{s,min}$ should be used

Detailing of reinforcement

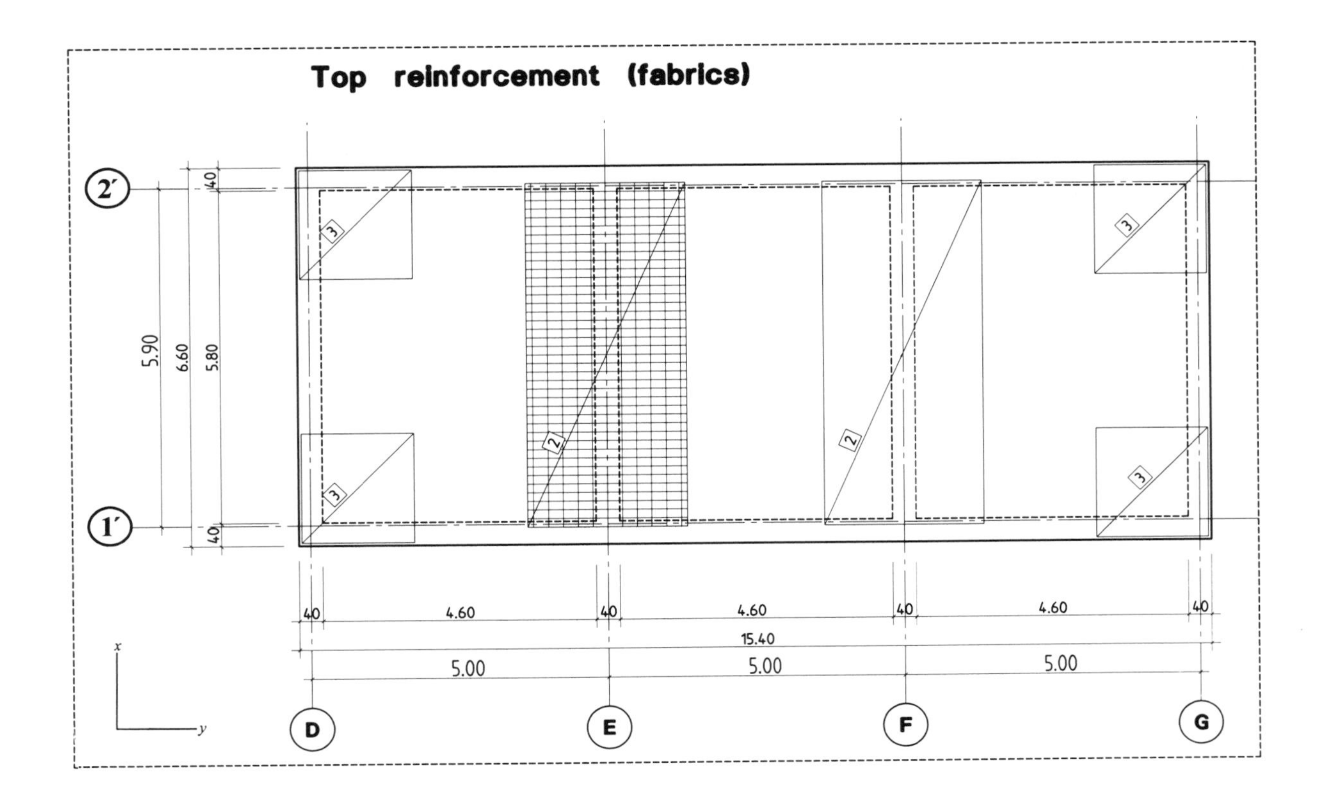

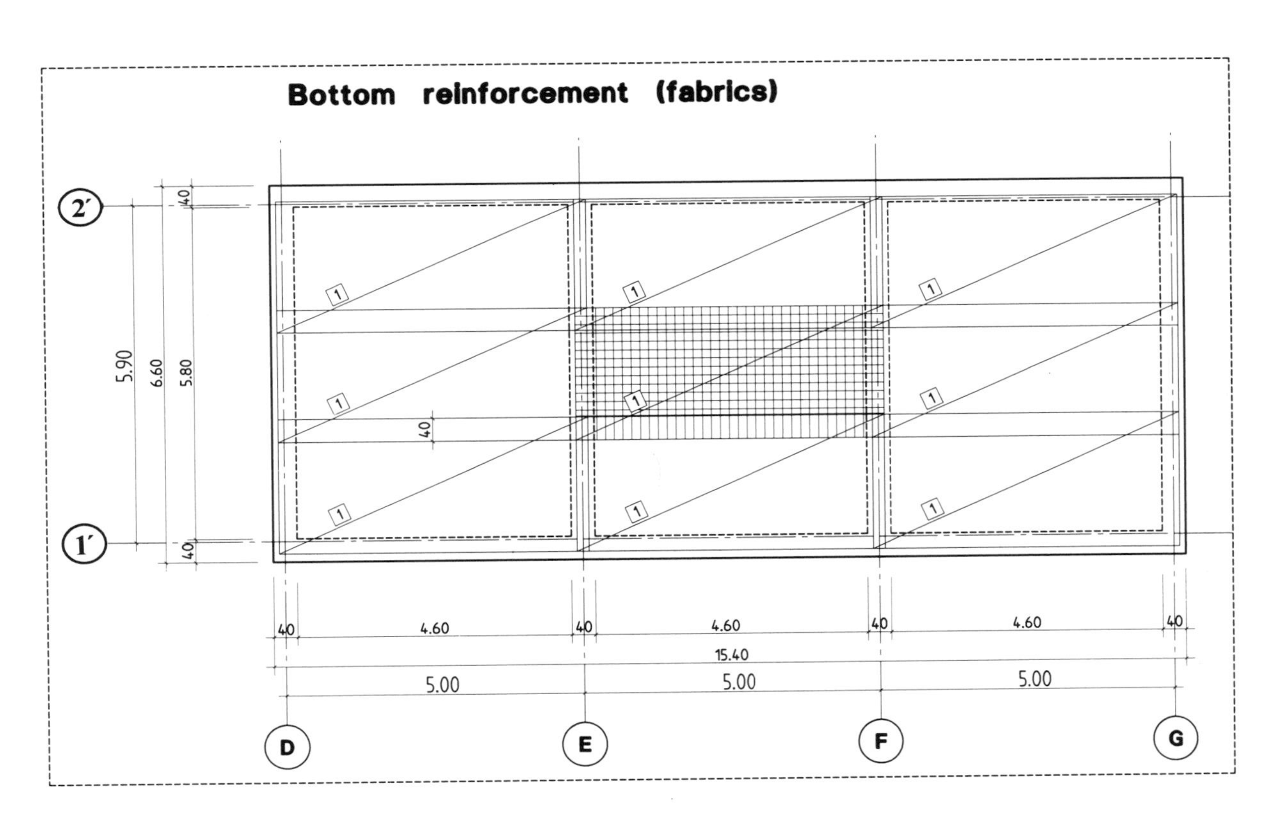

15.4.3 Continuous edge beam (end span)

Reference
see floor plan below

15.4.3.1 Structural system

Section A-A

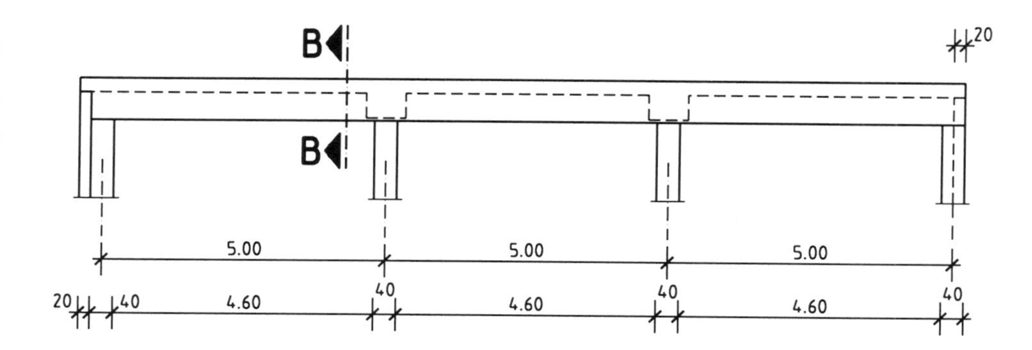

Section B-B

25
50
75
50

15.4.3.2 Actions

see 15.4.2 above

The beam is subjected to the following actions (see sketch below):

(a) permanent actions
- self-weight, $G_{k,1}$
- self-weight of parapets, $G_{k,2}$
- self-weight of supported slab, $G_{k,3}$
- self-weight of supported facade elements, $G_{k,4}$ to $G_{k,9}$
- concentrated forces due to permanent load, $G_{k,10}$ and $G_{k,11}$

(b) variable actions
- imposed load of the adjacent slab, $Q_{k,1}$
- variable actions transmitted by the facade elements, $Q_{k,2}$ to $Q_{k,7}$
- concentrated variable loads, $Q_{k,8}$ and $Q_{k,9}$.

In the following it is assumed that neither the permanent nor the variable actions are dependent upon each other.

Reference

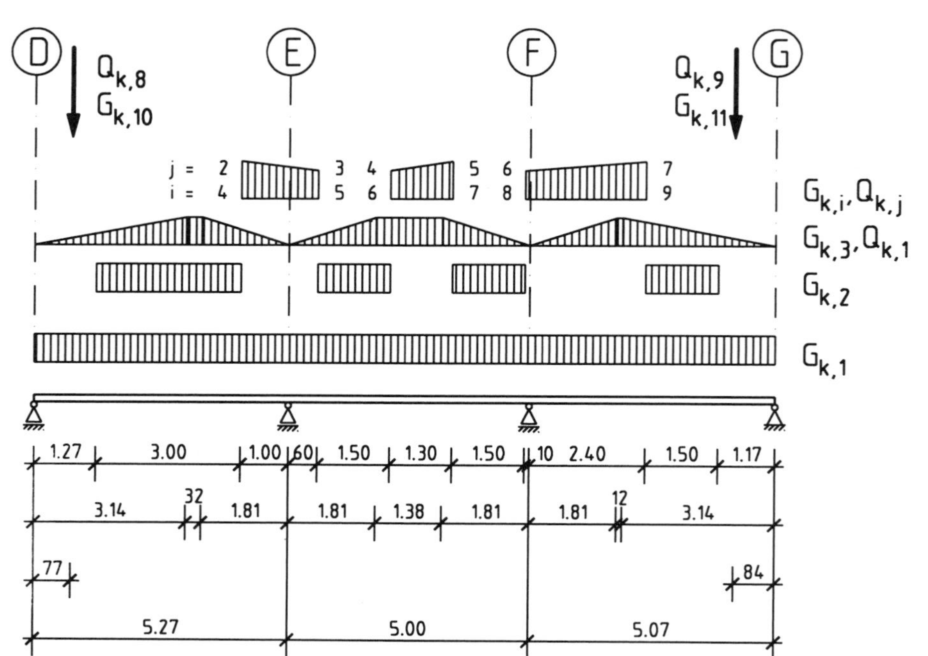

Table 15.3: Permanent, $G_{k,i}$, and variable actions, $Q_{k,j}$, acting on the beam

Action		Magnitude of the actions (kN/m; kN)										
subscript i =		1	2	3	4	5	6	7	8	9	10	11
subscript j =		-	-	1	2	3	4	5	6	7	8	9
$G_{k,i}$	γ_G=1.0	6.25	41.20	24.32	198.60	19.40	105.50	154.10	44.60	109.00	152.86	86.70
	γ_G=1.35	8.44	55.62	32.84	268.11	26.19	208.04	208.04	60.12	147.15	206.36	117.05
$Q_{k,j}$	γ_Q=1.0	-	-	6.28	20.60	2.00	16.50	24.00	3.40	8.40	18.36	-3.90
	γ_Q=1.5	-	-	9.42	30.90	3.00	24.75	36.00	5.10	12.60	27.54	-5.85

15.4.3.3 Structural analysis

(a) Linear analysis without redistribution

EC2, 2.5.3.4.2

The action effects resulting from a linear analysis without redistribution are summarized below.

Schematic shear and moment diagram

Reference

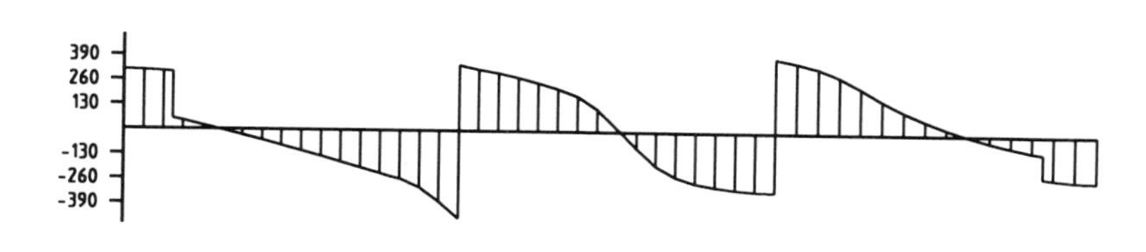

shear diagram

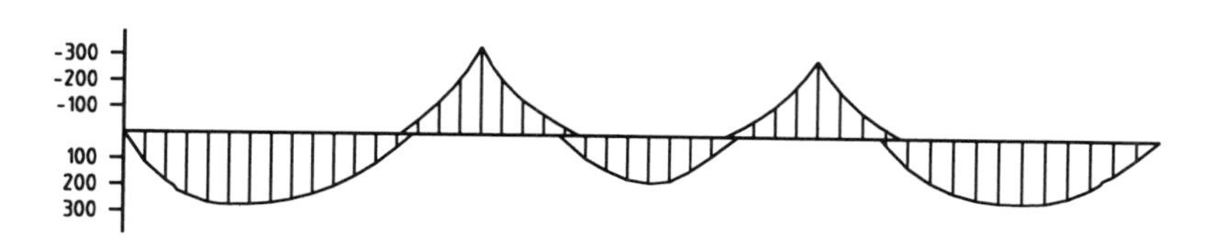

moment diagram

Support reactions

Support	Support reaction (kN) due to $G_{k,l}$	max $Q_{k,J}$	min $Q_{k,J}$	$G_{k,l} + Q_{k,J}$
D	276.05	310.43	271.71	306.10
E	717.47	807.24	711.78	801.55
F	668.51	749.30	660.05	740.84
G	212.23	226.48	202.75	217.00

Bending moments and shear forces in spans 1 to 3

Span	x (m)	max V_{Sd} (kN)	min V_{Sd} (kN)	max M_{Sd} (kNm)	min M_{Sd}
1	0.00	310.43	271.71	0.00	0.00
	0.77				205.91
	1.32	51.09	42.51	268.41	233.61
	1.91			283.85	
	2.64	-63.90	-73.58	259.60	218.08
	3.95	-188.03	-209.11	79.70	44.81
	5.27	-425.06	-472.84	-332.74	-383.43
2	0.00	334.40	286.72	-332.74	-383.43
	1.25	221.59	182.91	-23.95	-58.02
	2.50	42.99	25.87	157.08	97.69
	2.65			159.91	100.12
	3.75	-198.25	-242.77	8.30	-28.24
	5.00	-293.13	-342.25	-323.75	-374.64
3	0.00	407.05	366.92	-323.77	-374.66
	2.86			265.26	
	4.23				166.29
	5.07	-202.76	-226.49	-0.04	-

Calculation						Reference
(b) Linear analysis with redistribution						**Reference**
The cross-section over support E will be designed for the design bending moment						EC2, 2.5.3.4.2
M_{sd}	=			=	-333 kNm	see Table above
This corresponds to a distribution factor δ of						EC2, 2.5.3.4.2(3)
δ	=	332.74/383.43		=	0.867	

15.4.3.4 Design of span 1 for bending

Design data:						
C 30/37			f_{cd}	=	20 N/mm²	
B 500 B			f_{yd}	=	435 N/mm²	
effective depth			d	=	0.71 m	for bar diameter ø 25
Design of the cross-section of support E:						
b_w	=			=	0.50 m	
μ_{Sds}	=	$0.333/(0.50 * 0.71^2 * 20)$		=	0.066	
ω	=			=	0.070	[1], Table 7.1a
x/d	=			=	0.139	
$A_{s,req}$	=	0.070 * 500 * 710 * 20/435		=	$11.5 * 10^2$ mm²	
δ_{perm}	=	0.44 + 1.25 * 0.139		=	0.62	
				<	0.867	permissible coefficient δ

Selected 4 ø 20; $A_{s,prov}$ = 12.56 * 10² mm²

Design for maximum span moment:					
effective width					
b_{eff}	=	0.5 + 0.1 * 0.85 * 5.27	=	0.95 m	EC2, 2.5.2.2.1 for an L-beam
M_{Sd}	=	284 kNm			see Table above
μ_{SdS}	=	$0.284/(0.95 * 0.71^2 * 20)$	=	0.03	
ω	=		=	0.031	[1], Table 7.1a
$A_{s,req}$	=	0.031 * 950 * 710 * 20/435	=	$9.61 * 10^2$ mm²	

Selected 2 ø 25; $A_{s,prov}$ = 9.81 * 10² mm²

					Reference
15.4.3.5 Design for shear					EC2, 4.3.2
max V_{sd}	=		=	473 kN	see Table above
Design shear at the distance d from the face of the support:					
V_{Sd}	$\leq$	max V_{Sd} - d $G_{k,1}$			
	=	473 - 0.71 * 8.44	=	467 kN	see diagram of actions above; the opposite formula is a conservative assumption
The variable strut-inclination method is used; assumption:					EC2, 4.3.2.4.4
cot θ	=		=	1.25	$\theta \approx 40°$
ν	=	0.7 - 30/200	=	0.55	
V_{Rd2}	=	0.50 * 0.9 * 0.71 * 0.55 * 20/2.05 * 10^3	=	1714 kN	EC2, Eq.(4.26)
			>	V_{Sd}	
$(A_{sw}/s)_{req}$	=	467 * 10^3 /(0.9 * 0.71 * 435 * 1.25)	=	13.44*10^2 mm²/m	α = 90°

Selected stirrups ø 12 - spacing s = 150 mm

$(A_{sw}/s)_{prov}$	=		=	15.07*10^2 mm²/m	
maximum spacing:					EC2, 5.4.2.2(7)
V_{sd}/V_{Rd2}	=	467/1714	=	0.273	
s_{max}	=		=	300 mm	
			>	150 mm	
$(A_{sw}/s)_{min}$	=	0.0011 * 500 * 1 * 1000	=	5.5 * 10^2 mm²/m	EC2, Table 5.5
15.4.3.6 Control of cracking					EC2, 4.4.2
Cracking is controlled by limiting the bar diameter ø. The steel stress σ_s is estimated as					EC2, Table 4.11
σ_s	=	f_{yd} $A_{s,req}/A_{s,prov}$ $(1/\gamma_F)$			
	=	435 * 9.61/9.81 * (1/1.5)	≈	280 N/mm²	in span
From Table 4.11 in EC2 for reinforced concrete:					
$ø^*_s$	=		=	16 mm	
$ø_s$	=	16 * 71/(10*4)	=	28 mm	
			>	25 mm	
15.4.3.7 Detailing of reinforcement					EC2, 5
Basic anchorage length					
l_b	=	0.25 * √2 * 20 * 10^{-3} * 435/2.8	=	1.10 m	for ø = 20 mm
l_b	=	0.25 * 25 * 10^{-3} * 435/2.8	=	0.97 m	for ø = 25 mm

					Reference
Anchorage of bottom reinforcement					
- intermediate support					
$l_{b,net}$	=	$10 * 25 * 10^{-3}$	=	0.25 m	EC2, 5.4.2.1.5
- end support					
V_{Sd}	=		=	311 kN	see Table above
a_l	=	0.9 * 0.71 * 1.25/2	=	0.40 m	EC2, 5.4.2.1.3(1)
F_s	=	311 * 0.4/0.71	=	175 kN	EC2, 5.4.2.1.4(2)
$A_{s,req}$	=	$175 * 10^3/435$	=	$4.0 * 10^2$ mm²	
$l_{b,net}$	=	1.0 * 0.97 * 4.0/9.81	=	0.396 m	EC2, Eq.(5.4)
2/3 $l_{b,net}$	=	2/3 * 0.396	=	0.26 m	for straight bars
Anchorage of the top reinforcement					
$l_{b,net}$	=	0.3 * 1.10	=	0.33 m	EC2, Eq.(5.5)
or	=	d	=	0.71 m	EC2, 5.4.2.1.3(2)

Detailing of reinforcement

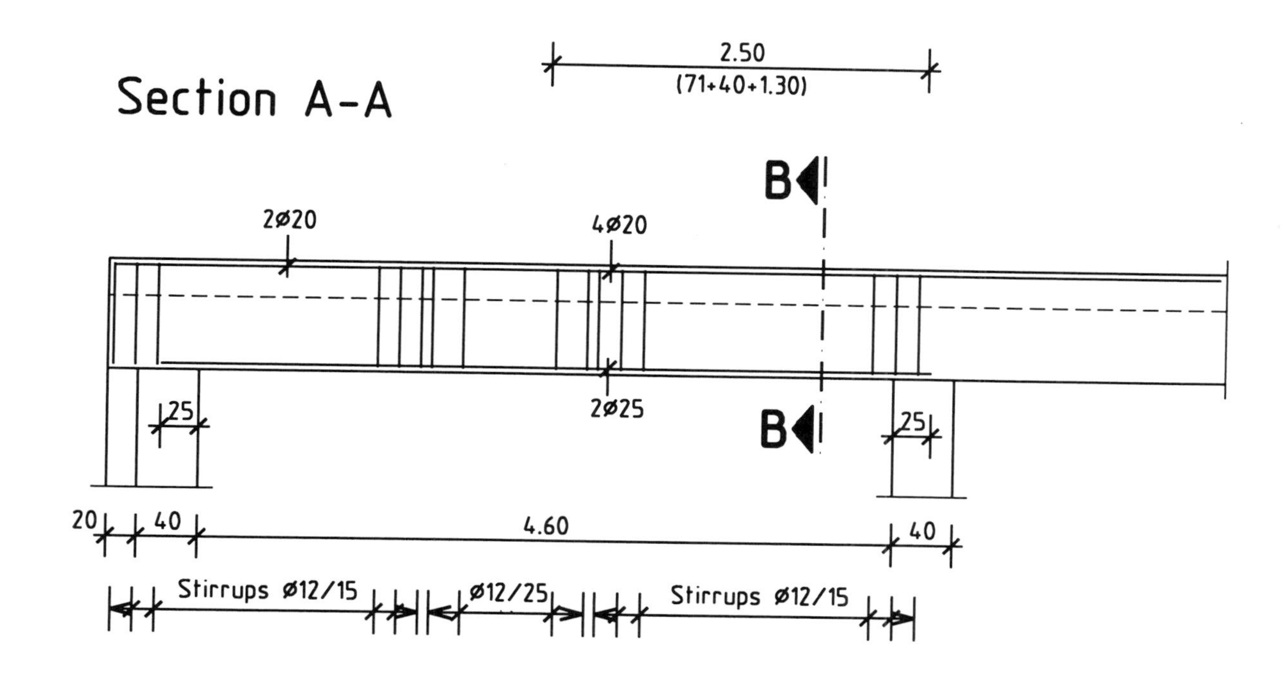

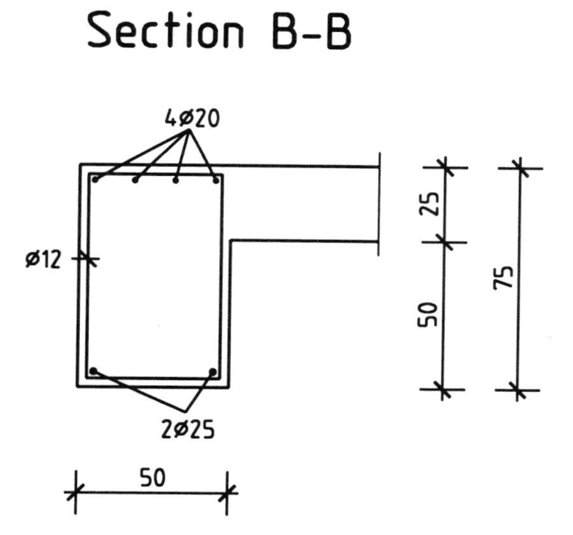

Reference

15.4.4 Braced transverse frame in axis E

15.4.4.1 Structural system; cross-sectional dimensions

see floor plan in 15.4.2.1

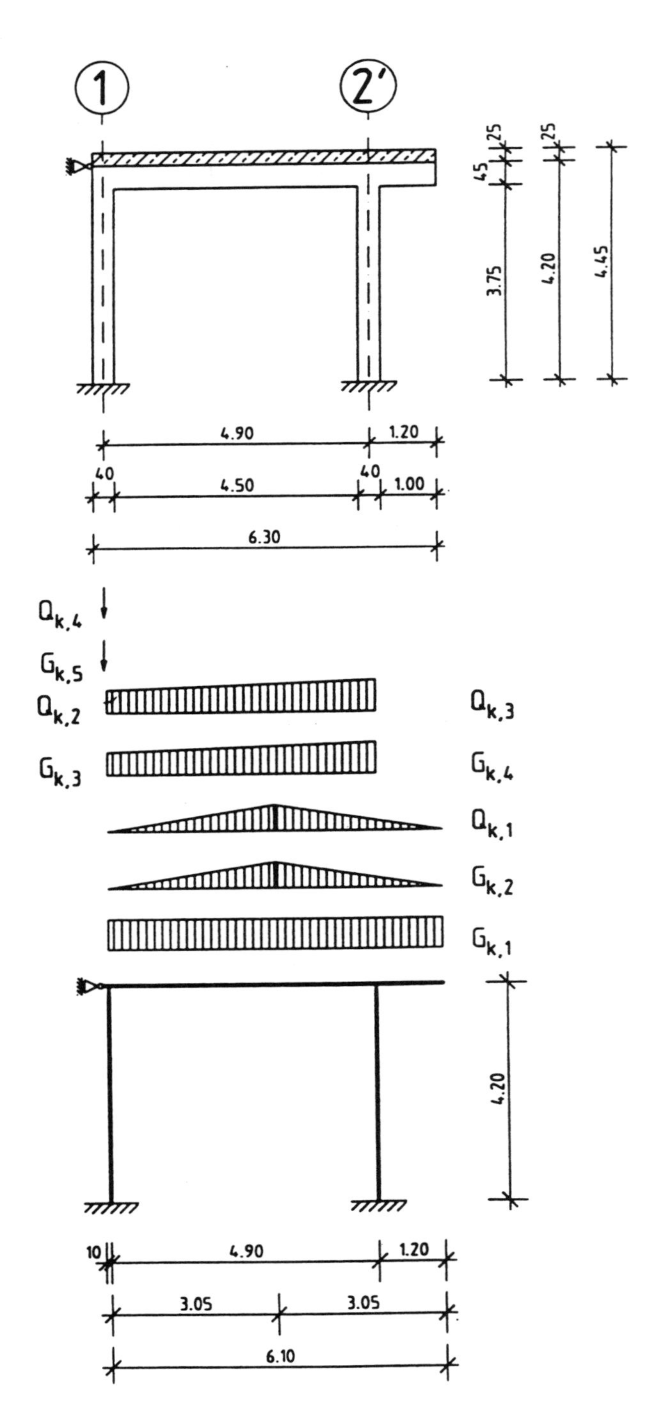

15.4.4.2 Actions

The frame is subjected to the following actions (see Figure above):

(a) permanent actions

- self-weight of beam, $G_{k,1}$
- self-weight of supported slab, $G_{k,2}$

Reference

- self-weight of supported slab, $G_{k,3}$; $G_{k,4}$
- support reaction of continuous beam, $G_{k,5}$.

see 15.4.3 above

(b) variable actions
- imposed load of supported slab, $Q_{k,1}$
- imposed load of supported slab, $Q_{k,2}$,; $Q_{k,3}$
- support reaction of continuous beam, $Q_{k,4}$.

Action		Magnitude of the actions (kN/m; kN)				
subscript i =		1	2	3	4	5
subscript j =		-	1	2	3	4
$G_{k,i}$	γ_G=1.0	7.85	28.1	30.0	227.0	540.4
	γ_G=1.35	10.60	37.9	40.5	306.5	729.5
$Q_{k,j}$	γ_Q=1.0	-	7.3	3.0	70.0	63.2
	γ_Q=1.5	-	10.9	4.5	105.0	94.8

It is assumed that all permanent actions and all imposed loads act simultaneously.

15.4.4.3 Structural analysis

EC2, 2.5

For the purposes of structural analysis, the frame is subdivided into elements and nodes as shown below.

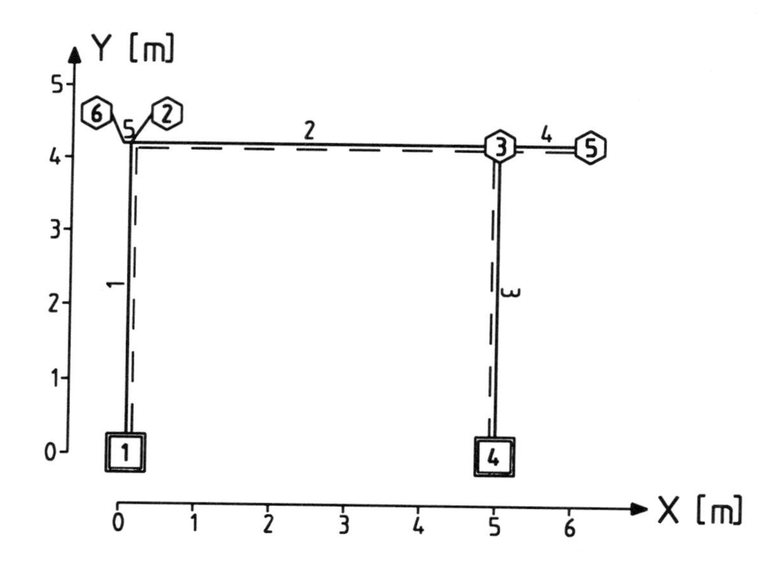

Reference

Node	Coordinates		Support conditions		
	x (m)	*y* (m)	*x*	*y*	*m*
1	0.10	0.00	1	1	1
2	0.10	4.20	0	0	0
3	5.00	4.20	0	0	0
4	5.00	0.00	1	1	1
5	6.20	4.20	0	0	0
6	0.00	4.20	0	0	0

x: free in *x*-direction
y: free in *y*-direction
m: free rotation
1: no freedom
0: freedom

Element	Defined by node		*A*
	left	right	(m^2)
1	1	2	0.160
2	2	3	0.677
3	3	4	0.160
4	3	5	0.677
5	2	6	0.677

Reference

Action effects due to permanent actions				
Element No.	x **(m)**	N_{Sd} **(kN)**	V_{Sd} **(kN)**	M_{Sd} **(kNm)**
1	0.00	-1146.09	-39.60	-15.58
	0.70	-1146.09	-39.60	-43.30
	1.40	-1146.09	-39.60	-71.02
	2.10	-1146.09	-39.60	-98.73
	2.80	-1146.09	-39.60	-126.45
	3.50	-1146.09	-39.60	-154.17
	4.20	-1146.09	-39.60	-181.88
2	0.00	-39.60	415.55	-254.88
	0.65	-39.60	368.58	
	0.82	-39.60	351.52	61.50
	1.63	-39.60	243.13	307.13
	2.45	-39.60	90.11	446.29
	2.85	-39.60		464.46
	3.27	-39.60	-106.85	442.29
	4.08	-39.60	-335.93	263.53
	4.70	-39.60	-526.38	
	4.90	-39.60	-593.22	-114.13
3	0.00	-614.92	39.60	-102.90
	0.70	-614.92	39.60	-75.18
	1.40	-614.92	39.60	-47.46
	2.10	-614.92	39.60	-19.75
	2.60	-614.92	39.60	
	2.80	-614.92	39.60	7.97
	3.50	-614.92	39.60	35.69
	4.20	-614.92	39.60	63.40
4	0.00	0.00	21.70	-11.23
	0.20	0.00	16.84	-7.38
	0.40	0.00	12.48	-4.46
	0.60	0.00	8.61	-2.36
	0.80	0.00	5.25	-0.98
	1.00	0.00	2.37	-0.23
	1.20	0.00		
5	0.00	0.00	-730.53	73.00
	0.02	0.00	-730.35	60.58
	0.03	0.00	-730.18	48.90
	0.05	0.00	-730.00	36.49
	0.07	0.00	-729.82	24.08
	0.08	0.00	-729.65	12.40
	0.10	0.00	-729.47	

Qualitative presentation of action effects due to permanent actions

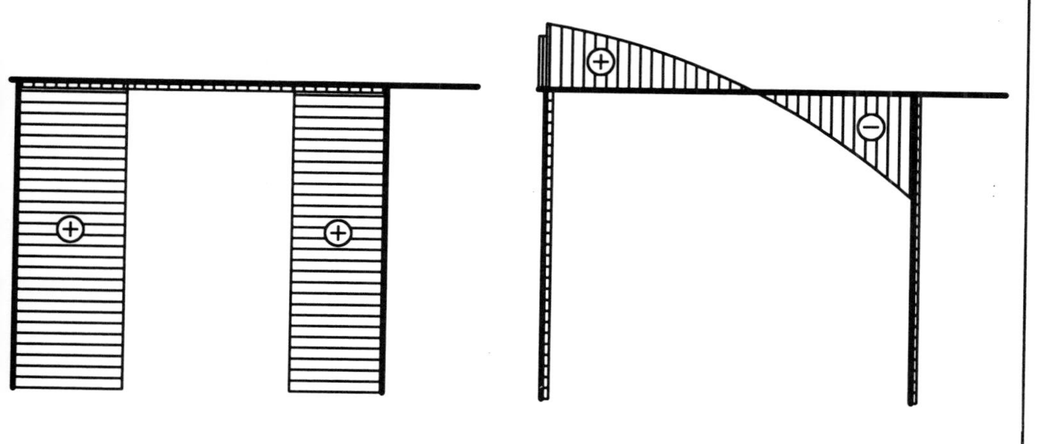

Reference

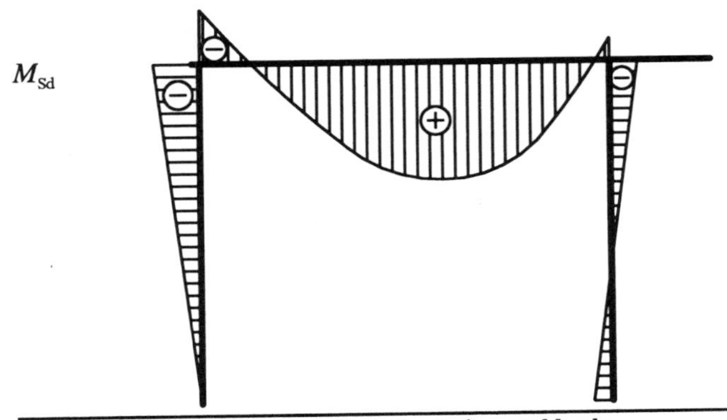

Action effects due to imposed loads				
Element	***x* (m)**	**N_{Sd} (kN)**	**V_{Sd} (kN)**	**M_{Sd} (kNm)**
	0.00	-208.12	-12.32	-8.34
	0.70	-208.12	-12.32	-16.97
1	1.40	-208.12	-12.32	-25.59
	2.10	-208.12	-12.32	-34.21
	2.80	-208.12	-12.32	-42.84
	3.50	-208.12	-12.32	-51.46
	4.20	-208.12	-12.32	-60.09
	0.00	-12.32	113.35	-69.56
	0.65	-12.32	105.73	
	0.82	-12.32	101.64	19.35
	1.63	-12.32	73.91	92.07
2	2.45	-12.32	30.08	135.64
	2.85	-12.32		142.39
	3.27	-12.32	-29.64	136.86
	4.08	-12.32	-101.87	83.97
	4.70	-12.32	-165.44	
	4.90	-12.32	-185.48	-32.64
	0.00	-188.04	12.32	-31.62
	0.70	-188.04	12.32	-23.00
	1.40	-188.04	12.32	-14.37
3	2.10	-188.04	12.32	-5.75
	2.60	-188.04	12.32	
	2.80	-188.04	12.32	2.88
	3.50	-188.04	12.32	11.50
	4.20	-188.04	12.32	20.12
	0.00	0.00	2.56	-1.02
	0.20	0.00	1.78	-0.59
	0.40	0.00	1.14	-0.30
4	0.60	0.00	0.64	-0.13
	0.80	0.00	0.28	-0.04
	1.00	0.00	0.07	0.00
	1.20	0.00		
	0.00	0.00	-94.77	9.48
	0.02	0.00	-94.77	7.87
	0.03	0.00	-94.77	6.35
5	0.05	0.00	-94.77	4.74
	0.07	0.00	-94.77	3.13
	0.08	0.00	-94.77	1.61
	0.10	0.00	-94.77	

	Reference
15.4.4.4 Design for the ultimate limit states	EC2, 4.3
15.4.4.4.1 Basic data	
Concrete C 30/37 f_{cd} = 20 N/mm^2 Steel B 500 f_{yd} = 435 N/mm^2	see 15.4.1 above
15.4.4.4.2 Design of the beam for the ultimate limit states of bending and longitudinal force	Element No. 2
(a) Design value of the acting bending moment M_{Sd} in node 2 M_{Sd} = -254.88 - 69.56 = -324.44 kNm	see Table in 15.4.4.3 above
Bending moment at the face of the support: M'_{Sd} = -324.44 + (415.55 + 113.35) * 0.2 = -219 kNm d = = 0.645 m	EC2, 2.5.3.3(5)
μ_{Sds} = 0.219/(0.7 * 0.645^2 * 20) = 0.038	
ω = = 0.04 $A_{s, req}$ = 0.04 * 700 * 645 * 20/435 = 8.30 * 10^2 mm^2	[1], Table 7.1a
Selected 5 ø 16; $A_{s, prov}$ = 10.1 * 10^2 mm^2	
(b) Design in mid-span M_{sd} = 464.46 + 142.39 = 607 kNm d = = 0.645 m	
Effective width for a T-beam:	EC2, 2.5.2.2.1
l_0 = 0.7 * 4.90 = 3.43 m b_{eff} = 0.7 + 0.2 * 3.43 = 1.39 m μ_{Sds} = 0.607/(1.39 * 0.645^2 * 20) = 0.053 ω = = 0.056 x/d = = 0.13 x = 0.13 * 0.645 = 0.084 m < 0.25 m $A_{s, req}$ = 0.056 * 1390 * 645 * 20/435 = 23.1 * 10^2 mm^2	[1], Table 7.1a
Selected 5 ø 20; $A_{s, prov}$ = 25.5 * 10^2 mm^2	
15.4.4.3 Design of the beam for shear	EC2, 4.3.2
max V_{Sd} = 593.22 + 185.48 = 778.7 kN	
For the design, the variable strut inclination method is used.	EC2, 4.3.2.4.4
Design shear force at the distance d from the face of the support:	
V'_{Sd} ≈ max V_{sd} - (0.2 + d) ($G_{k,1}$ + $G_{k,4}$ + $Q_{k,3}$) = 778.7 - 0.845(10.6 + 260 + 87) = 477 kN	$G_{k,4}$; $Q_{k,3}$ coordinates at x = 0.845 m
cot θ = 1.25; α = 90°	[1]

					Reference
A_{sw}/s	=	$477 * 10^3 /(0.9 * 0.645 * 435 * 1.25)$	=	$15.1 * 10^2$ mm²/m	EC2, Eq.(4.27)
V_{Rd2}	=	$0.7 * 0.9 * 0.645 * 0.55 * 20/2.05 * 10^3$	=	2180 kN	
			>	V'_{Sd}	
V'_{Sd}/V_{Rd2}	=	477/2180	=	0.22	

Maximum longitudinal spacing of stirrups: — EC2, 5.4.2.2(7)

max s_w	=	0.6 * 645	=	387 mm
			>	300 mm

Maximum transverse spacing of legs: — EC2, 5.4.2.2(9)

max $s_{w,t}$	=	=	300 mm

Selected shear links with four legs
ø 12 - 300 mm
$(A_{sw}/s)_{prov} = 15.08 * 10^2$ mm²/m

In mid-span:

Shear links with two legs
ø 12 - 300 mm
$(A_{sw}/s)_{prov} = 7.54 * 10^2$ mm²/m

ρ_w	=	$7.54 * 10^{-4} /(0.70 * 1.0 * 1.0)$	=	0.0011	
			=	min ρ_w	EC2, Table 5.5, for C 30/37 and B 500

15.4.4.4.4 Design for the ultimate limit states induced by structural deformations (buckling)

EC2, 4.3.5

In this example, only element No. 1 is designed to EC2.
Design action effects: — see 15.4.4.3 above

N_{sd}	=	-1146.09 - 208.12	=	-1354 kN	

Bending moment in node 1:

$M_{Sd,1}$	=	-15.58 - 8.34	=	-24 kNm	e_0 = 0.018 m

Bending moment in node 2:

$M_{Sd,2}$	=	-181.88 - 60.09	=	-242 kNm	e_0 = 0.179 m

Cross-sectional dimensions: b/h = 400/400 mm
Slenderness ratio in the plane of the frame: — EC2, 4.3.5.3.5; in the transverse direction, buckling is prevented by structural members

β	≈	0.7			
l_0	=	0.7 * 4.20	=	2.94 m	
λ	=	2.94/(0.289 * 0.40)	=	25.5	
λ_{lim}	=	$15/\sqrt{\nu_u}$			EC2, 4.3.5.3.5(2)
ν_u	=	$1.354/(0.4^2 * 20)$	=	0.423	
λ_{lim}	=	$15/\sqrt{(0.423)}$	=	23.0	not relevant here
λ_{crit}	=	25 * (2 - 0.018/0.179)	=	47.5	EC2, Eq.(4.62)

Check for second order effects is not necessary.

M_{Rd}	=	$N_{Sd} * h/20$ = 1354 * 0.4/20	=	27.1 kNm	EC2, Eq.(4.64)
			<	242 kNm	

						Reference
d	=			=	0.355 m	assumption
Design of the column in node 2 using the tables in [2]						
ν_{Sd}	=	-1.354/(0.16 * 20)		=	- 0.423	
μ_{Sd}	=	0.242/(0.16 * 0.4 * 20)		=	0.20	
ω	≈	0.28				[2], page 64, Table 6.4 b:
$A_{s,tot}$	=	$0.28 * 400^2 * 20/435$		=	$20.6 * 10^2$ mm²	

Selected 2 * 5 = 10 ø 16
$A_{s,prov} = 20.1 * 10^2$ mm²

In element 3, 4 ø 16 are provided on each side.

15.4.4.5 Detailing of reinforcement

15.4.4.5.1 Columns

						Reference
						elements 1 and 3
Bar diameters provided:			ø	=	16 mm	
				>	12 mm	EC2, 5.4.1.2.1
Minimum reinforcement areas:						
min A_s	=	$0.15 * 1354 * 10^3 /435$		=	$4.7 * 10^2$ mm²	EC2, 5.4.1.2.1(2)
or	=	$0.003 * 400^2$		=	$4.8 * 10^2$ mm²	
				<	$20.1 * 10^2$ mm²	
Transverse reinforcement (links)			$ø_w$	=	10 mm	EC2, 5.4.1.2.2(1)
				>	6 mm	
Spacing:						
$s_{w,max}$	=	12 * 16		=	192 mm	EC2,5.4.1.2.2(3), (4) relevant here
$0.6s_{w,max}$	=	0.6 * 192		=	115 mm	

15.4.4.5.2 Beam

						Reference
Minimum reinforcement area to avoid brittle failure:						
$A_{s,min}$	=	0.0015 * 700 * 645		=	$6.8 * 10^2$ mm²	EC2, Eq.(5.14)
				<	$A_{s,prov}$	
Anchorage of bottom reinforcement						
l_b	=	10 ø	$= 10 * 25 * 10^{-3}$	=	0.25 m	supports in nodes 2 and 3 are considered as restrained
Basic anchorage length of bars with ø = 16 mm						EC2, Eq.(5.3)
l_b	=	$0.25 * \sqrt{2} *16 * 10^{-3} * 435/3.0$		=	0.82 m	EC2, Table 5.3, for poor bond conditions
Lap length of the bars ø 16 in node 2:						
l_s	=	$\alpha_1\ l_{b,net}$				EC2, 5.2.4.1.3(1)
	=	2.0 * 1.0 * 0.82 * 8.30/10.1		=	1.36 m	EC2, Eq.(5.7)
$l_{s,min}$	=	0.3 * 1.0 * 2.0 * 0.82		=	0.50 m	EC2, 5.2.4.1.3(1) EC2, Eq.(5.8)

Calculation for a residential building

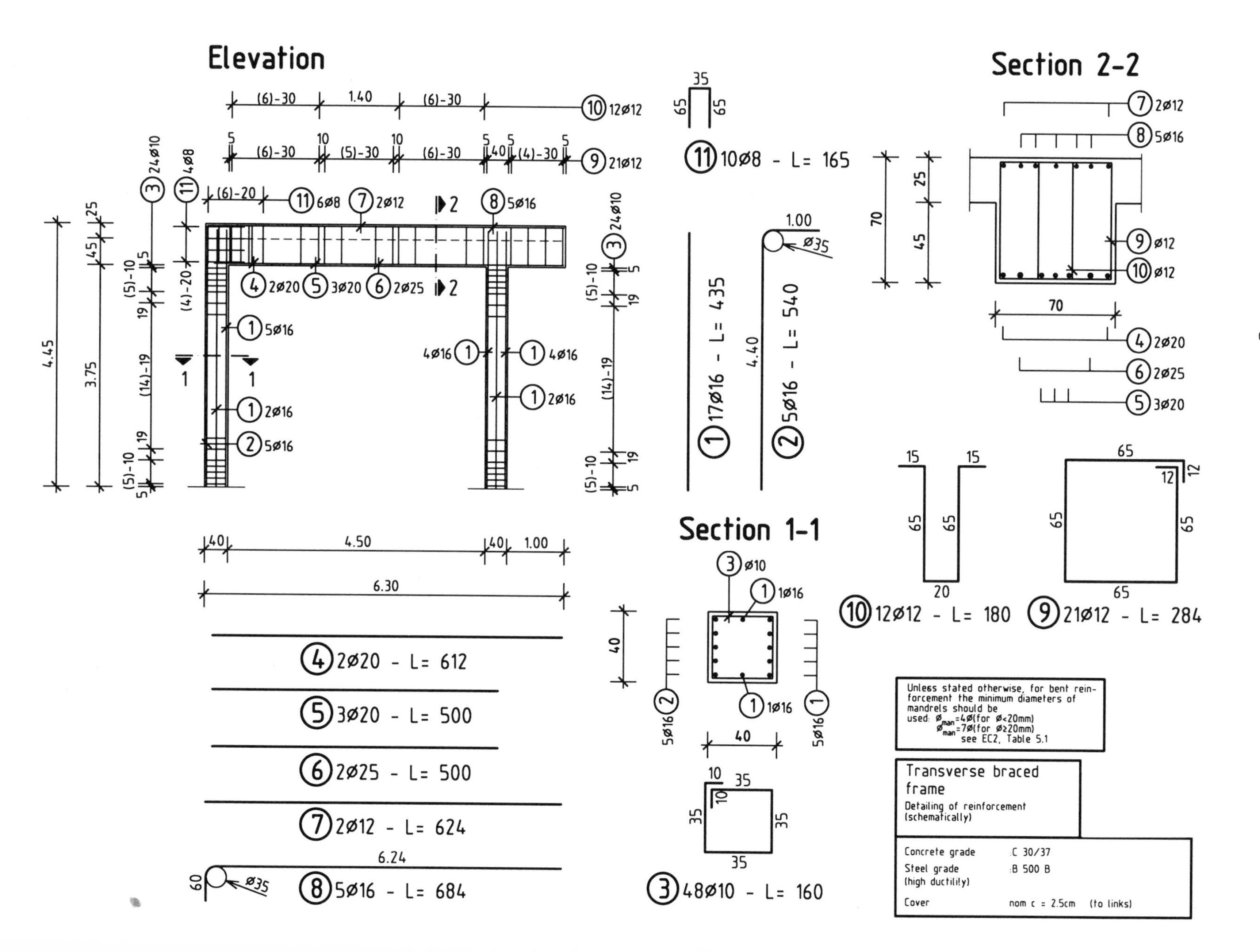

15.5.1 Floor plan; elevation

Reference

Floor plan

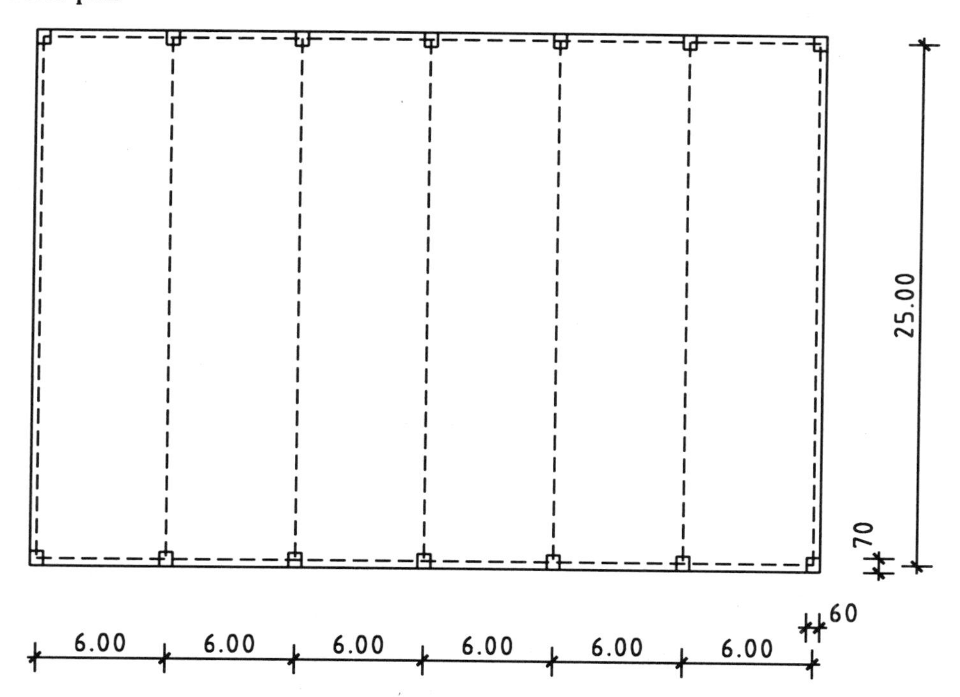

Elevation

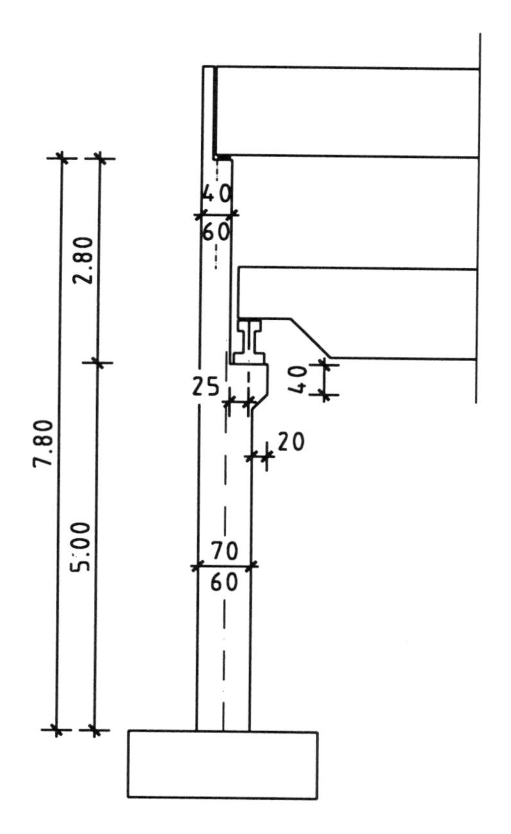

15.5.2 Calculation of prestressed concrete beam

Reference

15.5.2.1 Basic data

Structural system; cross-sectional dimensions

Elevation

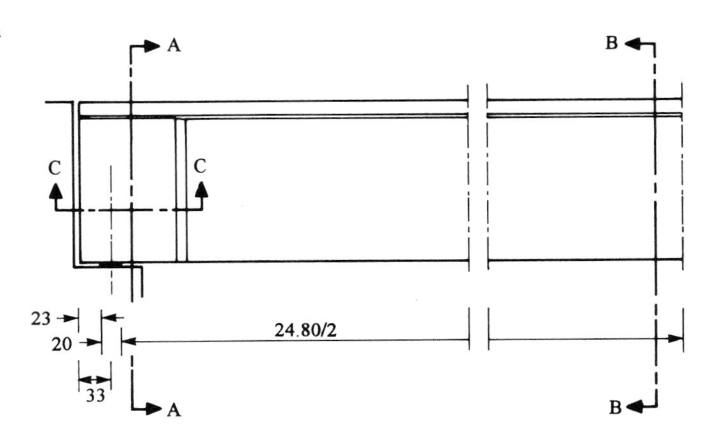

SECTION A-A SECTION B-B SECTION C-C

Exposure class:
Class 1 (indoor conditions) — EC2, Table 4.1

Materials:

		Reference
Concrete grade	C 35/45	EC2, Table 3.1
Steel grade	B 500	ENV 10 080

Tendons:

				Reference
7-wire strands	$f_{p0.1,k}/f_{pk}$	=	1500/1770 N/mm^2	
Modulus of elasticity	E_s	=	200 000 N/mm^2	
Relaxation class 2				EC2, 4.2.3.4.1(2)
Diameter of sheathing	$\varnothing_{duct}$	=	60 mm	
Cross-sectional area:	A_p	=	7.0 * 10^2 mm^2	

Reference

Coefficient of friction:	μ	=	0.22
Anchorage slip	Δl_{sl}	=	3.0 mm
Unintentional displacement	k	=	0.005

Cover to reinforcement:

- links:	nom c_w	=	25 mm	EC2, 4.1.3.3
- tendons:	nom c_p	=	65 mm	

Geometric data of the beam in mid-span section:

$\varnothing_{duct}$	=		=	60 mm	
A_{p1}	=	A_{p2}	=	$7.0 * 10^2$ mm^2	
α_e	=	200 000/33 500	=	5.97	modular ratio

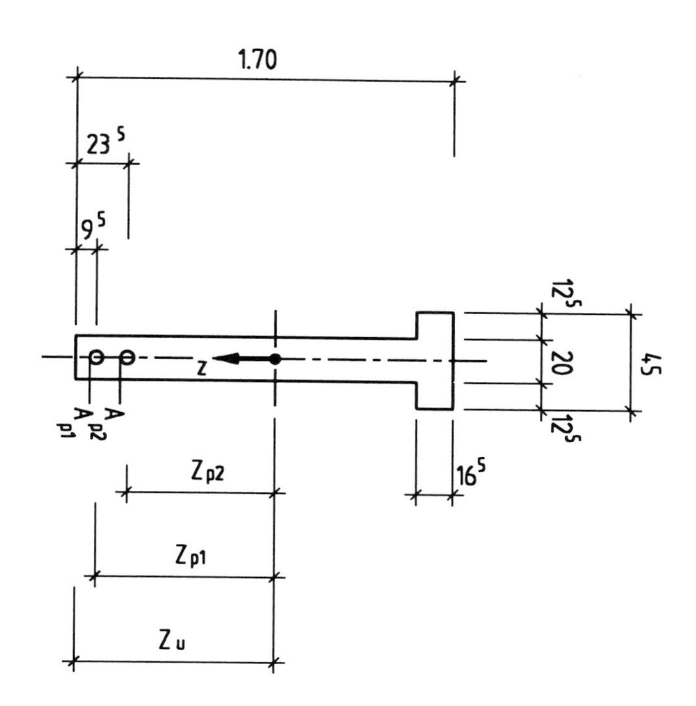

Cross-section	A_c; A_{ci} (m²)	I_c; I_{ci} (m⁴)	Z_u (m)	Z_{p1} (m)	Z_{p2} (m)
A_c	0.381	0.104	0.933	0.838	0.698
$A_{c,net}$	0.376	0.100	0.945	0.850	0.710
A_{ci}	0.406	0.122	0.927	0.832	0.692

Tendon profile

Reference

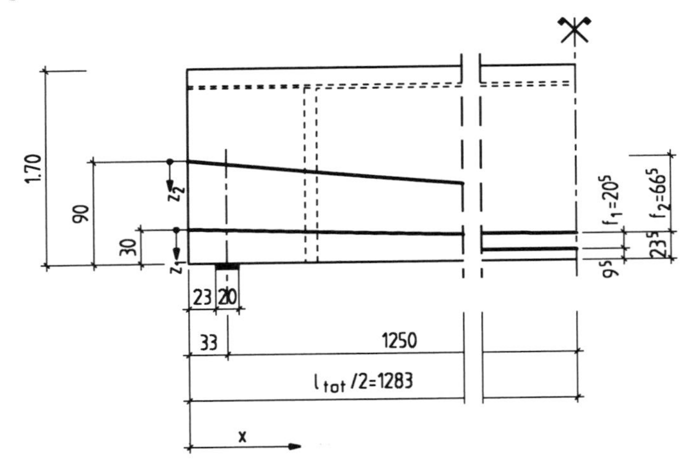

Description of the tendon profile:

Tendon 1:

$Z_1(x) \quad = \quad 4 * 0.205 * [x/l_{tot} - (x/l_{tot})^2]$

l_{tot} = 25.66 m

Tendon 2:

$Z_2(x) \quad = \quad 4 * 0.665 * [x/l_{tot} - (x/l_{tot})^2]$

15.5.2.2 Actions

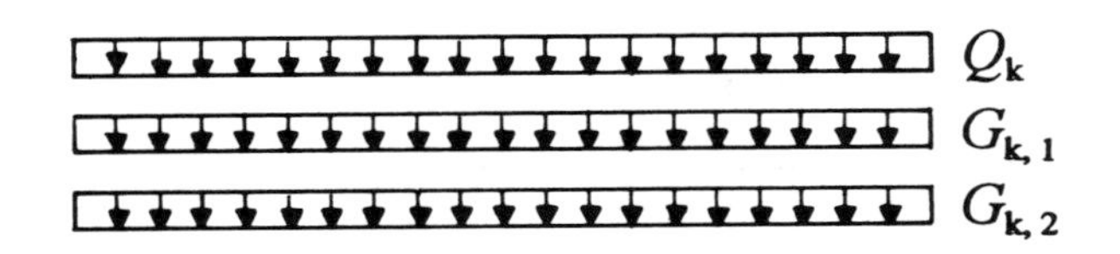

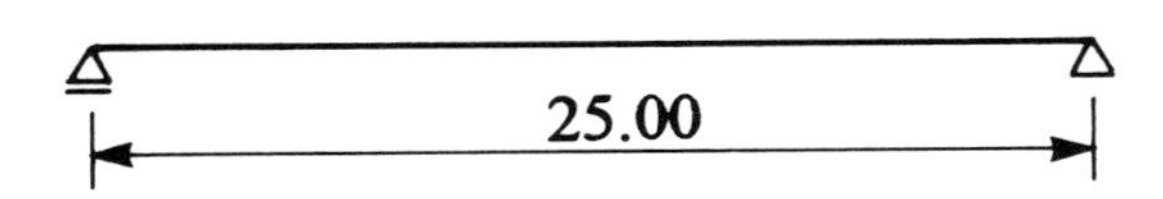

$G_{k,1}$	=	=	9.5 kN/m	self-weight of beam
$G_{k,2}$	=	=	10.0 kN/m	roofing
Qk	=	=	4.8 kN/m	snow

15.5.2.3 Action effects due to $G_{k,1}$, $G_{k,2}$ and Q_k

max M_{sd} = $[1.35 * 19.5 + 1.5 * 4.8] * 25.0^2/8$ = 2620 kNm

max V_{sd} = $[1.35 * 19.5 + 1.5 * 4.8] * 25.0/2$ = 420 kN

15.5.2.4 Action effects due to prestress

Reference

Stresses σ_{pm0} in the tendons at $t = 0$ allowing for friction, anchorage slip and unintentional angular displacement

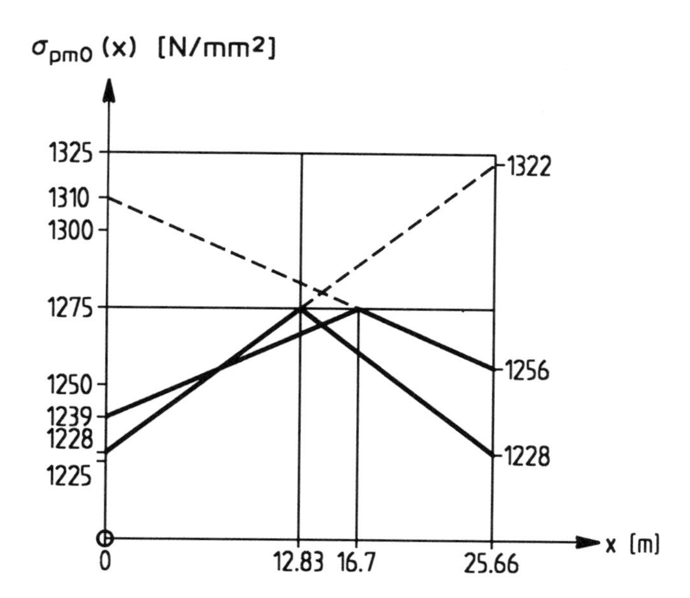

Action effects N_p, M_p and V_p due to prestressing at the serviceability limit states

Location	Action effects at					
	$t = 0$			$t = \infty$		
	N_p (kN)	M_p (kNm)	V_p (kN)	N_p (kN)	M_p (kNm)	V_p (kN)
Left support	-1727.2	-483.6	-117.3	-1452.8	-406.8	-98.6
Mid-span	-1779.4	-1387.9	0	-1505.0	-1146.8	0
Right support	-1738.8	-486.9	-118.1	-1464.4	-410.1	-99.4

15.5.2.5 Design for the ultimate limit states for bending and longitudinal force

Reference
EC2, 4.3.1

(a) Material data; design values of material strength

Concrete C 35/45 f_{ck} = 35 N/mm²

f_{cd} = f_{ck}/γ_c = 35/1.5 = 23.33 N/mm²

Reinforcing steel B 500 f_{yk} = 500 N/mm²

f_{yd} = f_{yk}/γ_s = 500/1.15 = 435 N/mm²

Prestressing steel 1500/1770 f_{pk} = 1770 N/mm²

f_{pd} = $0.9 f_{pk}/\gamma_s$ = 0.9 * 1770/1.15 = 1385 N/mm²

(b) Design at mid-span

max M_{Sd} = = 2620 kNm (see 15.5.2.3 above)

Effective depth at mid-span:

d_m = 1.70 - [(4 * 2.0/26.0) * 4.1 + (2 * 2.0/26.0) * 7.7 + (14.0/26.0) *16.5] * 10^{-2} = 1.58 m

Related bending moment:

μ_{Sds} = max $M_{Sd}/(b_f d^2_m f_{cd})$ = 2.620/(0.45 * 1.58² * 35/1.5) = 0.10

with

h_f/d = 0.165/1.58 ≈ 0.1 ([2], p. 59)

b_f/b_w = 45/20 = 2.25

The mechanical reinforcement ratio ω is given as:

ω = 108/1000 = 0.108 ([2], p. 59, Table 6.3a)

$A_{s, req}$ = $(1/f_{yd})(\omega\, b_f\, d_m f_{cd} - A_p\, \sigma_{pd})$

where

σ_{pd} = $(\gamma_p\, \epsilon_{pm} + \Delta\epsilon_p) E_p \le f_{pd}$ = 1385 N/mm²

A trial calculation has shown that

$(\gamma_p * \epsilon_{pm} + \Delta\epsilon_p) > \epsilon_{p0.1k}$,

so that

σ_{pd} = f_{pd} = 1385 N/mm²

$A_{s, req}$ = (1/435)(0.108 * 0.45 * 1.58 * 23.33 - 14.0 * 10^{-4} * 1385)10^4 < 0

i.e., for the resistance of max M_{Sd}, no reinforcement is necessary.

Reference

Minimum reinforcement area required to avoid brittle failure: — EC2, 5.4.2.1.1(1)

$A_{s,\,min} = 0.0015 * 200 * 1580 = 4.74 * 10^2\ mm^2$

Selected reinforcing steel B 500
6 bars ø 16; $A_{s,\,prov} = 12 * 10^2\ mm^2$

(c) Check of the pre-compressed tensile zone

It needs to be checked that the resistance of the pre-compressed tensile zone subjected to the combination of the permanent load $G_{k,1}$ and prestress is not exceeded.

Design value of bending moment due to $G_{k,1}$:

$M_{Sd,G} = \gamma_G\ G_{k,1}\ l^2_{eff} / 8 = 1.0 * 9.5 * 25.0^2 / 8 = 742\ kNm$

Characteristic value of prestress: — see 15.5.2.4 above

$P_k = \gamma_p\ P_{m0} = 1.0 * 1780 = 1780\ kN$

Bending moment due to prestress:

$M_k = \gamma_p\ M_p = -1.0 * 1388 = -1388\ kNm$

The cross-section in mid-span needs to be designed for the combination of

$N_{Sd} = -\gamma_p\ P_k = -1780\ kN$

$M_{Sd} = M_{Sd,G} + \gamma_p\ M_k = 742 - 1388 = -646\ kNm$

Distance z_s of the reinforcement in the flange from the centre of gravity:

$z_s = h - z_u - h_f /2 = 1.70 - 0.945 - 0.165/2 = 0.67\ m$

$d = h - h_f /2 = 1.70 - 0.165/2 = 1.60\ m$

$\mu_{Sds} = (0.646 - 1780 * 0.67) / (0.2 * 1.60^2 * 23.33) = 0.15 < 0.40$ — no compression reinforcement necessary

$\omega = 0.167;\quad \sigma_{sd} = f_{yd} = 435\ N/mm^2$ — [1], Table 7.1

$A_{s,\,req} = (1/435)\ (0.167 * 0.2 * 1.6 * 23.33 - 1.78)10^4 < 0$

No reinforcement in the flange is necessary.

15.5.2.6 Design for shear

Design value of the applied shear force: — EC2, 4.3.2.4.6

$V_{Sd} = V_{od} - V_{pd}$ — EC2, Eq.(4.32)

Design value V'_{od} at a distance d from the face of the support: — EC2, 4.3.2.2(10)

$V'_{od} = (\gamma_G\ G_k + \gamma_Q\ Q_k)\ (l_{eff}/2 - a_L/2 - d) =$

$= (1.35 * 19.5 + 1.5 * 4.8)\ (12.5 - 0.1 - 1.65) = 361\ kN$

Force component V_{pd} due to the inclined tendons:

$V_{pd} = \gamma_p\ \sigma_{pm,t}\ A_p\ \tan \alpha_i$

Reference

The stress $\sigma_{pm,\infty}$ for $t = \infty$ was calculated as:

$\sigma_{pm,\infty}$ = = 1040 N/mm^2

see 15.5.2.1 above

tan α_i is given by:

- for tendons 1:

$\tan \alpha_1$ = 4 * 0.205[1/25.66 - 2 (1.65 + 0.43)/25.66^2] = 0.0265

- for tendons 2:

$\tan \alpha_2$ = 4 * 0.665[1/25.66 - 2(2.08)/25.66^2] = 0.0868

V_{pd} = 0.9 * 1040 * 7 * 10^{-1} (0.0268 + 0.0868) = 74 kN

V_{Sd} = 361 - 74 = 287 kN

Selected stirrups ø 8 - 200
A_{sw}/s_w = 5.0 * 10^2 mm^2/m

Design shear resistance V_{Rd3} using the variable strut inclination method and assuming α = 90° and cot θ = 1.25: — EC2, 4.3.2.4.4

V_{Rd3} = 5.0 * 10^2 * 0.9 * 1.65 * 435 * 1.25 * 10^{-3} = 403.7 kN

Design shear resistance of the compression struts

$b_{w,net}$ = b_w - ø$_{duct}$/2 = 0.20 - 0.006/2 = 0.17 m

ν = 0.7 - 35/200 = 0.525

V_{Rd2} = 0.17 * 0.9 * 1.65 * 0.525 * 23.33/2.05 * 10^3 = 1508 kN — EC2, Eq.(4.26)

Minimum shear reinforcement: — EC2, 5.4.2.2(5)

$(A_{sw}/s_w)_{min}$ = 0.0011 * 200 * 1000 = 2.2 * 10^2 mm^2/m — EC2, Eq.(5.16)

maximum longitudinal spacing $s_{w,max}$: — EC2, 5.4.2.2(7)

V_{sd}/V_{Rd2} = 287/1508 = 0.19 < 0.20

$s_{w,max}$ = = 300 mm — EC2, Eq.(5.17)

Detailing of reinforcement

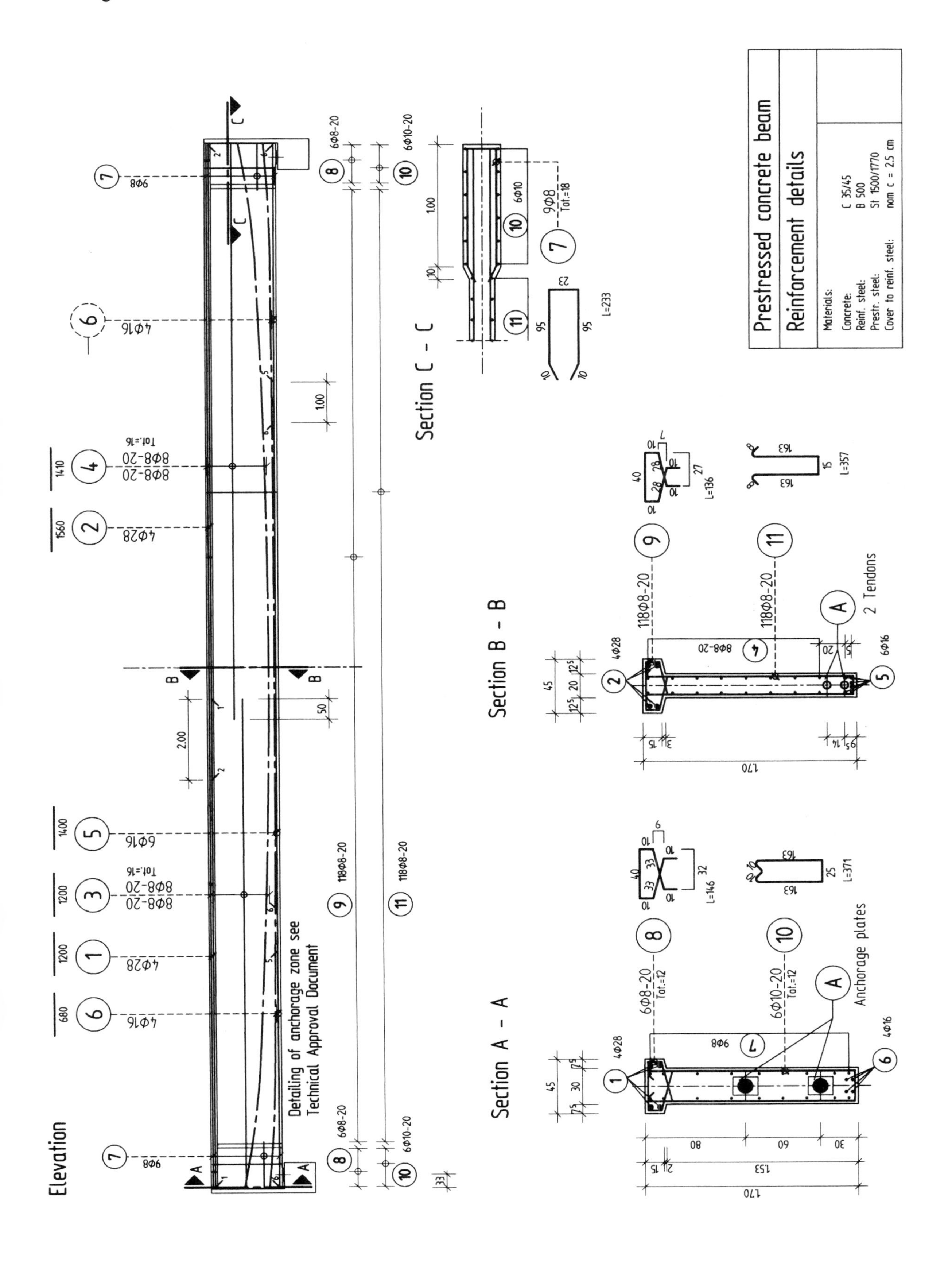

15.5.3 Calculation of edge column subjected to crane-induced actions

15.5.3.1 Basic data and design value of actions

Structural system; cross-sectional dimensions

Elevation

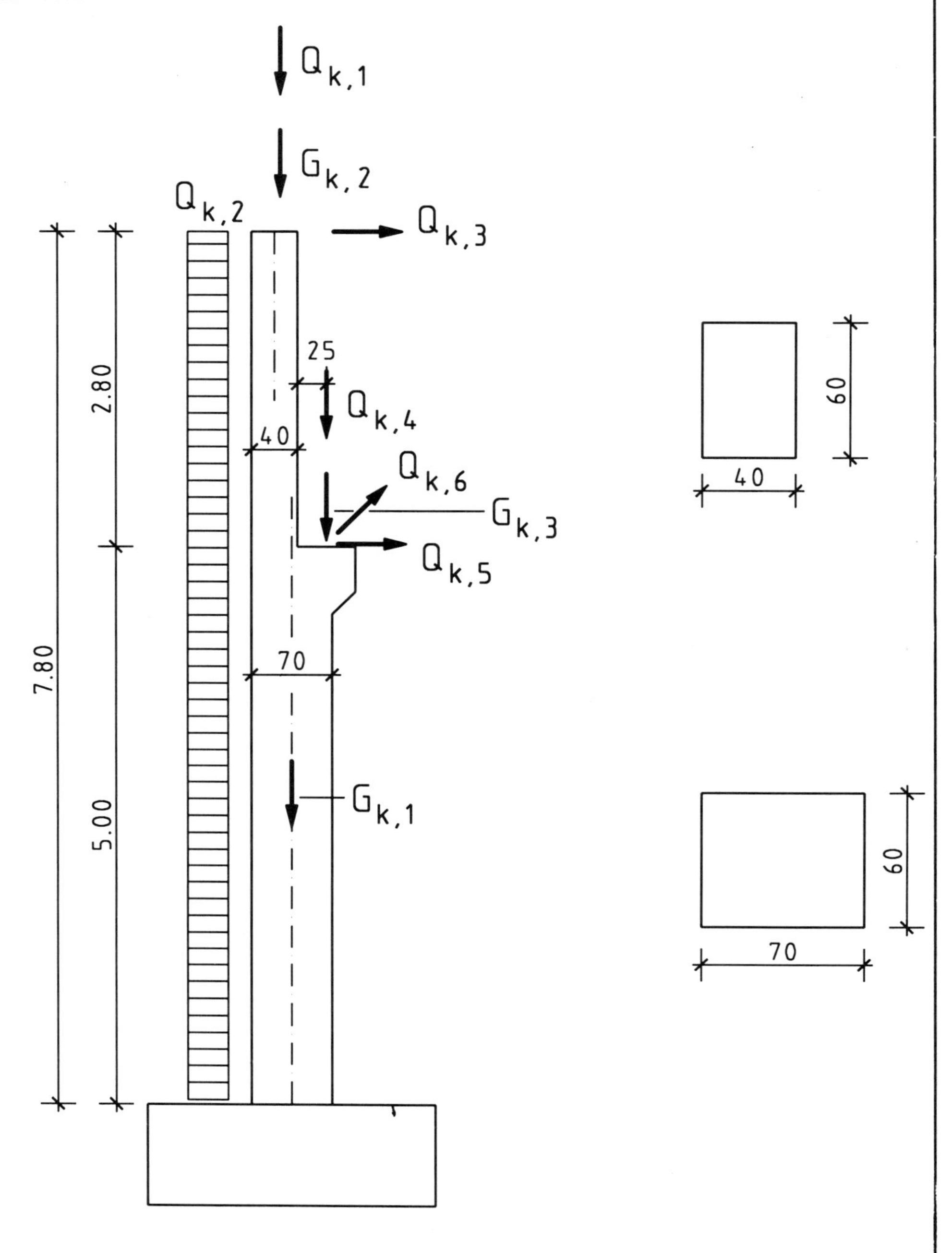

Reference
i.e. fatigue verification to EC2-2 is performed

Exposure class:
Class 2a (humid environment without frost)

Reference: EC2, Table 4.1

Materials:
Concrete grade C 45/55

EC2, Table 3.1

For the second order analysis of the column (see 15.5.3.2 below), the general stress-strain diagram acc. to Figure 4.1 in EC2 will be used. For the design of the cross-section, the parabolic-rectangular diagram will be applied.

EC2, 4.2.1.3.3(a), (5)
EC2, Eq. 4.2

Steel grade B 500
For structural analysis and the design of cross-sections, the bi-linear diagram with a horizontal top branch will be used.

ENV 10 080
EC2, 4.2.2.3.2
EC2, Fig. 4.5

Cover to reinforcement (stirrups)

min c_w = = 20 mm
nom c_w = = 25 mm

EC2, Table 4.2, for exposure class 2a

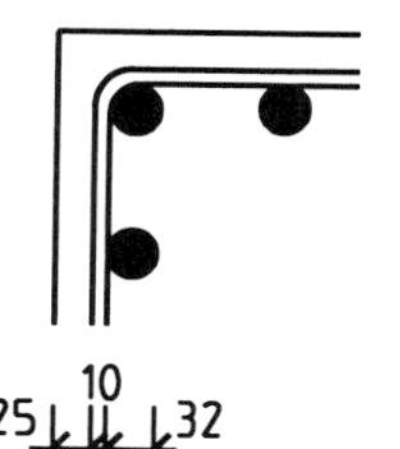

Actions

Permanent actions (self-weight)

$G_{k,1}$ = 25.0 kN/m^3
$G_{k,2}$ = 244.0 kN (prestressed beam)
$G_{k,3}$ = 42.0 kN (crane girder)

see 15.5.2.1 above

Crane-induced variable actions

$Q_{k,1v}$ = 551.0 kN (vertical)
$Q_{k,1t}$ = 114.0 kN (transverse action)
$Q_{k,1b}$ = ± 70.0 kN (braking force)

EC1-1, 9.4.4

Variable actions except crane-induced actions

$Q_{k,2}$ = 60.0 kN (snow)
$Q_{k,3}$ = 3.6 kN/m (wind)
$Q_{k,4}$ = 16.0 kN (sliding force)

see 15.5.2.1 above

Combination coefficients

- for crane-induced actions ψ_{0c} = 1.0
- for snow ψ_{0s} = 0.6
- for wind ψ_{0w} = 0.6
- for sliding force ψ_{0sl} = 0.6

15.5.3.2 Design values of actions

EC2, 2.2.2.4

(a) Permanent actions (γ_G = 1.35)

The combination with γ_G = 1.0 is not relevant in this example

$\gamma_G G_{k,1}$ = 1.35 * 25.0 = 33.8 kN/m^3
$\gamma_G G_{k,2}$ = 1.35 * 244.0 = 329.4 kN
$\gamma_G G_{k,3}$ = 1.35 * 42.0 = 56.7 kN

(b) Variable actions ($\gamma_Q = 1.50$)

Reference

					Reference
$\gamma_Q\, Q_{k,1v}$	=	1.5 * 551.0	=	826.5 kN	vertical crane load
$\gamma_Q\, Q_{k,1t}$	=	1.5 * 114.0	=	171.0 kN	transverse action
$\gamma_Q\, Q_{k,1b}$	=	±1.5 * 70.0	=	±105.0 kN	braking force
$\gamma_Q\, \psi_{0.2}\, Q_{k,2}$	=	1.5 * 0.6 * 60.0	=	54.0 kN	snow
$\gamma_Q\, \psi_{0.3}\, Q_{k,3}$	=	1.5 * 0.6 * 3.6	=	3.24 kN/m	wind
$\gamma_Q\, \psi_{0.4}\, Q_{k,4}$	=	1.5 * 0.6 * 16.0	=	14.4 kN	sliding force

(c) Fundamental combination of actions

EC2, Eq.(2.7a); the accidental combination of actions is not considered

$$\Sigma(\gamma_{G,j}\, G_{k,j}) + \gamma_{Q,1}\, Q_{k,1} + \Sigma(\gamma_{Q,i}\, \psi_{0,i}\, Q_{k,i})$$
$$= 1.35(G_{k,1} + G_{k,2} + G_{k,3}) + 1.5(Q_{k,1v} + Q_{k,1t} + Q_{k,1b}) + 1.5 * 0.6(Q_{k,2} + Q_{k,3} + Q_{k,4})$$

Crane-induced actions are the main variable actions.

15.5.3.3 Design of the column for the ultimate limit states induced by structural deformations

EC2, 4.3.5

(a) General

For the design of the column at the ultimate limit states induced by structural deformations, a rigorous computer-based second-order analysis is carried out. The design model is shown below. In this program, the steel reinforcement, $A_{s,req}$, required in the individual cross-sections is calculated automatically.

see 15.5.3.3 (e) below

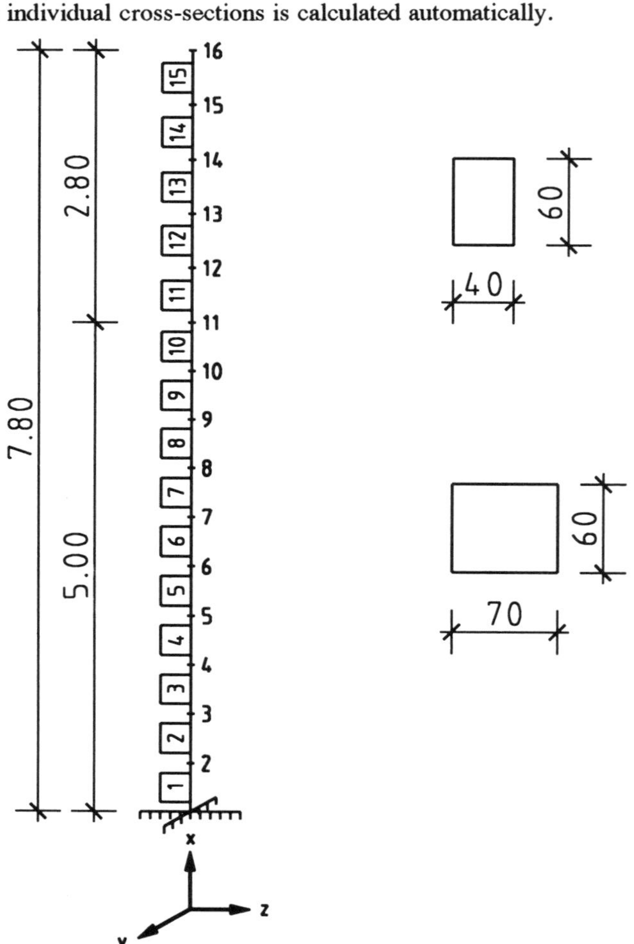

Reference

(b) Imperfections
For structural analysis, an inclination of

$\nu \quad = \quad 1/200$

in the direction of the theoretical failure plane is assumed.

EC2, 4.3.5.4 and 2.5.1.3, Eq.(2.10)

introduced by iteration

(c) Creep
Allowance for creep deformations is made by using the simplified method and Appendix 3 proposed by QUAST in [2], i.e. to multiply the second order eccentricities e_2 by a coefficient

EC2, 4.3.5.5.3

$$f_\psi = (1 + M_{Sd,c}/M_{Sd})$$

[2], p. 85, 9.4.3

where

$M_{Sd,c}$ is the factored bending moment due to quasi-permanent actions
M_{Sd} is the bending moment due to the relevant combination of permanent and variable actions

(d) Design actions in the nodes of the design model

Node	$F_{Sd,x}$ (kN)	$F_{Sd,y}$ (kN)	$F_{Sd,z}$ (kN)	$M_{Sd,y}$ (kNm)
16	14.4	0	383.4	-28.8
11	171.0	-105.0	883.2	-204.0

Wind:

$q_{Sd,x} = 3.24$ kN/m

Rotation due to imperfections and creep

$\nu_x = -4.24 * 10^{-3}$
$\nu_y = -7.57 * 10^{-3}$

(e) Summary of design results
Internal forces and moments

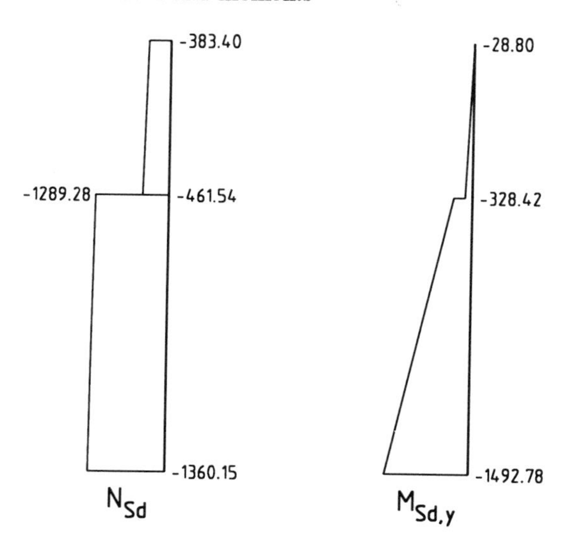

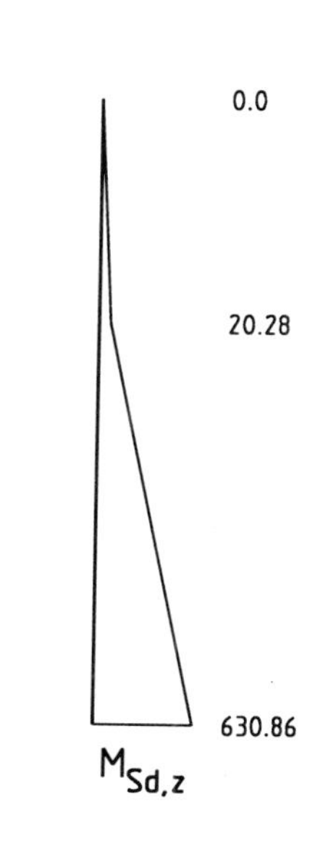

0.0
20.28
630.86
$M_{Sd,z}$

Reference

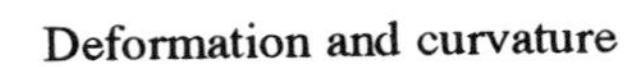

Deformation and curvature

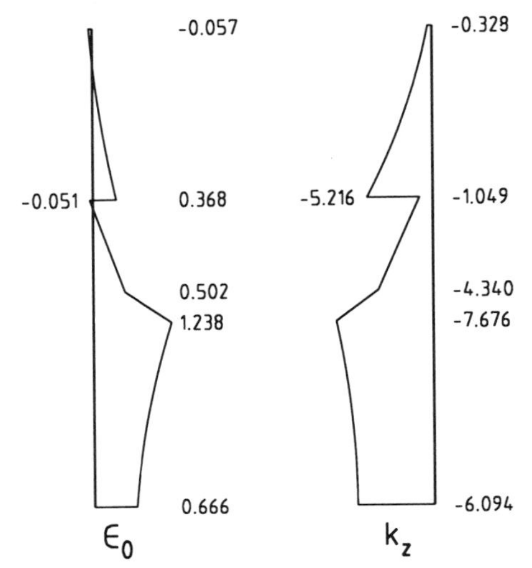

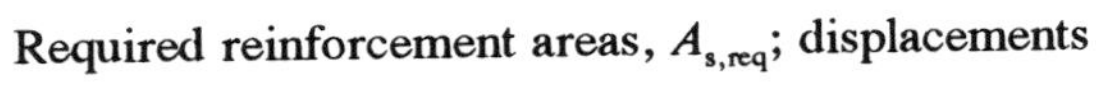

Required reinforcement areas, $A_{s,req}$; displacements

$A_{s,req}$ in mm² * 10²

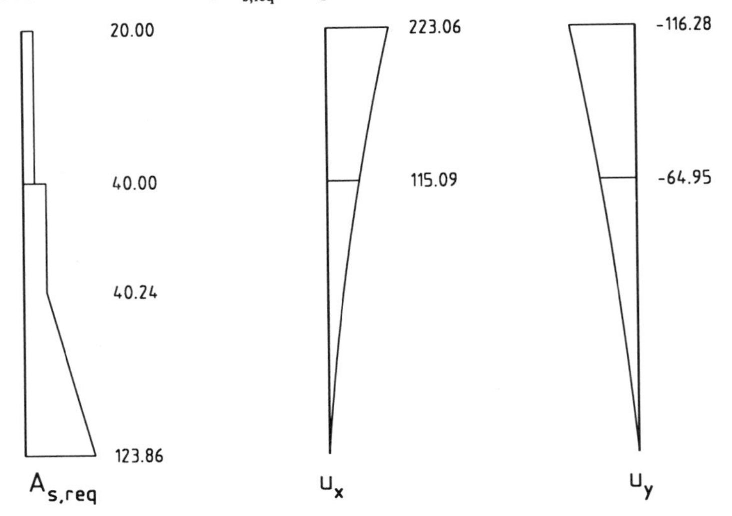

15.5.3.4 Design of the column; detailing of reinforcement

Required reinforcement area, $A_{s,req}$ at the restrained cross-section:

$A_{s,req}$ = = 123.86 * 10² mm² — see 15.5.3.3 (e), above

Selected steel B 500 2 * 8 = 16 ø 32 **$A_{s,prov}$ = 128.68 * 10² mm²**

Detailing of reinforcement: — EC2,5
see Figure below

Detailing of reinforcement

Elevation

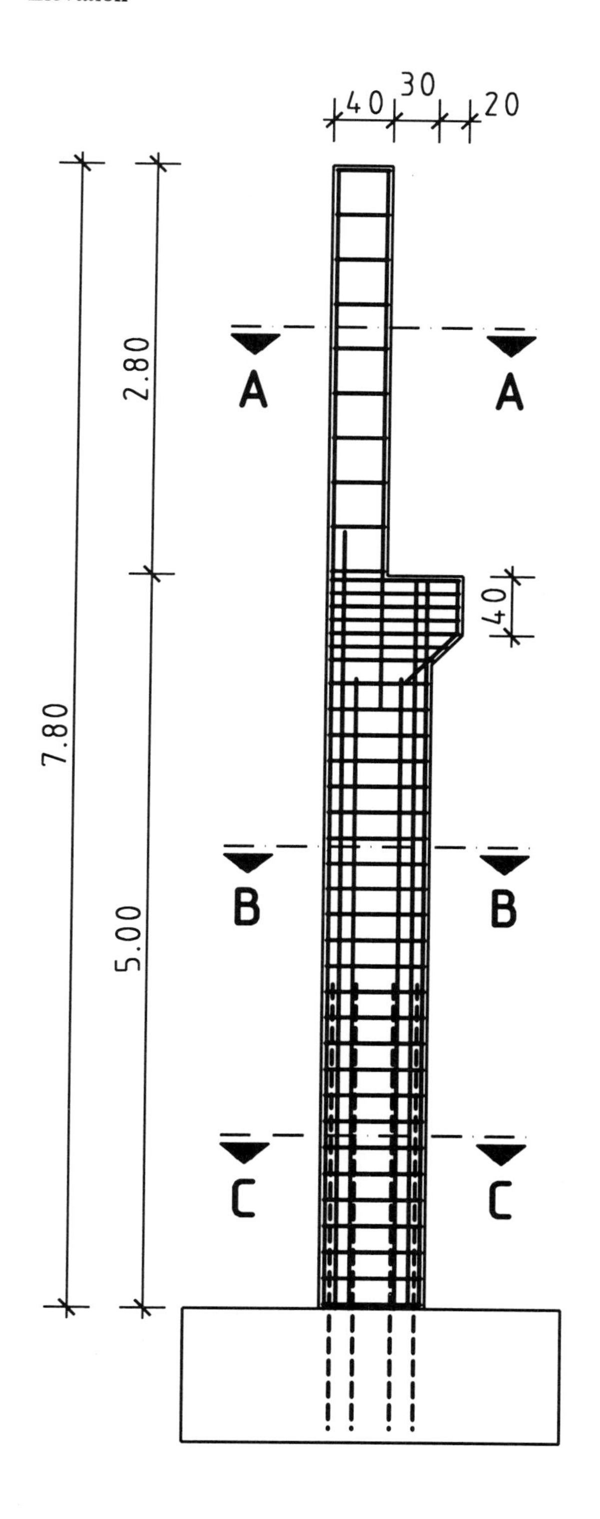

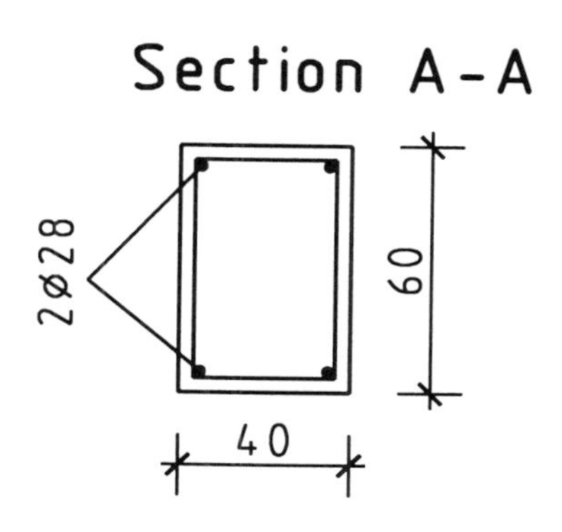

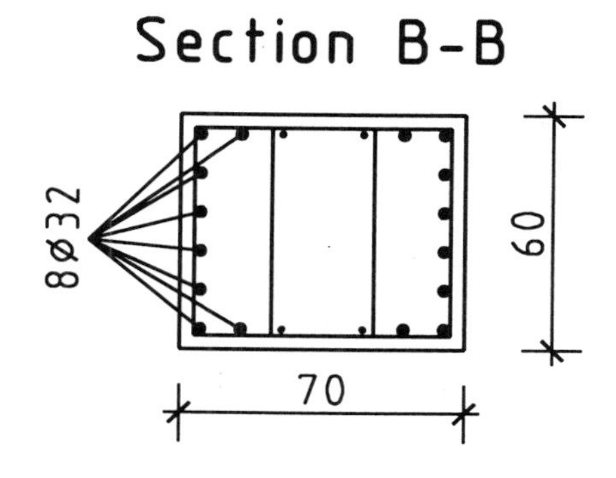

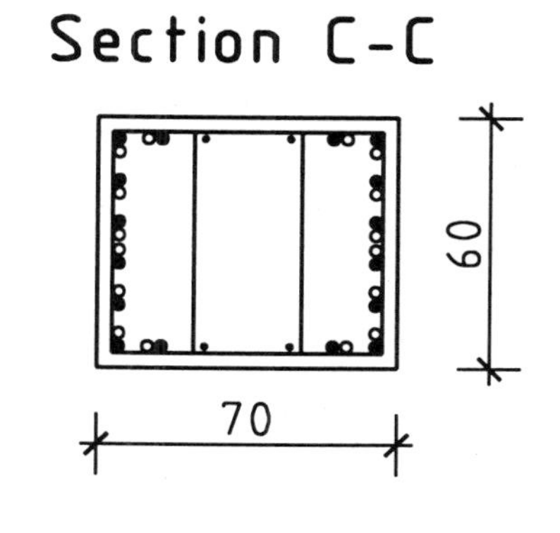

15.5.3.5 Ultimate limit state of fatigue

Reference

EC2-2, 4.3.7

15.5.3.5.1 General

The edge column is subjected to crane-induced actions. It needs therefore to be checked for the ultimate limit state for fatigue. In this ultimate limit state, see 15.5.3.0 above, it shall be verified that

$$D_{sd} = \sum_i (n_i/N_i) \leq 1$$

EC2-2, 4.3.7.5

where

D_{Sd} is the design value of the fatigue damage factor calculated using the PALMGREN-MINER summation

n_i denotes the number of acting stress cycles associated with the stress range for steel and the actual stress levels for concrete

N_i denotes the number of resisting stress cycles

For the above verification, the stress calculation shall be based on the assumption of cracked cross-sections neglecting the tensile stength of concrete but satisfying compatibility of strains.

EC2-2, 4.3.7.3

The fatigue strength of reinforcing steel and concrete are given by EC2-2, 4.3.7.8 and 4.3.7.9 respectively.

15.5.3.5.2 Combination of actions

In the present example, fatigue verification will be performed under the frequent combination of actions using the partial safety factors

EC2-2, 4.3.7.2

γ_F = 1.0 for actions

γ_{Sd} = 1.0 for model uncertainties

$\gamma_{c,fat}$ = 1.5 for concrete

$\gamma_{s,fat}$ = 1.15 for reinforcing steel

Therefore the relevant combination of actions is given by:

$$E_d = \sum (1.0 * G_{k,j}) + 1.0 * \psi_{1,1} (Q_{k,1v} + Q_{k,1b}) + \sum (1.0 * \psi_{2,i} Q_{k,i})$$

see 15.5.3.1 above
i = 1t, 2, 3, 4

where

$Q_{k,1b}$ = the component of the braking force $Q_{k,1b}$ that is relevant for fatigue verification. In this example, it is assumed that $Q_{k,1b}$ = ± 59.7 kN

see 15.5.3.1 above

For the verification, the following combination coefficients $\psi_{1,1}$ and $\psi_{2,i}$ are asssumed:

$\psi_{1,1}$ = 1.0 (for crane-induced actions)

$\psi_{2,2}$ = 0 (for snow loads)

$\psi_{2,3}$ = 0 (for wind)

$\psi_{2,i}$ = 0 (for all other variable actions $Q_{k,i}$)

Design values of actions:

(a) Vertical actions

$\sum G_{d,j}$ = 25.0 + 244.0 + 42.0 = 311.0 kN

$Q_{d,1v}$ = = 551.0 kN

				Reference
(b) Horizontal actions				
$Q_{d,1b}$	=		= ± 59.7 kN	
$Q_{d,4}$	=		= 16.0 kN	

15.5.3.5.3 Damage factor D_{Sd}

For the calculation of the damage factor D_{Sd}, the spectrum of actions S2 in DIN 15 018 Part 1 is assumed. This approach is based on a linear relationship between actions and stresses assuming cracked cross-sections.

15.5.3.5.4 Calculation of the stress range $\Delta\sigma_s$

Reference: for reinforcing steel

A trial calculation shows that the most unfavourable stress range $\Delta\sigma_s$ occurs at the coordinate x = 3.0 m above the level of restraint. A rigorous second order analysis leads to the following stresses:

$\sigma_{s,max}$	=		= +112.2 N/mm²	Tension
$\sigma_{s,min}$	=		= -24.3 N/mm²	Compression
$\Delta\sigma_s$	=	112.2 - (-24.3)	= 136.5 N/mm²	

Since both tensile and compressive stresses in the reinforcing steel occur, fatigue verification is necessary. *(EC2-2, 4.3.7.1)*

15.5.3.5.5 Calculation of the stress range $\Delta\sigma_c$

Reference: for concrete

The extreme concrete stresses occur at the level of restraint (i.e. x = 0). A rigorous second order analysis leads to he following values:

$\sigma_{c,max}$	=		= -11.3 N/mm²	Compression
$\sigma_{c,min}$	=		= -2.9 N/mm²	Compression
$\Delta\sigma_c$	=	11.3 - 2.9	= 8.4 N/mm²	

15.5.3.5.6 Verification of the fatigue strength of the reinforcing steel

Reference: EC2-2, 4.3.7.5

The fatigue requirement for reinforcing steel will be met if the following expression is satisfied:

$$\gamma_F \, \gamma_{Sd} \, \Delta\sigma_{s,equ} \leq \Delta\sigma_{Rsk}(N^*)/\gamma_{s,fat}$$

where

$\Delta\sigma_{Rsk}(N^*)$	=	stress range at N^* cycles from the appropriate S-N lines	
$\Delta\sigma_{s,equ}$	=	the damage equivalent stress range which is the stress range of a constant stress spectrum with N^* = 10^6 stress cycles which results in the same damage as the spectrum of stress ranges caused by flowing traffic loads	
γ_F	=	1.0	
γ_{Sd}	=	1.0	see 15.5.3.4.2 above
$\gamma_{s,fat}$	=	1.15	

For bars with diameter, ø > 25 mm, the stress range, $\Delta\sigma_{Rsk}$, at N^* = 10^6 cycles is given as:

$\Delta\sigma_{Rsk}(N^*)$	=		= 195 N/mm²	EC2-2, 4.3.7.8

The shape of the relevant S-N curve is defined by the coefficients

k_1 = 5.0

k_2 = 9.0

Reference

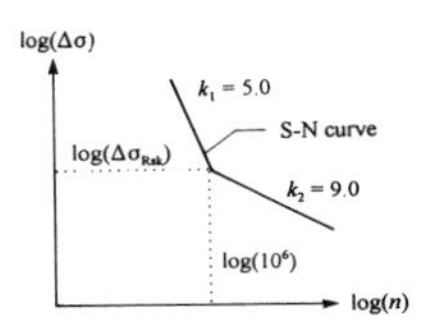

For $N + N^*$ cycles, the damage equivalent stress range $\Delta\sigma_{s,equ}$ is given by:

$\Delta\sigma_{s,equ}$ = $\alpha_p\ \kappa\ \Delta\sigma_s$

For the calculation of the coefficients α_p and κ, the following assumptions have been used:

- spectrum of actions S2 according to DIN 15 018 Part 1 — see 15.5.3.4.3 above

- $\Delta\sigma_s$ = 136.5 N/mm² — see 15.5.3.4.4 above

- number of cycles: n
  design lifetime: 50 years
  working time: 10 hours/per day
  one cycle/minute

n = 50 * 365 * 10 * 60 * 1 = $1.1 \cdot 10^7$

From the α_p-diagram below:

α_p = = 1.0

Annex A gives more details on the fatigue verification

Coefficient A:

A = $\gamma_{Sd}\ \Delta\sigma_{s,max}\ \gamma_s/\Delta\sigma_{Rsk}$

= 1.0 * 136.5 * 1.15/195 = 0.81

< 1.0

From the κ-diagram below:

A = = 1.0

Thus:

$\Delta\sigma_{s,equ}$ = 1.0 * 1.0 * 136.5 < 195/1.15 = 169.5 N/mm²

Requirements for reinforcing steel are met.

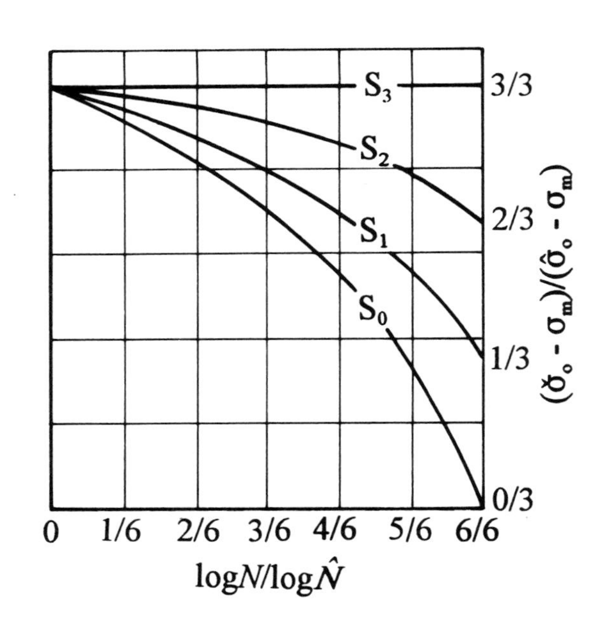

Diagrams for reinforcing steel

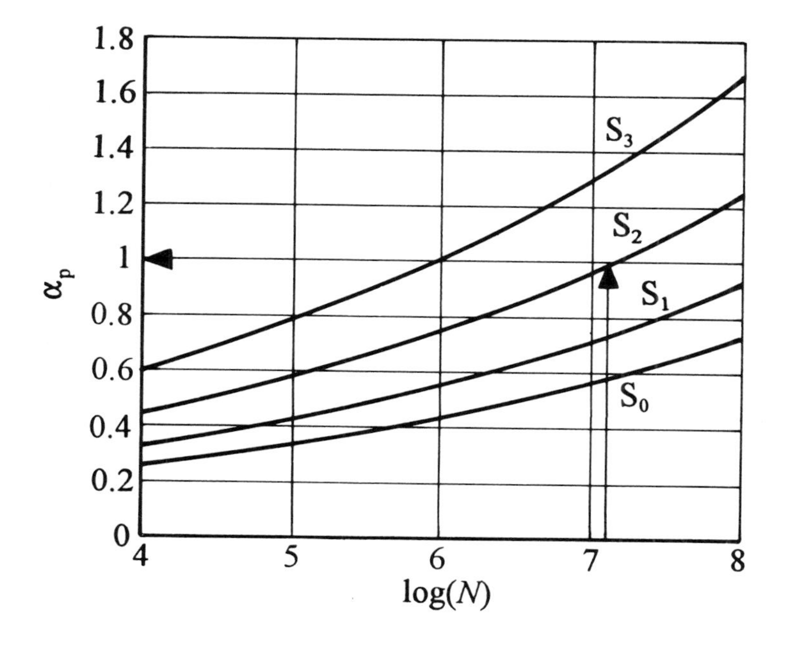

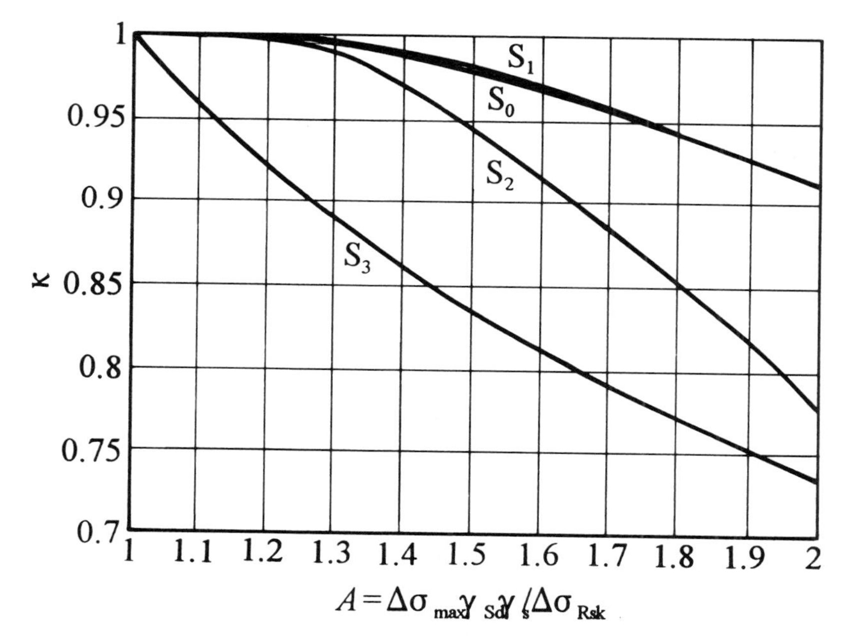

15.5.3.5.7 Verification of the fatigue strength of concrete

Reference
EC2-2, 4.3.7.4

The design fatigue strength of concrete is given by the S-N curve according to:

$$\log(N) = 14 \frac{1 - S_{\text{cd,max}}}{\sqrt{1 - R}}$$

$$S_{\text{cd,max}} = \left|\sigma_{\text{c,max}}\right| \frac{1}{f_{\text{cd,fat}}}$$

$$S_{\text{cd,min}} = \left|\sigma_{\text{c,min}}\right| \frac{1}{f_{\text{cd,fat}}}$$

$$R = \frac{S_{\text{cd,min}}}{S_{\text{cd,max}}}$$

$$f_{\text{cd,fat}} = \beta_{\text{cc}}(t) \frac{f_{\text{ck}}}{\gamma_{\text{c}}}$$

$$\beta_{\text{cc}} = \exp\left(s\left[1 - \sqrt{\frac{28}{t_0/t_1}}\right]\right)$$

where

N	=	number of cycles to failure
$\beta_{cc}(t_0)$	=	coefficient which depends on the age of concrete t_0 in days when fatigue loading starts. If no information is available β_{cc} can be taken as 1.0
t_1	=	1 day

In the present case, the basic data are given by:

Concrete strength class C 45/55:			f_{yk}	=	45 N/mm²
$f_{\text{cd,fat}}$	=	45/1.5		=	30 N/mm²
$S_{\text{cd,max}}$	=	11.3/30		=	0.38
$S_{\text{cd,min}}$	=	2.9/30		=	0.10
				≈	0

see 15.5.3.4.5 above

from the diagram below:

N = $7.9 * 10^9$ > $1.1 * 10^7$

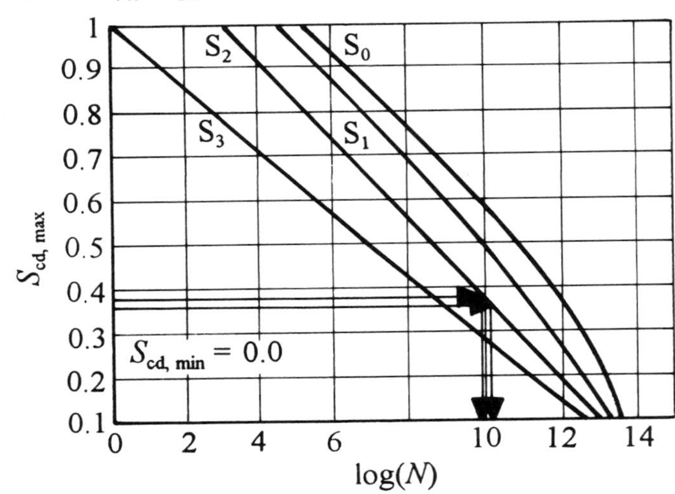

ANNEX A

15.6 Guidance for the calculation of the equivalent stress range $\Delta\sigma_{s,equ}$ for reinforcing steel and of the S-N curve for concrete in compression using the single load level method

15.6.1 Reinforcing steel

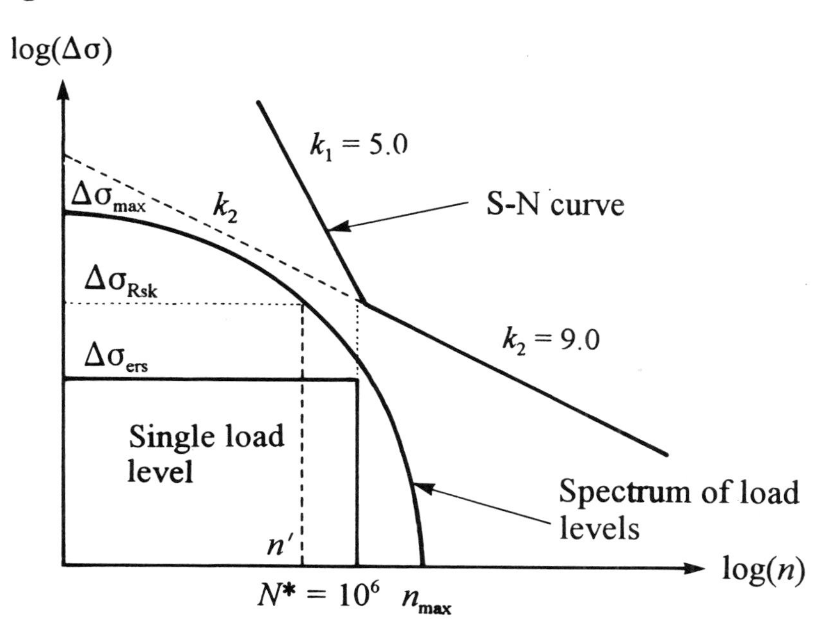

Figure A1 Graphical presentation of the design concept for reinforcing steel.

Design value of the fatigue damage D_d, using the PALMGREN-MINER summation:

$$D_d = \sum \frac{n_i}{N_i} \tag{1}$$

where

n_i denotes the number of acting stress cycles associated with the stress range for steel and the actual stress levels for concrete

N_i denotes the number of resisting stress cycles

The shape of the S-N curve is given by:

$$\Delta\sigma^{k_j} N^* = \Delta\sigma_i^{k_j} N_i \; ; \; k_j = 1 \text{ or } 2 \tag{2}$$

$$\Delta\sigma^* = \frac{\Delta\sigma_{Rsk}}{\gamma_s}$$

$$\Delta\sigma_i = \Delta\sigma_{max}\, \eta\, \gamma_{Sd} \tag{3}$$

where

η = coefficient describing the spectrum of load levels

$\Delta\sigma_{max}$ = maximum acting stress range

γ_s, γ_{Sd} = partial safety factors

Equation 2 may be written as:

$$N_i = \left(\frac{\Delta\sigma^*}{\Delta\sigma_i}\right)^{k_j} N^* \tag{4}$$

Equations 1 and 4 lead to the following expression for E_d:

$$D_d = \sum_{0}^{n'} \left[\frac{n_i}{\left(\frac{\Delta\sigma^*}{\Delta\sigma_i}\right)^{k_1} N^*}\right] + \sum_{n'}^{n_{max}} \left[\frac{n_i}{\left(\frac{\Delta\sigma^*}{\Delta\sigma_i}\right)^{k_2} N^*}\right] \tag{5}$$

$$D_d = \sum_{0}^{n'} \left[\frac{n_i}{\left(\frac{\Delta\sigma^*}{\Delta\sigma_i}\right)^{k_1} N^*}\right] + \sum_{0}^{n_{max}} \left[\frac{n_i}{\left(\frac{\Delta\sigma^*}{\Delta\sigma_i}\right)^{k_2} N^*}\right] - \sum_{0}^{n'} \left[\frac{n_i}{\left(\frac{\Delta\sigma^*}{\Delta\sigma_i}\right)^{k_2} N^*}\right] \tag{6}$$

or

$$D_d = \frac{1}{N^*}\left(\frac{1}{\Delta\sigma^*}\right)^{k_2} \left[\sum_{0}^{n_{max}} \left(\Delta\sigma_i^{k_2} n_i\right) + \sigma^{*(k_2 - k_1)} \sum_{0}^{n'} \left(\Delta\sigma_i^{k_1} n_i\right) - \sum_{0}^{n'} \left(\Delta\sigma_i^{k_2} n_i\right)\right] \tag{7}$$

An equivalent single load level with N^* cycles shall satisfy the condition:

$$D_d = D_{equ} = \frac{N^*}{N_{equ}} \tag{8}$$

Using equation 2 for the S-N curve and equation 7 for the equivalent damage factor, the equivalent steel stress $\Delta\sigma_{s,equ}$ may be calculated as:

S-N curve

$$\Delta\sigma^{*k_2} N^* = \Delta\sigma_{s,equ}^{\;k_2} N_{s,equ} \tag{9}$$

Equivalent number of cycles

$$N_{s,equ} = \left(\frac{\Delta\sigma^*}{\Delta\sigma_{s,equ}}\right)^{k_2} N^* \tag{10}$$

Equivalent damage factor

$$D_d = D_{s,equ} = \left[\frac{\Delta\sigma_{s,equ}}{\Delta\sigma^*}\right]^{k_2} \tag{11}$$

Equivalent steel stress

$$\Delta\sigma_{s,equ} = \Delta\sigma^* \sqrt[k_2]{D_d} \tag{12}$$

From equation 7, it follows that

$$\Delta\sigma_{s,equ} = \Delta\sigma^* \sqrt[k_2]{\frac{1}{N^*}\left(\frac{1}{\Delta\sigma^*}\right)^{k_2}\left[\sum_0^{n_{max}}\left(\Delta\sigma_i^{k_2} n_i\right) + \sigma^{*(k_2-k_1)}\sum_0^{n'}\left(\Delta\sigma_i^{k_1} n_i\right) - \sum_0^{n'}\left(\Delta\sigma_i^{k_2} n_i\right)\right]} \qquad (13)$$

The equivalent steel stress $\Delta\sigma_{s,equ}$ may be expressed by:

$$\Delta\sigma_{s,equ} = \Delta\sigma_{max}\,\gamma_{Sd}\,\alpha_p\,\kappa \qquad (14)$$

The coefficients α_p and κ are defined as:

$$\alpha_p = \sqrt[k_2]{\frac{1}{N*}\sum_0^{n_{max}} \eta_i^{k_2} n_i} \qquad (15)$$

and

$$\kappa = \sqrt[k_2]{1 + \frac{\Delta\sigma^{*(k_2-k_1)}\sum_0^{n'}\left(\Delta\sigma_i^{k_1} n_i\right)}{\sum_0^{n_{max}}\left(\Delta\sigma_i^{k_2} n_i\right)} - \frac{\sum_0^{n'}\left(\Delta\sigma_i^{k_2} n_i\right)}{\sum_0^{n_{max}}\left(\Delta\sigma_i^{k_2} n_i\right)}} \qquad (16)$$

Format for fatigue verification:

$$\Delta\sigma_{s,equ} \le \frac{\Delta\sigma_{Rsk}}{\gamma_s}$$

$$\gamma_{Sd}\,\Delta\sigma_{max}\,\alpha_p\,\kappa \le \frac{\Delta\sigma_{Rsk}}{\gamma_s} \qquad (17)$$

The coefficients α_p and κ may be taken from Figure A2.

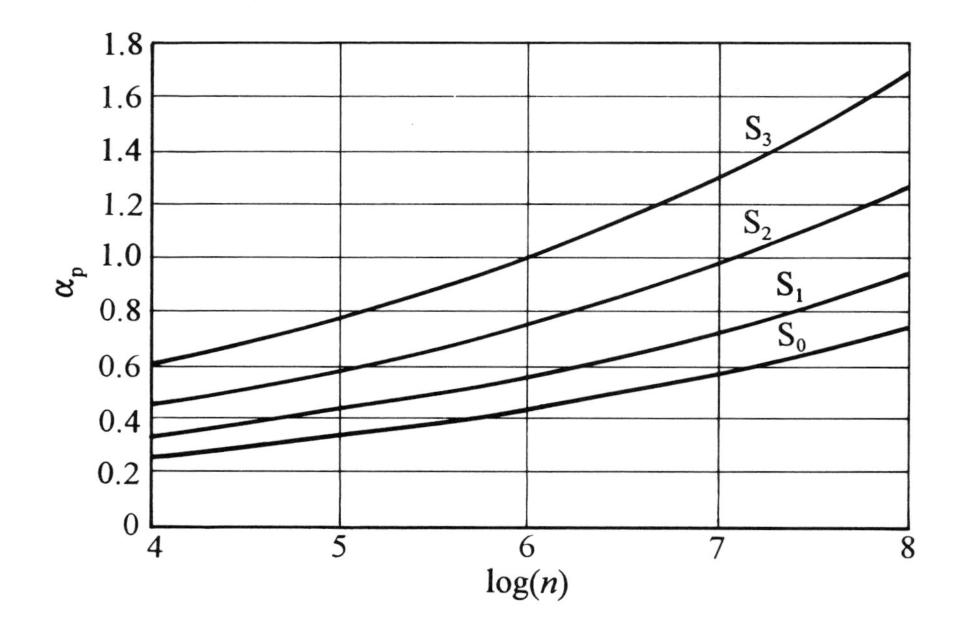

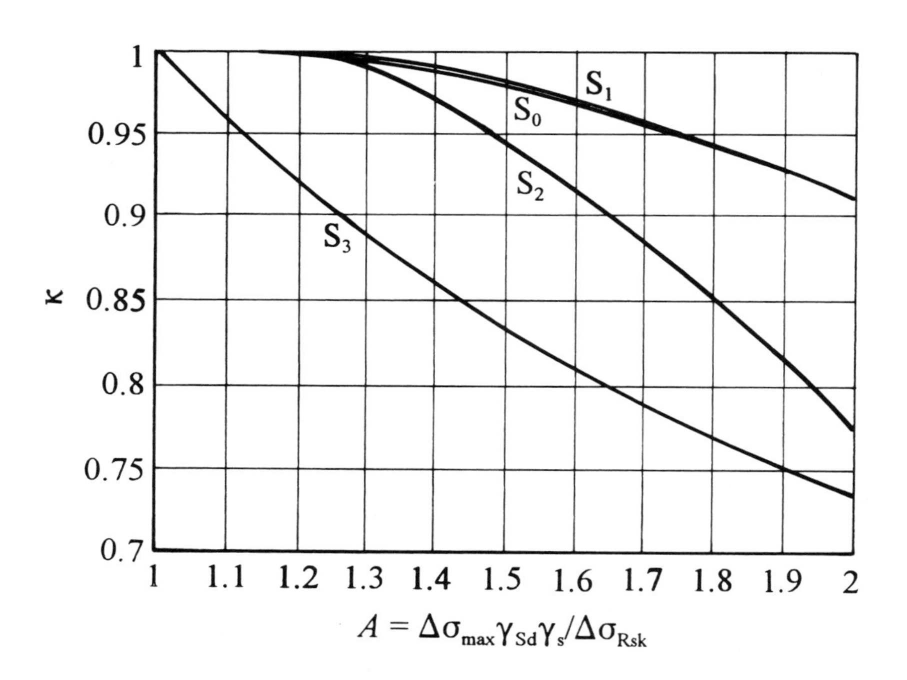

Figure A2 Coefficients for reinforcing steel. (a) α_p (b) κ.

15.6.2 Concrete

The fatigue verification of concrete is analogous to that for steel reinforcement. However, there are differences. The fatigue requirements under cyclic loading will be met if the required lifetime (number of cycles, n_{max}) is less than or equal to the number of cycles to failure (N_{equ}). In addition, the simplified S-N function given by equation 18 is used.

S-N curve of concrete:

$$\log(N) = 14 \frac{1 - S_{cd,max}}{\sqrt{1 - R}} \tag{18}$$

The calculation of the coefficients in equation 18 is based on EC2-2 (pr ENV 1992-2):

S-N curve of concrete, i.e. equation 1:

$$D_d = \sum \frac{n_i}{N_i} \tag{1}$$

Equivalent damage factor, D_{equ}:

$$D_d \le D_{equ} = \frac{n_{max}}{N_{equ}} \tag{19}$$

Equivalent spectrum of load levels:

N_{equ} is calculated on the assumption that $D_{sd} = 1.0$ for a given spectrum of load levels and for given parameters $S_{cd,max}$ and R of the relevant S-N curve for concrete.

$$D_d \le 1$$

Verification format:

$$n_{max} \le N_{equ} \tag{20}$$

N_{equ} should be taken from Figure A3.

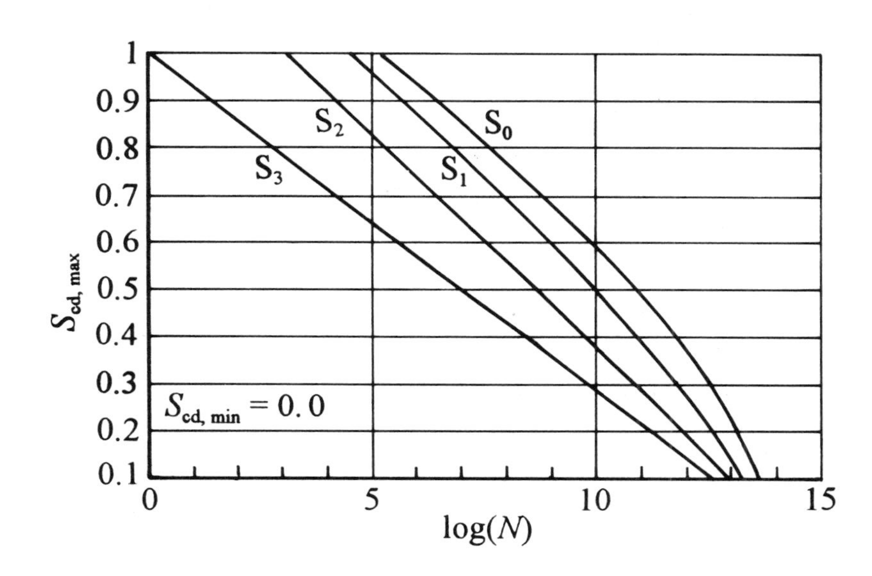

Figure A3 Relationship between N_{equ} and $S_{cd,\,max}$ for different values of $S_{cd,\,min}$.

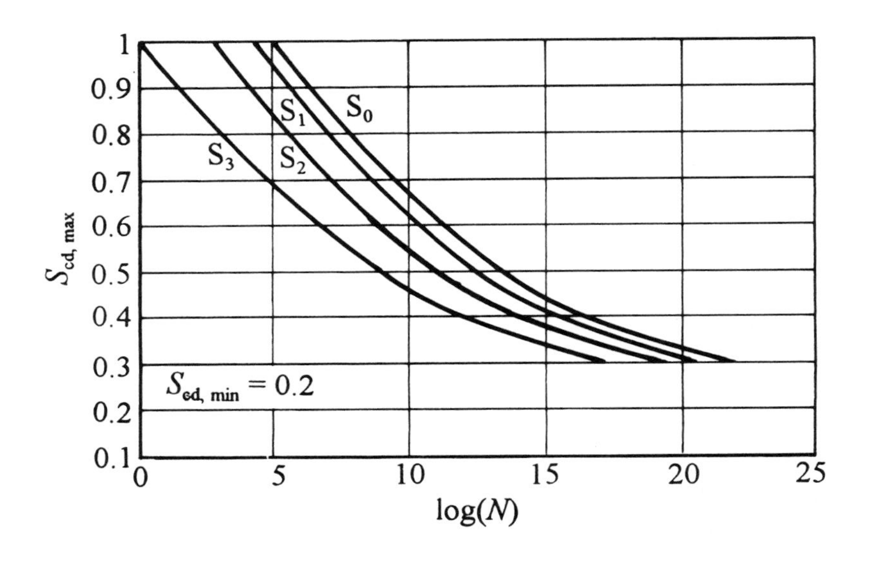

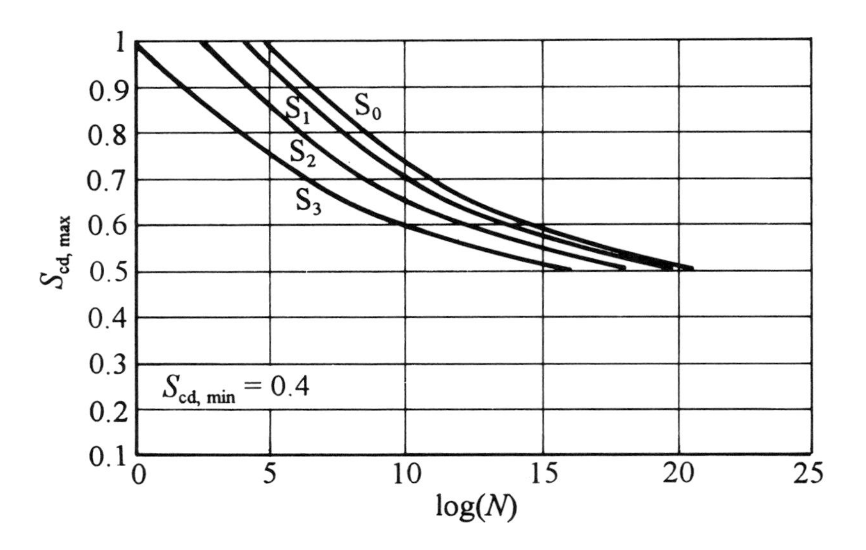

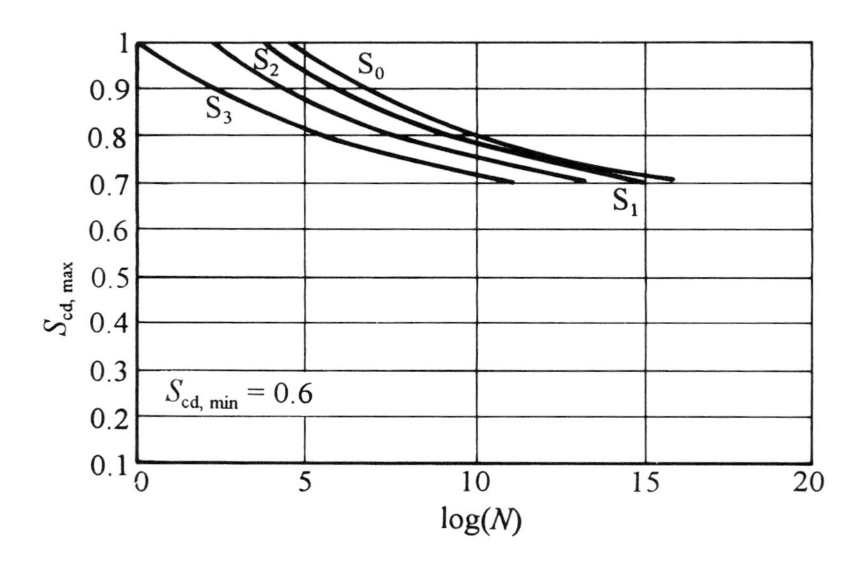

ANNEX B

15.7 Design of purpose-made fabrics

In the present design examples, purpose-made fabrics as defined in ENV 10 080 *Steels for the reinforcement of concrete; Weldable ribbed reinforcing steel B 500* have been chosen. The graphical representation is shown in Figure B1 for the top reinforcement of a continuous slab and in Figure B2 for the respective bottom reinforcement.

Each individual fabric is characterized by a position number, i.e. ① to ③ in Figures B1 and B2. Their characteristics are described graphically in Figures B3 to B5 by means of the diameters and spacing of both the longitudinal and transverse bars. The total number of bars and their lengths lead to the total weight of the fabric.

The presentation of the fabrics corresponds to ISO 3766 - 1977(E) *Building and civil engineering drawings - Symbols for concrete reinforcement*, particularly clause 2.3.1. Each fabric is characterized by a rectangular frame (see, for example, Figure B2), the diagonal connected to the position number denoting the direction of the main bars. In Figure B2, the lap length (s_t = 400 mm) of the transverse bars is also defined.

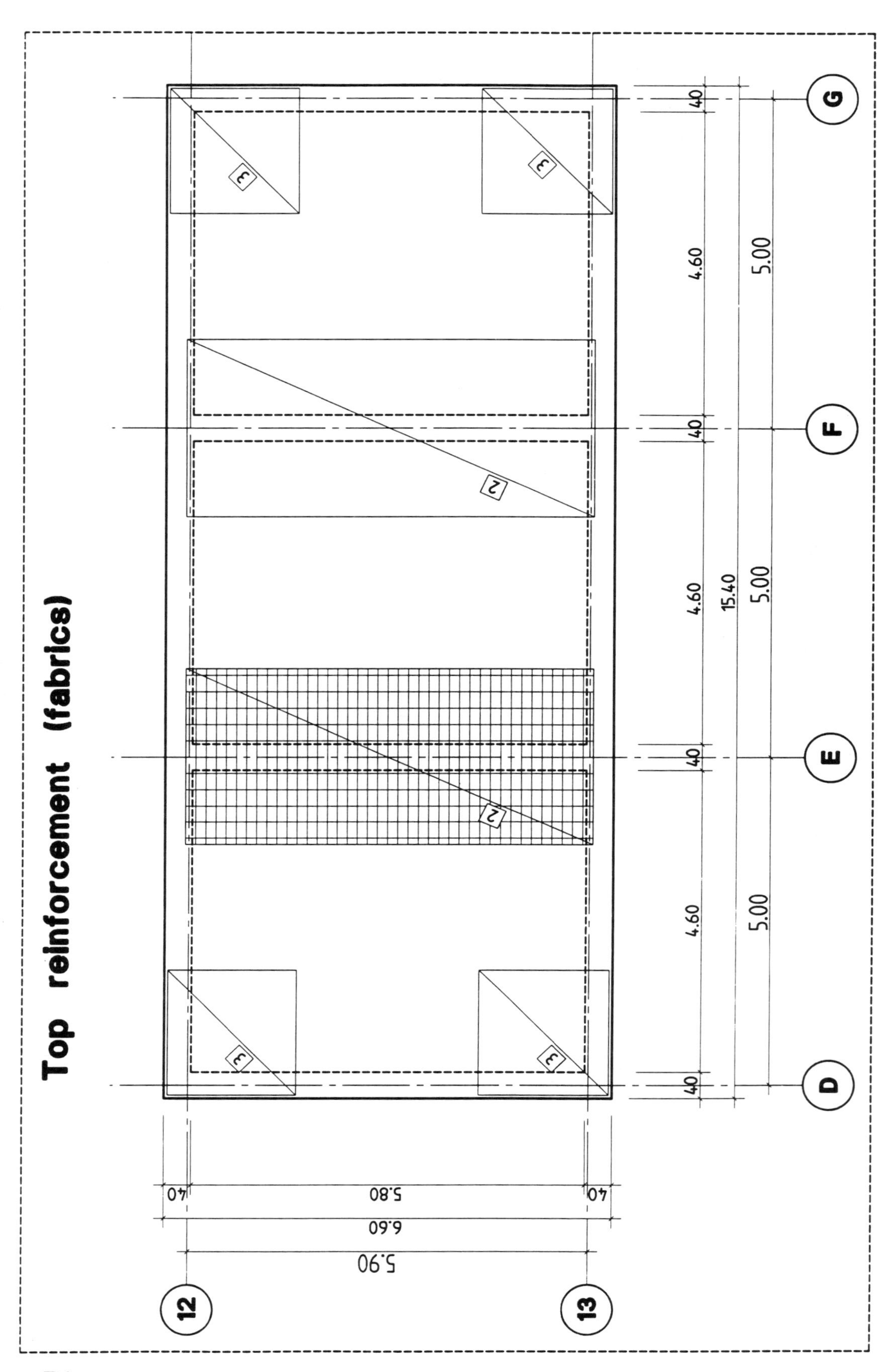
Top reinforcement (fabrics)
G
F
E
D
12
13
40
4.60
5.00
15.40
5.80
6.60
5.90
2
3

Figure B1

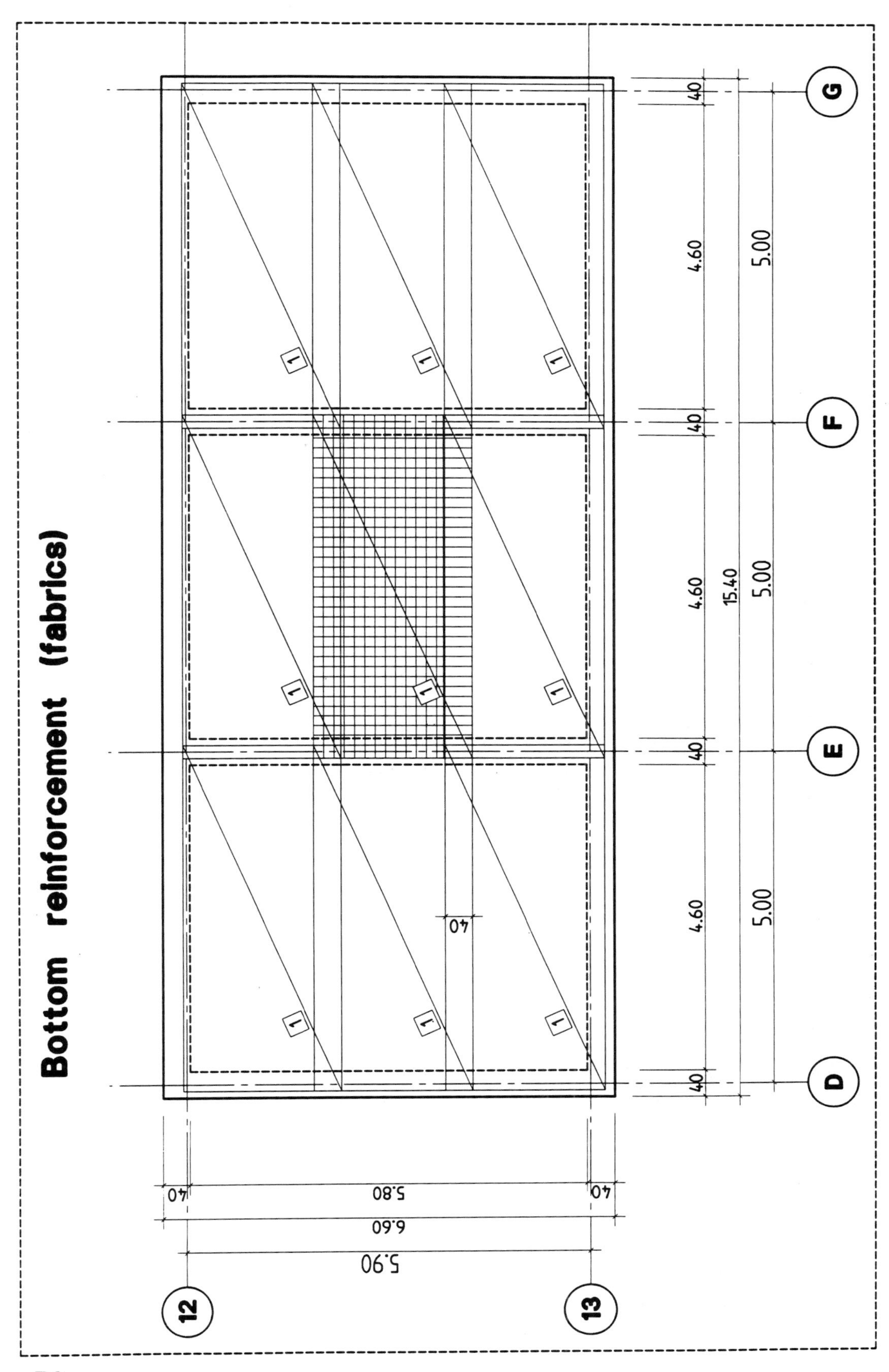

Bottom reinforcement (fabrics)
1
40
4.60
5.00
15.40
5.80
6.60
5.90
D
E
F
G
12
13

Figure B2

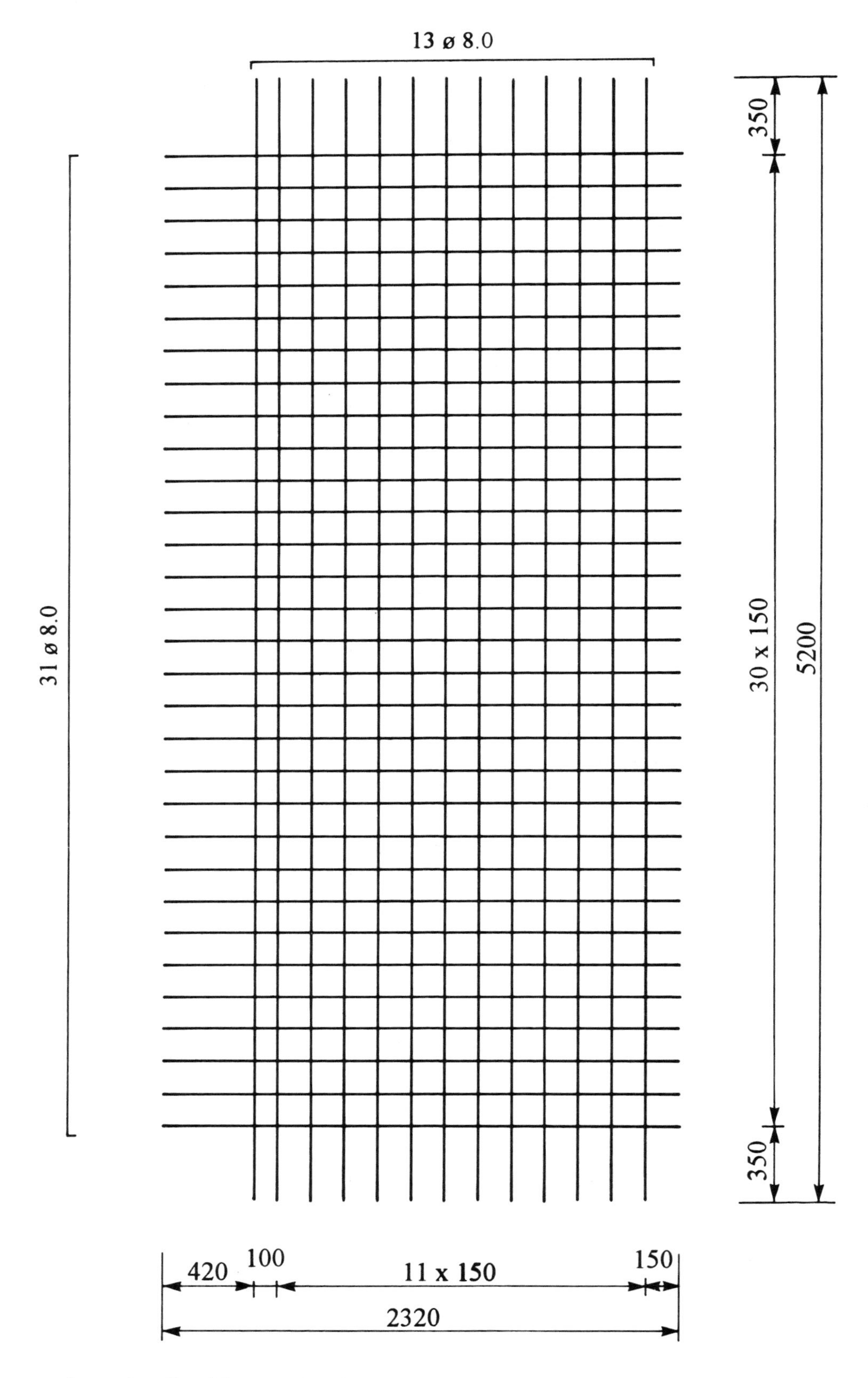

Longitudinal bars 13 ø 8.0 x 5200 = 26.702 **kg**
Transverse bars 31 ø 8.0 x 2320 = 28.408 kg
Total weight 58.110 kg

Figure B3

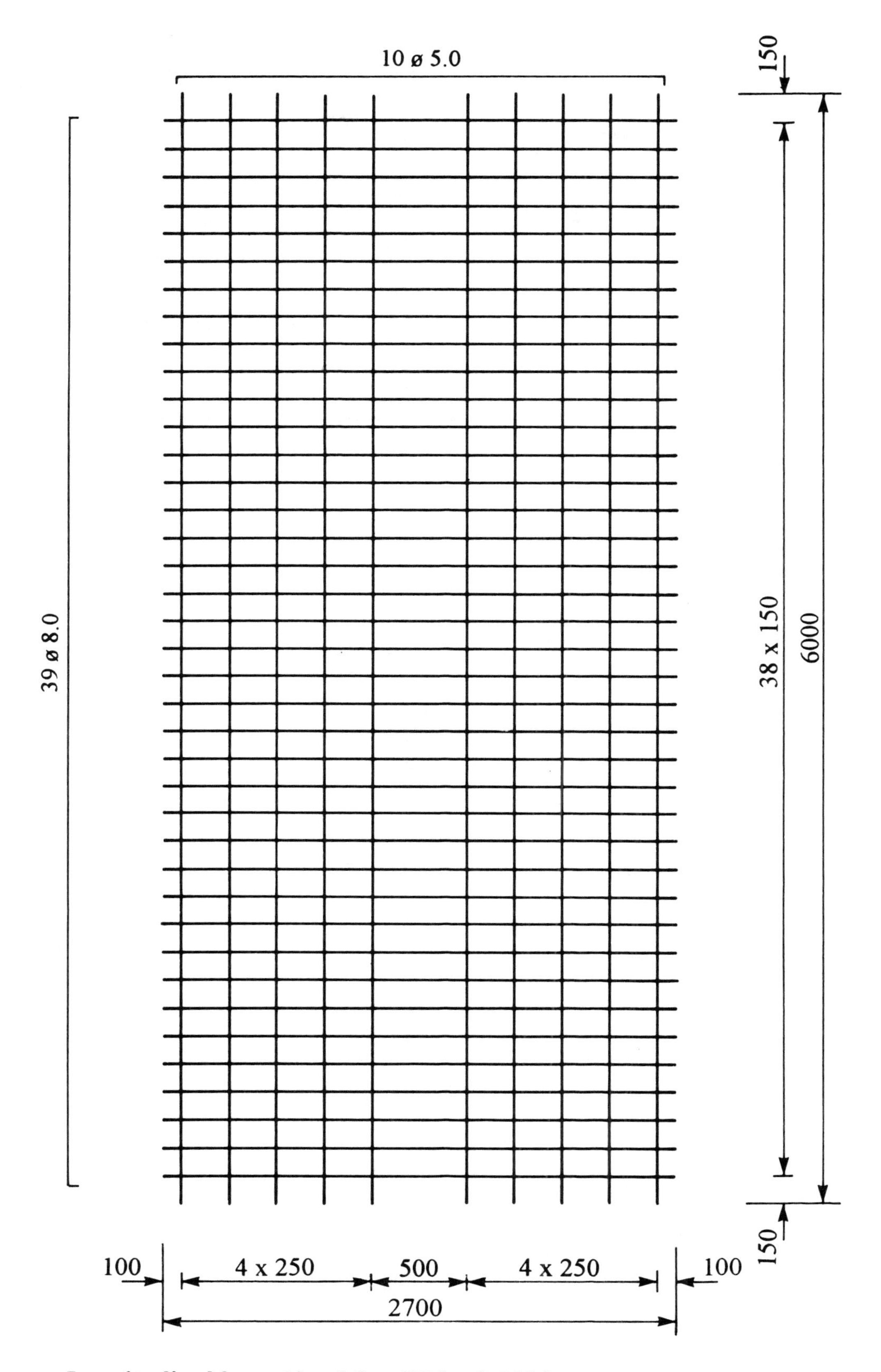

Longitudinal bars 10 ø 5.0 x 6000 = 9.240 **kg**
Transverse bars 39 ø 8.0 x 2700 = 41.594 kg
Total weight 50.834 kg

Figure B4

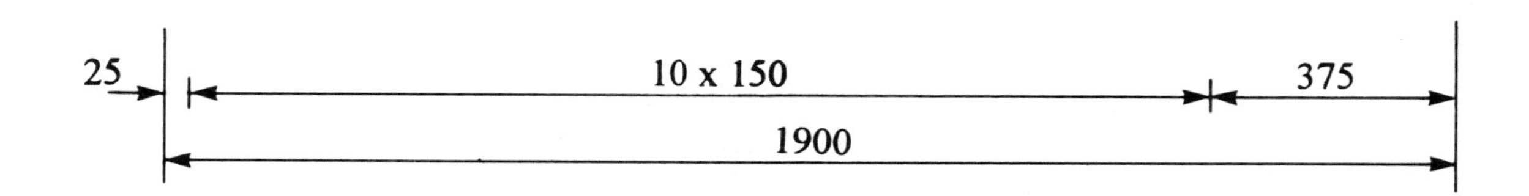

Longitudinal bars 11 ø 7.0 x 1900 = 6.312 kg
Transverse bars 11 ø 7.0 x 1900 = 6.312 kg
Total weight 12.624 kg

Figure B5

Index